FIGHTING MULTIDRUG RESISTANCE WITH HERBAL EXTRACTS, ESSENTIAL OILS AND THEIR COMPONENTS

FIGHTING MULTIDRUG RESISTANCE WITH HERBAL EXTRACTS, ESSENTIAL OILS AND THEIR COMPONENTS

Edited by

MAHENDRA KUMAR RAI MSc, PhD
Sant Gadge Baba Amravati University, Amravati, Maharashtra, India
Department of Chemical Biology, Institute of Chemistry, University of Campinas, Campinas SP, Brazil

KATERYNA VOLODYMYRIVNA KON MD, PhD
Kharkiv National Medical University, Kharkiv, Ukraine

ELSEVIER

AMSTERDAM • BOSTON • HEIDELBERG • LONDON
NEW YORK • OXFORD • PARIS • SAN DIEGO
SAN FRANCISCO • SINGAPORE • SYDNEY • TOKYO
Academic Press is an Imprint of Elsevier

Academic Press is an imprint of Elsevier
32 Jamestown Road, London NW1 7BY, UK
225 Wyman Street, Waltham, MA 02451, USA
525 B Street, Suite 1800, San Diego, CA 92101-4495, USA

First edition

Notice

No responsibility is assumed by the publisher for any injury and/or damage to persons or property as a matter of products liability, negligence or otherwise, or from any use or operation of any methods, products, instructions or ideas contained in the material herein. Because of rapid advances in the medical sciences, in particular, independent verification of diagnoses and drug dosages should be made

British Library Cataloguing-in-Publication Data
A catalogue record for this book is available from the British Library

Library of Congress Cataloging-in-Publication Data
A catalog record for this book is available from the Library of Congress

ISBN: 978-0-12-398539-2

For information on all Academic Press publications
visit our website at elsevierdirect.com

Typeset by TNQ Books and Journals

Working together
to grow libraries in
developing countries

www.elsevier.com • www.bookaid.org

Contents

Preface

Mahendra Kumar Rai

Kateryna Volodymyrivna Kon

Drug resistance is now documented for all types of infectious agents. Furthermore, multidrug-resistant microorganisms have been found not only in nosocomial infections but also as community-associated etiologic agents. The most notorious multidrug-resistant bacterium, methicillin-resistant *Staphylococcus aureus*, was first described in 1961 and nowadays its prevalence has reached 50% among invasive isolates in some countries. In addition, resistance to the classical antimalarial, chloroquine, in *Plasmodium falciparum* strains has now reached more than 50%. Likewise, one of the commonly used antimycotics, fluconazole, has recently become less effective owing to the emergence of fluconazole-resistant *Candida albicans* strains and intrinsically resistant species such as *Candida glabrata* and *Candida krusei*.

Widespread drug resistance is especially alarming because in moderate and severe infections treatment is usually prescribed empirically before the results of sensitivity tests are known. This has led to an increasing probability of inappropriate initial treatments for infections, which many studies have demonstrated can adversely impact the outcome, resulting in increased morbidity and mortality.

Resistance to all classes of antimicrobials has emerged, leading to an urgent need for the development of new approaches to cope with this problem. The synthesis of new antibiotic, antimycotic, or antiprotozoal drugs is restricted by the challenges involved in identifying new substances that are both effective against microorganisms and nontoxic to mammalian cells. The use of herbal products to combat infections has been popular for many centuries, although the discovery of the first antibiotics and the great successes of antibiotic treatments diminished the role of antimicrobial plant products. However, the decreasing effectiveness of antimicrobial drugs and difficulties in producing effective replacements has led to a renewed interest in the antimicrobial properties of herbs, as indicated by the many recent publications on this topic.

This book explores the approaches used to combat different multidrug-resistant microorganisms, including bacteria, fungi, protozoa, and viruses, as well as multidrug-resistant tumor cells, using herbal extracts, essential oils, and isolated plant components. The advantages and disadvantages of herbal products, their mechanisms of action, and current trends in their use, both alone and in combination therapies, are also discussed.

Potential readers include researchers in applied microbiology, pharmacology, infection control, and antibiotic resistance, and clinicians in the areas of dermatology, gynecology, infectious diseases, and surgery.

The editors would like to thank Andy Albrecht, the editorial project manager at Elsevier, San Diego, California, USA for his help and advice.

List of Figures

List of Tables

Contributors

María José Abad Department of Pharmacology, Faculty of Pharmacy, University Complutense, Ciudad Universitaria, Madrid, Spain

Emrah Sefik Abamor Department of Bioengineering, Yildiz Technical University, Istanbul, Turkey

B. A. Adeniyi Department of Pharmaceutical Microbiology, University of Ibadan, Ibadan, Nigeria

Adil M. Allahverdiyev Department of Bioengineering, Yildiz Technical University, Istanbul, Turkey

Sezen Canim Ates Department of Bioengineering, Yildiz Technical University, Istanbul, Turkey

Melahat Bagirova Department of Bioengineering, Yildiz Technical University, Istanbul, Turkey

Serap Yesilkir Baydar Department of Bioengineering, Yildiz Technical University, Istanbul, Turkey

Luis Miguel Bedoya Department of Pharmacology, Faculty of Pharmacy, University Complutense, Ciudad Universitaria, Madrid, Spain

Paulina Bermejo Department of Pharmacology, Faculty of Pharmacy, University Complutense, Ciudad Universitaria, Madrid, Spain

Maria Evangelina Carezzano Universidad Nacional de Rio Cuarto, Córdoba, Argentina

Iracilda Zeppone Carlos São Paulo State University, Faculdade de Ciências Farmacêuticas, Araraquara, São Paulo, Brazil

Sumitra V. Chanda Phytochemical, Pharmacological and Microbiological Laboratory, Department of Biosciences, Saurashtra University, Gujarat, India

Ademar A. Da Silva Filhoa Department of Pharmaceutical Sciences, Faculty of Pharmacy, Federal University of Juiz de Fora, Juiz de Fora, Minas Gerais State, Brazil

Camila Bernardes de Andrade Carli São Paulo State University, Faculdade de Ciências Farmacêuticas, Araraquara, São Paulo, Brazil

Clarissa Campos Barbosa de Castro Department of Pharmaceutical Sciences, Faculty of Pharmacy, Federal University of Juiz de Fora, Juiz de Fora, Minas Gerais State, Brazil

Maria de las Mercedes Oliva Universidad Nacional de Rio Cuarto, Córdoba, Argentina

Túlio Pessoa de Rezende Department of Pharmaceutical Sciences, Faculty of Pharmacy, Federal University of Juiz de Fora, Juiz de Fora, Minas Gerais State, Brazil

Mirta Susana Demo Universidad Nacional de Rio Cuarto, Córdoba, Argentina

Mirna Meana Diasa Department of Pharmaceutical Sciences, Faculty of Pharmacy, Federal University of Juiz de Fora, Juiz de Fora, Minas Gerais State, Brazil

Luiz Felipe Domingues Passero Laboratório de Patologia de Moléstias Infecciosas, Departamento de Patologia, Faculdade de Medicina da Universidade de São Paulo, Brazil

B. J. Doyle Department of Biology and Department of Biochemistry, Alma College, Alma, Michigan, USA

Valerie Edwards-Jones School of Research, Enterprise and Innovation, Faculty of Science and Engineering, Manchester Metropolitan University, Manchester, United Kingdom

Serhat Elcicek Department of Bioengineering, Firat University, Elazig, Turkey

M. L. Faleiro Centro de Biomedicina Molecular e Estrutural, Instituto de Biotecnologia e Bioengenharia, Faculdade de Ciências e Tecnologia, Universidade do Algarve, Campus de Gambelas, Faro, Portugal

Mauro Nicolás Gallucci Universidad Nacional de Rio Cuarto, Córdoba, Argentina

Ameenah Gurib-Fakim Center for Phytotherapy Research, Ebene, Mauritius

Amir Reza Jassbi Medicinal & Natural Products Chemistry Research Center, Shiraz University of Medical Science, Shiraz, Iran

Mital J. Kaneria Phytochemical, Pharmacological and Microbiological Laboratory, Department of Biosciences, Saurashtra University, Gujarat, India

Philip G. Kerr School of Biomedical Sciences, Charles Sturt University, WaggaWagga, New South Wales, Australia

Rabia Cakir Koc Department of Biomedical Engineering, Yeni Yuzyil University, Istanbul, Turkey

Kateryna Volodymyrivna Kon Kharkiv National Medical University, Kharkiv, Ukraine

Victor Kuete Department of Biochemistry, Faculty of Science, University of Dschang, Dschang, Cameroon

João Henrique G. Lago Laboratório de Química Bioorgânica Prof. Otto R. Gottlieb. Instituto de Ciências Ambientais, Químicas e Farmacêuticas, Universidade Federal de São Paulo, Brazil

Márcia D. Laurenti Laboratório de Patologia de Moléstias Infecciosas, Departamento de Patologia, Faculdade de Medicina da Universidade de São Paulo, Brazil

T. O. Lawal Department of Pharmaceutical Microbiology, University of Ibadan, Ibadan, Nigeria

Lizandra Guidi Magalhães Núcleo de Pesquisas em Ciências Exatas e Tecnológicas, University of Franca, Franca, São Paulo State, Brazil

G. B. Mahady Department of Pharmacy Practice and Department of Medicinal Chemistry and Pharmacognosy, PAHO/WHO Collaborating Centre for Traditional Medicine, College of Pharmacy, University of Illinois at Chicago, Chicago, Illinois 60612, USA

M. Fawzi Mahomoodally Department of Health Sciences, Faculty of Science, University of Mauritius, Réduit, Mauritius

Lyndy McGaw Phytomedicine Programme, Department of Paraclinical Sciences, Faculty of Veterinary Science, University of Pretoria, Onderstepoort, South Africa

M. G. Miguel Centro Biotecnologia Vegetal, Instituto de Biotecnologia e Bioengenharia, Faculdade Ciências Lisboa, Universidade de Lisboa Portugal, and Departamento de Química e Farmácia, Faculdade de Ciências e Tecnologia, Universidade do Algarve, Portugal

Olga Nehir Oztel Department of Bioengineering, Yildiz Technical University, Istanbul, Turkey

Ali Parsaeimehr G. S. Davtyan Institute of Hydroponics Problems, National Academy of Sciences Republic of Armenia, Yerevan, Armenia

Marcela Bassi Quilles São Paulo State University, Faculdade de Ciências Farmacêuticas, Araraquara, São Paulo, Brazil

Mahendra Kumar Rai Sant Gadge Baba Amravati University, Amravati, Maharashtra, India; Department of Chemical Biology, Institute of Chemistry, University of Campinas, Campinas, São Paulo, Brazil

Kalpna D. Rakholiya Phytochemical, Pharmacological and Microbiological Laboratory, Department of Biosciences, Saurashtra University, Gujarat, India

Román Yesid Ramírez Rueda Universidad de Boyacá, Campus Universitario, Tunja (Boy), Colombia

Gabriela Santos-Gomes Unidade de Ensino e Investigação de Parasitologia Médica, Centro de Malária e outras Doenças Tropicais, Instituto de Higiene e Medicina Tropical, Universidade Nova de Lisboa, Lisboa, Portugal

Elmira Sargsyan G. S. Davtyan Institute of Hydroponics Problems, National Academy of Sciences Republic of Armenia, Yerevan, Armenia

Patrícia Sartorelli Laboratório de Química Bioorgânica Prof. Otto R. Gottlieb. Instituto de Ciências Ambientais, Químicas e Farmacêuticas, Universidade Federal de São Paulo, Brazil

Bruno Luiz Soares Campos Laboratório de Patologia de Moléstias Infecciosas, Departamento de Patologia. Faculdade de Medicina da Universidade de São Paulo, Brazil

K.K. Soni Department of Pharmacy Practice and Department of Medicinal Chemistry and Pharmacognosy, PAHO/WHO Collaborating Centre for Traditional Medicine, College of Pharmacy, University of Illinois at Chicago, Chicago, Illinois, USA

Serkan Yaman Department of Bioengineering, Yildiz Technical University, Istanbul, Turkey

1

Alternative Antimicrobial Approaches to Fighting Multidrug-Resistant Infections

Valerie Edwards-Jones

School of Research, Enterprise and Innovation, Faculty of Science and Engineering, Manchester Metropolitan University, Manchester, United Kingdom

BACKGROUND

The increasing prevalence of multidrug-resistant bacteria in the healthcare environment has left clinicians with fewer treatment options, often resulting in the use of more costly treatments. In the United Kingdom, there has been a 26% reduction in the number of antibiotics available for prescription compared to 20 years ago [1]. The reason for this is threefold. First, many common pathogens have developed resistance to first-generation antimicrobials, thus making them unsuitable for treatment. Second, many large pharmaceutical companies have changed the focus of their research and development (R&D) programs to meet the treatment needs for lifestyle-related illnesses such as diabetes and obesity. Third, the cost of development of a single antibiotic, estimated at £300 million −£550 million, can be cost prohibitive and the associated scientific and regulatory

challenges make this type of R&D riskier than other drug-related research.

Since 2009, only two new antibiotics have been approved in the United States (the fifth-generation cephalosporin, ceftaroline (Teflaro), in 2010 and macrocyclic fidaxomicin (Dificid) in 2011) for the treatment of *Clostridium difficile* and, in 2011, nine intravenous compounds active against resistant Gram-negative bacilli were reported to be in Phase II or Phase III clinical trials, of which two had a novel mechanism of action [2]. It is expected that the number of new antibiotics brought onto the market in the near future will continue to decline, causing a number of organizations to highlight this plight to their governments, and with some challenging both the governments and drug companies to develop 10 new antibiotics by 2020 [3,4].This may be difficult, as greater profits are generated for pharmaceutical companies from lifestyle drugs, which many

Fighting Multidrug Resistance with Herbal Extracts,
Essential Oils and Their Components
http://dx.doi.org/10.1016/B978-0-12-398539-2.00001-X

patients take for life, compared to a short treatment duration (typically 7–14 days) for antibiotics.

Antibiotic resistance is seen in a range of pathogenic and nonpathogenic bacteria; however, methicillin-resistant *Staphylococcus aureus* (MRSA) is the most commonly reported. Multiple antibiotic resistance is also increasing in Gram-negative bacteria (e.g., *Acinetobacter* spp. and *Klebsiella* spp.) associated with the production of extended spectrum β-lactamases (ESBLs) [5]. Treating other communicable diseases such as gonorrhea and tuberculosis is becoming more difficult, with many strains of gonorrhea now resistant to penicillin and multiple resistant strains of tuberculosis persisting globally. There are very few treatment options left for the clinician, with no new drugs on the horizon. This problem is not just confined to the hospital or healthcare environment. Nowadays, many antibiotic-resistant strains are seen in the community and also among animals (both domestic and farmed). This has resulted in a number of changes in practice and a tightening up of the guidance made available to those able to prescribe antibiotics.

In 2001, a committee of experts met in the UK to reduce the inappropriate use of antibiotics; six priority action areas were targeted to minimize further development including prudent antimicrobial use in humans (in both the community and hospitals) and animals, improved infection control, and improved education, information dissemination, research, and surveillance. The UK Antimicrobial Resistance Action Plan was fully implemented in 2005 [6]. In 2011, a task force was established in the USA to address problems with antimicrobial resistance in the USA, as well as in Europe [7]. This Transatlantic Taskforce on Antimicrobial Resistance (TATFAR) was established by Presidential declaration in 2009 at the annual summit between the EU and US presidencies; three key areas were to be addressed with

some urgency, along with 17 recommendations. The three key areas were:

1. Appropriate therapeutic use of antimicrobial drugs in the medical and veterinary communities;
2. Prevention of both healthcare- and community-associated drug-resistant infections; and
3. Strategies for improving the pipeline of new antimicrobial drugs.

The scientific community has raised concerns over antimicrobial resistance globally, and recently Dr. Chan from the World Health Organization has said that we face a return to the pre-antibiotic era unless something is done to limit the speed of further antimicrobial resistance developing. Although increased resources have been made available for research into antimicrobial resistance, the development of new antibiotics and the associated regulation make implementation slow. There have been calls to have some of the new antibiotics fast-tracked to market, which would allow their availability in a shorter time frame. However, if this is approved, the rigor of safety checking through the various clinical trial stages must not be compromised.

To challenge the problems of treating antibiotic-resistant bacterial infections in healthcare environments, many pharmacist-facilitated restriction programs have been introduced into hospitals and primary care trusts in the UK. In addition, different antibiotic administrative techniques are now being recommended such as combination therapy, the use of narrow spectrum drugs, and in some cases continuous versus intermittent antimicrobial therapy. Targeted release of antibiotics is also being developed and trialed to help minimize the further development of resistance and ensure more appropriate treatment regimes.

Owing to these problems, there has been a renewed interest in research into alternative antimicrobials and more targeted treatment strategies, including therapies used before the advent of antibiotics, with a return to traditional remedies and medicines. Some of these alternative therapies can be administered systemically but there are many more that can be administered topically. These new treatments may complement antibiotic therapy and help reduce the further spread and development of resistance in the future. Some examples of these are listed in Table 1-1 and described briefly below:

Bacteriocins are ribosomally synthesized peptides produced by Gram-positive and Gram-negative bacteria. They are classified according to their mechanisms of killing, resistance mechanisms, and the producing strain; a new classification scheme was adopted in 2006 [8]. The bacteriocins of lactic acid-fermenting bacteria are known as *lantibiotics*; they possess unique intramolecular ring structures formed by the thioether amino acids lanthionine or methyllanthionine. These provide structural rigidity for binding to their cellular

TABLE 1-1 Some Examples of Alternative Antimicrobial Therapies: Systemic and Topical Administration

Alternative Therapy	Possible Administration Route
Bacterial interference/ probiotics/prebiotics	Topical and systemic
Bacteriocins—lantibiotics	Topical and systemic
Essential oils	Topical
Herbs and plant extracts	Topical and systemic
Honey	Topical
Maggots	Topical
Metal ions	Topical
Peptide antibiotics	Topical and systemic
Phage therapy	Topical and systemic
Vaccine therapy	Systemic

targets and provide resistance against protease degradation. Lantibiotics are produced by, and act primarily on, Gram-positive bacteria: nisin (the best known) has been being used in the dairy industry since the early 1980s.

PEPTIDE ANTIBIOTICS

There are two classes of peptide antibiotics: nonribosomally and ribosomally synthesized. Nonribosomally synthesized peptides e.g., bacitracins, gramicidins, and polymyxins are produced predominantly by bacteria and have a narrow spectrum of activity. Many of these antibiotics were superseded in the 1970s as other new classes of antibiotics were produced with broader spectrums of activity. Recently, there has been renewed interest in these antibiotics, following chemical modification of insoluble streptogramins and the production of two water-soluble antibiotics, dalfopristin and quinupristin (Synercid) [2].

Ribosomally synthesized cationic peptides are known as Nature's antibiotics and are produced by all types of organisms (including bacteria) as part of the innate host defense mechanism: mammalian, amphibian, insect, and plant bacteriocins have been identified. Frog skin has been used as an antimicrobial agent for centuries in South America and is still used today: it contains a 24-amino-acid peptide, bombinin, which is hemolytic. Melittin, another antimicrobial cationic peptide in bee venom, has not been exploited owing to the toxic nature of this defensin [9].

BACTERIAL INTERFERENCE

Bacterial interference relates to interactions between probiotics/prebiotics and bacteria. Understanding the interaction of one species with another is an area of intense research as we try to fully understand the pathogenic process and find alternative methods of preventing infection. Understanding the signaling processes used in microbial communication and involved in adhesion, toxin production, and biofilm formation may help prevent and even treat infectious diseases in the future. The use of probiotics (predominantly *Lactobacillus* spp.) with or without prebiotics to prevent and treat infection, especially gastrointestinal infections, is very promising. There have been a number of reports in which probiotics used prophylactically with or without prebiotics (synbiotics) have led to a reduction in the number of surgical site infections or the incidence of gastrointestinal disease [10]. Whether this is solely due to the production of extracellular products that can suppress the growth of the pathogen or a case of preferential colonization is an area of much debate.

Researchers are also studying bacterial interactions at a cellular level, by investigating the interactive effects of genetic regulatory molecules among different species. It is known that the role of effector molecules (oligopeptides in Gram-positive bacteria and *N*-acyl homoserine lactones in Gram-negatives) and their accumulation in a population controls the production of virulence factors, as well as biofilm formation. How they accumulate, facilitate cell-cell communication, and alter gene expression may eventually lead to the production of a range of new molecules and potential vaccine targets for future treatments [11]. This area is especially promising for the future regulation of virulence factors and for increasing our understanding of the infectious process in mixed populations. The possibility of adding certain known molecules (or the avirulent bacteria that produce them) to prevent adhesion, colonization, and the production of virulence factors that promote overt infection in areas such as wound care, post-surgery, and gastrointestinal infection is a fascinating, if controversial, therapeutic opportunity.

VACCINE THERAPY

Vaccine development has been a very successful strategy in the prevention of highly contagious or serious disease. An excellent example is that of smallpox, which is no longer a threat to human health. The development of vaccines for all infectious agents would be an ideal, but sometimes there are difficulties with immunogenicity and the protective nature of vaccines. The development of vaccines for some non-life threatening diseases is not always cost effective and antibiotic treatment may be preferred (e.g., *S. aureus*). However, with antimicrobial resistance becoming prevalent in common pathogens, vaccines are now being developed. Most invasive strains of *S. aureus* have either type 5 or type 8 capsules; therefore, this component has been used in staphylococcal vaccine development for a number of years (e.g., StaphVax). However, Nabi Biopharmaceuticals intend to add additional virulence determinants such as T336, Panton-Valentine leukocidin (PVL), and alpha toxin to improve efficacy in the next generation of StaphVAX [12]. New ideas for staphylococcal vaccines have led to the development of combinatory vaccines that target microbial surface components recognizing adhesive matrix molecules (MSCRAMMs), a family of bacterial proteins that bind to human extracellular matrix components [13]. A person receiving this type of vaccine could be protected from staphylococcal infection through inhibition of the bacteria binding to the host.

BACTERIOPHAGE THERAPY

Lytic phage therapy has been effectively used in Russia for decades (e.g., Stalin therapy) and can be effective against a number of pathogens. Treatment is based on the administration of either a bacteriophage specific for the offending pathogen (monotherapy) or a cocktail of bacteriophages (polytherapy). This treatment is nontoxic to the host but lethal to the bacteria. Bacteriophages can be administered orally, intramuscularly, or topically and, because the bacteriophages reproduce within the susceptible bacteria, a second dosage should be unnecessary. Bacteriophage therapy is therefore a cheap, effective, and specific way of treating a range of diseases. Bacteriophages tend to minimally disrupt normal flora, are equally effective against antibiotic-sensitive and antibiotic-resistant bacteria, are often easily discovered, seem to be capable of disrupting bacterial biofilms, and can have low inherent toxicities [14]. A number of different preparations are available (cream, drops, liquids, and impregnated solids), especially in the veterinary field. Moreover, because of their relatively low cost, much research is currently being undertaken to expand their application. These therapies may have huge benefits in areas such as dentistry and wound care, where biofilms become established and their removal is difficult because of the complex interactions that occur between species. At the time of writing, several articles provide a comprehensive overview of the history, advantages and disadvantages, and applications of bacteriophage therapy to the treatment of human infections [14–16].

ANTIMICROBIAL METALS

A number of metals are used as antimicrobials, with silver, copper, and titanium being the most common. They all have a variety of applications, but are predominantly incorporated into nanocomposites, medical devices, or surfaces to prevent contamination or infection. A recent study in which the antimicrobial effect of copper alloys in furnishings in the clinical environment was investigated showed lower microbial counts on copper alloy surfaces compared to standard materials and significantly

lower numbers of vancomycin-resistant enterococci, *S. aureus*, and coliforms [17].

More literature is available on the antimicrobial effect of silver ions. Silver has been used as an antimicrobial in medicine and the environment since antiquity and is incorporated into a number of materials and applications (e.g., catheters, dressings, washing machines, and computer keyboards). The most common use of silver in health care is the management of heavily colonized or infected wounds, especially those colonized with multiply resistant antibiotic strains such as MRSA. Nanocrystalline silver appears to have superior antimicrobial activity over normal crystalline silver and other silver salts (*in vitro*), due to the smaller particle size (20–120 nm) and a silver level of 70–100 ppm [18,19].

MEDICINAL GRADE HONEY DRESSINGS

Honey has been used as an antimicrobial agent for centuries, and is still used extensively in wound care. The high sugar content and the presence of both hydrogen peroxide [20] and propolis are known to confer antimicrobial properties. In addition, honey's osmotic effect helps remove excess exudate from wounds and it is also a very good desloughing agent (i.e., removes wound debris and necrotic tissue). Manuka honey is a monofloral honey produced from the nectar of the New Zealand Manuka tree (*Leptospermum scoparium*). The incorporation of this honey into dressings and lotions is claimed to provide added benefits over normal honey because it contains trace amounts of terpenoids, which are natural antimicrobials

HERB AND PLANT EXTRACTS

Herbal and plant remedies have been used for centuries and have a huge potential for future antibacterial therapies. In the fifth century BC, Hippocrates wrote a text containing the medicinal properties of over 350 medicinal plants. Similarly, the Greek physician Dioscorides wrote the informative reference book about medicinal plants, *De Materia Medica*, which has been used for over 1,600 years [21]. In addition, the Egyptians used herbs before 5000 BC for embalming their dead to prevent decomposition [22].

Many research groups have renewed their interest in these areas and have focused their research programs on investigating the antimicrobial activities of plants and their extracts against multidrug-resistant bacterial strains in the hope of discovering new antibiotics.

MAJOR GROUPS OF ANTIMICROBIAL COMPOUNDS FROM PLANTS

The use of gas chromatography mass spectrometry (GC-MS) analysis has shown that plant extracts consist of a complex mixture of multiple organic compounds, many of which are antimicrobial. These include alkaloids, coumarins, essential oils (EOs), flavonoids, flavonols, flavones, lectins, phenolics and polyphenols, polyacetylenes, polypeptides, quinines, tannins, and terpenoids [21]. The isolation and identification of plant secondary metabolites and their use as active compounds in medicinal preparations has recently attracted the attention of the pharmaceutical industry. It is now known that these compounds have a range of biologic, chemical, and physical activities, and that plants such as allspice, clove, cinnamon, garlic, and thyme contain numerous antimicrobial compounds, including alkaloids, catechins, furans, polyphenols, tannins, terpenoids, and terpenes [23].

The traditional use of each plant is usually guided by its availability and the ease of extract preparation. There are a variety of extraction methods using appropriate solvents and

extraction techniques, and the success of extraction is influenced by the length of the extraction period, solvent used, pH, and temperature [21,24]. Both fresh and dried plant material can be used as treatments, although most researchers tend to use dried material for extract studies. Water is an excellent solvent and traditional healers use water to extract compounds from dried plant material. Compounds such as tannins and terpenoids are water soluble and both of these compounds are known to have antimicrobial activity. However, many of the major components are not water soluble; to extract these, organic solvents of increasing polarity may be used, e.g., ether, petroleum ether, chloroform, ethyl acetate, and ethanol [25,26].

ESSENTIAL OILS

EOs are a complex mixture of the volatile organic components of fragrant plant matter that contribute to the flavor and fragrance of the plant. They can be extracted from any part of the plant by steam distillation, cold pressing, or using organic solvents; the oil yield is affected by environmental conditions and distillation processes. The yield can vary between batches from the same trees and between sites [23]. Significant differences in the chemical composition of oils from the same species of plant give rise to *chemotypes*, which are the result of biologic variations caused by the effects of different soils, temperature, weather conditions, and light. That is, chemical components can differ in plants that are botanically identical. For example, thyme oil (obtained from the common species *Thymus vulgaris)* normally contains 20–54% thymol along with additional compounds, such as carvacrol, *p*-cymene, linalool, and myrcene. However, a different chemotype of the same species of plant may have linalool as the predominant component [23].

EOs are an example of secondary metabolites that consist of hydrocarbon compounds based on isoprene (C_5H_8), or 2-methyl-1,3-butadiene. Polymers of this compound are collectively called terpenes and have the structure (C_5H_8)n. Monoterpenes have two units, ($C_{10}H_{16}$), sesquiterpenes three units, ($C_{15}H_{24}$), diterpenes four units ($C_{20}H_{32}$), etc., and terpenoids, which are terpene derivatives, include abscisic acid and gibberellin (plant growth substances) and the carotenoid and chlorophyll pigments [21]. EOs can also contain oxygenated terpenoids e.g., acids, alcohols, aldehydes, esters, furans, ketones, lactones, oxides, peroxides, and phenols. These compounds can all affect bacterial growth [27].

The mode of action of EOs is thought to be through disruption of the cell membrane, causing potassium leakage [28]. It has also been shown that tea tree oil can inhibit glucose-dependent respiration in *Candida albicans*, *Escherichia coli*, and *S. aureus*. In *C. albicans*, inhibition of ion transport processes and respiration causes increased membrane permeability [28]. Tea tree oil has been widely accepted globally as a topical antiseptic and is used in a wide range of applications in the cosmetic and medical industries, ranging from head lice shampoo to dressings for burns patients [29].

Several microbiologic studies have demonstrated that there is an improved effect when EOs are blended [30,31], and there are proven synergistic effects with known biocides [32] and antibiotics [33,34]. Studies undertaken in the United Kingdom using an *in vitro* model have shown that the addition of eucalyptus oil to the antiseptic chlorhexidine can enhance the penetration of chlorhexidine into skin compared to chlorhexidine alone [35]. In addition, using a biofilm model the same research group showed that eucalyptus oil in combination with tea tree oil and chlorhexidine can kill both sessile and planktonic cultures of *Staphylococcus epidermidis* [36].

It is possible that, when blended with other antimicrobial agents, the components of EOs can open cell membrane channels, thus allowing the entry of antimicrobial agents to their target sites, when otherwise entry is resisted; however, this has yet to be proven conclusively.

The vapors of EOs are also now studied extensively, as their antimicrobial effects may have potential usage for decontaminating environments, preserving foods, and in wound dressings [30,31,37,38].

The use of individual herbs and plant extracts in the treatment of infectious disease is a huge area of research and many examples are covered in the following chapters of this book. A review by Cowan in 1999 on the use of plants as antimicrobial agents provides an excellent overview of individual components and their activities [21]. Since then, the number of publications in this area has expanded exponentially and examples of their use both topically and systemically are now described.

CONCLUSIONS

While we are faced with increasing numbers of cases of infections with microorganisms that are multi-antibiotic resistant, we must continue to research into alternative therapies. Many of these are traditional medicines that were abandoned following the introduction of antibiotics and, without doubt, antibiotics have proven time and time again to be invaluable in the war on microbes. Antibiotics are currently the best treatment we have for systemic disease and the prospect of having none available for treatment in the future is extremely worrying. We must therefore continue to develop and accept alternatives for less life-threatening infections and save the remaining antibiotics for treating more serious infections.

References

[1] Cooke J. Infectious diseases—the need for new antibiotics. Hospital Pharmacist 2004;11:265–8.

[2] Freire-Morana L, Aronssona B, Manzb C, Gyssensc IC, Sob AD, Monnete DL, et al. Critical shortage of new antibiotics in development against multidrug-resistant bacteria—Time to react is now! Drug Resistance Updates 2011;14:118−24.

[3] Gilbert DN, Guidos RJ, Boucher HW, Talbot GH, Spellberg B, Edwards Jr JE, et al. The 10 x '20 Initiative: pursuing a global commitment to develop 10 new antibacterial drugs by 2020. Clin Infect Dis 2010;50:1081−3.

[4] Bulletin of the World Health Organization. 2011;89:88−89. doi:10.2471/BLT.11.030211.

[5] The Healthcare Associated Infections Report 2009. http://www.hpa.org.uk/webc/HPAwebFile/HPAweb_C/1252326222452; last [accessed 21.05.12].

[6] The Antimicrobial Resistance Plan 2002−2005, <http://www.dhsspsni.gov.uk/amrap-report-23jan02.pdf>; [last accessed 21.05.12].

[7] Transatlantic Taskforce on Antimicrobial Resistance. http://www.cdc.gov/drugresistance/pdf/tatfar-report.pdf; last [accessed 21.05.12].

[8] Nes IF, Yoon SS, Diep DB. Ribosomally synthesized antimicrobial peptides (bacteriocins) in lactic acid bacteria: a review. Food Sci Biotechnol 2007;16: 675−90.

[9] Hancock REW, Chapple DS. Peptide antibiotics. Antimicrob Agents Chemother 1999;43:1317−23.

[10] Jeppsson B, Mangell P, Thorlacius H. Use of probiotics as prophylaxis for postoperative infections. Nutrients 2011;3:604−12.

[11] McDowell P, Affas Z, Reynolds C, Holden MT, Wood SJ, Saint S, et al. Structure, activity and evolution of the group I thiolactone peptide quorum-sensing system of Staphylococcus aureus. Mol Microbial 2001;41:503−12.

[12] DeLeo FR, Otto M. An antidote for Staphylococcus aureus pneumonia? J Exp Med 2008;205:271−4.

[13] Otto M. Targeted immunotherapy for staphylococcal infections: focus on anti-MSCRAMM antibodies. BioDrugs 2008;22:27−36.

[14] Loc-Carrillo C, Abedon ST. Pros and cons of phage therapy. Bacteriophage 2011;1:111−4.

[15] Hanlon GW. Bacteriophages: an appraisal of their role in the treatment of bacterial infections. Int J Antimicrob Agents 2007;30:118−28.

[16] Abedon ST, Kuhl SJ, Blasde BG, Kutter EM. Phage treatment of human infections. Bacteriophage 2011;1:66−85.

[17] Karpanen TJ, Casey AL, Lambert PA, Cookson BD, Nightingale P, Miruszenko L, et al. The antimicrobial efficacy of copper alloy furnishing in the clinical environment: a crossover study. Infect Control Hosp Epidemiol 2012;33:3−9.

[18] Dunn K, Edwards-Jones V. The role of Acticoat with nanocrystalline silver in the management of burns. Burns 2004;30(Suppl. 1):S1−9.

[19] Edwards-Jones V. Antimicrobial and barrier effects of silver against methicillin-resistant Staphylococcus aureus. J Wound Care 2006;15:285−90.

[20] Brudzynski K. Effect of hydrogen peroxide on antibacterial activities of Canadian honeys. Can J Microbiol 2006;52:1228−37.

[21] Cowan MM. Plant products as antimicrobial agents. Clin Microbiol Rev 1999;12:564−82.

[22] Martin KW, Ernst E. Antiviral agents from plants and herbs: a systematic review. Focus on Alternative and Complementary Therapies 2003;8:285−99.

[23] Price S. Practical aromatherapy: how to use essential oils to restore vitality. London: HarperThorsons Publishing; 1987.

[24] Ncube NS. Anti-bacterial properties of the methanol extract of Helichrysum pedunculatum. MSc Thesis, University of Fort Hare, <http://hdl.handle.net/10353/44>; 2008 [last accessed 29.09.12].

[25] Eloff J. Which extractant should be used for the screening and isolation of antimicrobial components from plants? J Ethnopharmacol 1998;60:1−8.

[26] Parekh J, Jadeja D, Chanda S. Efficacy of aqueous and methanol extracts of some medicinal plants for potential antibacterial activity. Turk J Biol 2005;29:203−10.

[27] Janssen MA, Scheffer JJC, Svendsen AB. Antimicrobial activities of essential oils: a 1976−86 literature review on possible applications. Pharmaceutische Weekblad (Scientific Edition) 1987;9:193−7.

[28] Cox SD, Mann CM, Markham JL, Bell HC, Gustafson JE, Warmington JR, et al. The mode of antimicrobial action of the essential oil of Melaleuca alternifolia (tea tree oil). J Appl Microbiol 2000;88:170−5.

[29] Carson CF, Hammer KA, Riley TV. Melaleuca alternifolia (Tea Tree) oil: a review of antimicrobial and other medicinal properties. Clin Microbiol Rev 2006;19:50−62.

[30] Edwards-Jones V, Buck R, Shawcross SG, Dawson MM, Dunn K. The effect of essential oils on methicillin-resistant Staphylococcus aureus using a dressing model. Burns 2004;30:772−7.

[31] Doran AL, Morden WE, Dunn K, Edwards-Jones V. Vapour-phase activities of essential oils against antibiotic sensitive and resistant bacteria including MRSA. Lett Appl Microbiol 2009;48:387−92.

[32] Cannon RD, Tompkins GR. Improving the effectiveness of chlorhexidine. N Z Dent J 2006;102:17.

[33] Hemaiswarya S, Kruthiventi AK, Doble M. Synergism between natural products and antibiotics against infectious diseases. Phytomedicine 2008;15:639−52.

[34] Rosato A, Vitali C, De Laurentis N, Armenise D, Antonietta Milillo M. Antibacterial effect of some essential oils administered alone or in combination with Norfloxacin. Phytomedicine 2007;14:727−32.

[35] Karpanen TJ, Conway BR, Worthington T, Hilton AC, Elliott TS, Lambert PA. Enhanced chlorhexidine skin penetration with eucalyptus oil. BMC Infect Dis 2010;10:278.

[36] Karpanen TJ, Worthington T, Hendry ER, Conway BR, Lambert PA. Antimicrobial efficacy of chlorhexidine digluconate alone and in combination with eucalyptus oil, tea tree oil and thymol against planktonic and biofilm cultures of Staphylococcus epidermidis. J Antimicrob Chemother 2008;62:1031−6.

[37] Fisher K, Phillips CA. The effect of lemon, orange and bergamot essential oils and their components on the survival of Campylobacter jejuni, Escherichia coli 0157, Listeria monocytogenes, Bacillus cereus and Staphylococcus aureus *in vitro* and in food systems. J Appl Microbiol 2006;101:1232−40.

[38] Laird K, Phillips C. Vapour phase: a potential future use for essential oils as antimicrobials? Lett Appl Microbiol 2012;54:169−74.

2

Natural Plant Products Used against Methicillin-Resistant *Staphylococcus aureus*

Román Yesid Ramírez Rueda

Universidad de Boyacá, Campus Universitario Cra 2a este 64–169 Tunja (Boy), Colombia

INTRODUCTION

Following its discovery, penicillin began to be used as a treatment for *Staphylococcus aureus* infections in the 1940s, initially providing very good results. However, just 1 year after the implementation of this antibiotic, penicillin-resistant strains were identified [1]. Such resistance is caused by the production of plasmid-encoded β-lactamase. Since the 1950s, there has been a substantial and continuous increase in penicillin resistance. The use of other antistaphylococcal drugs, such as chloramphenicol, erythromycin, streptomycin, sulfonamides, and tetracyclines, led to resistance to these drugs in *S. aureus* in the 1950s. The introduction of methicillin (a semisynthetic penicillin) in 1960 marked a major advance in the treatment of staphylococcal infections resistant to penicillin; however, the first strain of methicillin-resistant *S. aureus* (MRSA) was described in 1961, just a few months after onset of clinical use of the antibiotic [2]. The main resistance mechanism of MRSA is the production of a penicillin-binding protein (PBP) with lower affinity for β-lactam, called PBP2a. This resistance is encoded by the *mecA* gene and is part of a mobile element located within a pathogenicity island known as staphylococcal chromosomal cassette mec (SCCmec) [3]. MRSA is resistant to all β-lactams (including carbapenems and cephalosporins) [4] and usually resistant to aminoglycosides, clindamycin, erythromycin, quinolones, rifampin, sulfonamides, and tetracyclines, while it is usually susceptible to glycopeptides [5]. In 1996 in Japan, a strain of MRSA was isolated with reduced susceptibility to vancomycin [6], exhibiting a minimum inhibitory concentration (MIC) of $8\,\mu g/mL$. Subsequently, isolates with the same characteristics were reported in other countries. In 2002 in the United States, the first MRSA strain with high resistance was detected (MIC $> 32\,\mu g/mL$) that had acquired a *vanA* gene identical to that of vancomycin-resistant enterococci [7].

MRSA infection often occurs endogenously in sites with previous disturbance of the

Fighting Multidrug Resistance with Herbal Extracts,
Essential Oils and Their Components
http://dx.doi.org/10.1016/B978-0-12-398539-2.00002-1

mucocutaneous barrier due to traumatic injury, surgery, intravenous drug abuse, skin diseases, and intravascular catheterization, among others. This causes further colonization with MRSA, resulting in infections of the skin and subcutaneous tissue (abscesses), wound infections, and intravascular or urinary catheter-related infection. The spread of bacteria by blood can cause septic shock and severe metastatic infections such as acute endocarditis, arthritis, meningitis, myocarditis, pericarditis, pneumonia, and osteomyelitis [8].

As described above, bacteria such as *S. aureus* develop resistance mechanisms easily over a short time; this resistance rate is much higher than the rate of creation or discovery of new antibiotics, which makes it important to investigate therapeutic alternatives derived from various sources. Plants represent an important source of antibacterial molecules, some of which are known and probably others that are unknown. Consequently, a fourfold increase in the number of publications since the late 2000s compared with the first 5 years of the 1990s reflects the increased research interest in this field. In the next sections, some of the most important results of research studies aimed at combating multidrug resistance in *S. aureus* with plant-derived compounds are shown.

PLANTS EXTRACTS WITH ANTI-METHICILLIN-RESISTANT *STAPHYLOCOCCUS AUREUS* EFFECTS

Antibiotic resistance of microorganisms is a growing problem in public health. It is considered as an emerging phenomenon worldwide that has led to the increased resistance of pathogenic microorganisms involved in the development of nosocomial and community acquired infections.

However, in relation to the total number of plant species in the world, the plants that have already been studied in order to evaluate their antimicrobial properties represent a tiny amount and a large number of scientists around the world are studying this phenomenon. This interest is mainly due to the development of bacterial strains with multiple resistance to antibiotics, a problem that is as old as the use of antibiotics itself. Thus, over time researchers have developed several techniques that enable the study of antimicrobial effects. In order to demonstrate an effect or activity of a plant against a living organism, it is necessary to separate the active principle(s) from the inert or inactive components. The most common way to do this is by exposing small amounts of plant tissues (fresh or dried) to water or organic solvent, depending on the nature of compound to be obtained (polar solvents for hydrophilic compounds and nonpolar for lipophilic compounds), or by successive extractions using different solvents to obtain different nature extracts. Another way is to obtain essential oil (EO) of a plant using techniques such as hydrodistillation [9].

Once the extracts or EOs are obtained, bioassays may be carried out. This involves exposing organisms to decreasing concentrations of extract or EO and determining the MIC. In this procedure, one should take into account that concentrations > 1 mg/mL (in the case of an extract) may produce false-positive results; therefore, candidates are those plants extracts that show antimicrobial activity at < 100 µg/mL [10]. Several methods that can be used to establish the MIC include agar dilution (plate dilution), microdilution (microtiter wells), macrodilution (in tubes), and indirectly by agar diffusion, among others. For these tests, the most commonly used culture medium is Mueller-Hinton in either agar or broth form (depending on the bioassay technique). This culture medium is accepted by the Clinical and Laboratory Standards Institute (CLSI) for evaluating the antimicrobial susceptibility of most bacterial strains. According to the selected technique, a specific bacterial concentration is used, but in general the standard reference is a concentration equivalent to tube number 0.5 in the McFarland scale: this represents a turbidity of 1.5×10^8 colony-forming units (CFU) per mL. This bacterial concentration is also used established to test antimicrobial susceptibility by the agar diffusion method. Results are frequently achieved after 18–24-h incubation at a temperature range of 35–37 °C (the temperatures at which most pathogenic bacteria thrive) [11].

The antimicrobial effect may be indicated by the MIC or the minimum bactericidal concentration (MBC), according to the technique used. Some researchers also report the IC50 (50% inhibitory concentration) or IC_{90} (90% inhibitory concentration) values. For extracts, the MIC is reported in units of mg/mL or smaller. For EOs, the MIC or MBC is usually expressed in units of μL/mL or % v/v. As mentioned above, several research groups are working on this issue worldwide. Table 2-1 lists some plants that have been investigated and shown to possess antimicrobial activities, either as an antibiotic adjuvant or by direct action against MRSA.

As shown in Table 2-1, various plant families have been shown to possess antimicrobial effect, indicating that their distribution is ubiquitous. This indicates that plants from all regions of the world may be natural sources of antimicrobials, which may therefore generate a wide field of research.

Furthermore, various different parts of the plants can be studied, ranging from the seed to the fruit, and even the entire plant. It is important to ensure that the plant resources used should be quickly renewable and that the maximum amount of material should be obtained from each plant for obtaining the extract or metabolite. The leaves are the parts of the plant that best fulfills these characteristics, because they are the most abundant elements of the plant and are quickly renewed after harvesting. Although leaves are the ideal source material, we should not ignore other parts of the plant, since both the distribution and concentration of metabolites can vary from one tissue to another.

Regarding the MIC of extract required to inhibit for MRSA, studies have reported concentrations ranging from 250 mg/mL to 1 μg/mL. However, it is important to remember that some authors believe that concentrations > 1 mg/mL (in extracts) are not significant, since the inhibitory effect may not relate to the activity of the antimicrobial or metabolite(s) but may instead be a result of other effects, such as osmotic pressure exerted by solutes on the bacterial cell walls [10].

MOLECULES WITH ANTI-METHICILLIN-RESISTANT *STAPHYLOCOCCUS AUREUS* ACTIVITY DERIVED FROM PLANTS

Once a plant extract is found to have an antimicrobial effect, the next logical step is to determine which compounds or molecules are involved in this process. This can be accomplished using separation techniques such as high performance liquid chromatography (HPLC) or gas chromatography-coupled mass spectrometry, among others. The basis of the technique is the same, but some adjustments are made in order to work with extracts. Sometimes, after identifying the antimicrobial molecule, modifications are performed to the purified compound in an attempt to increase its antimicrobial activity [29].

The wide diversity and large number of molecules that may be contained in a natural extract raises the possibility that the constituent molecules may react to generate effects such as synergism or antagonism. Thus, some individual molecules may have stronger effects than the whole extract (i.e., there is an antagonist within the extract) or a reduced effect (i.e., there is a synergistic factor in the extract) [30]. However, Ríos et al. [10] propose that individual molecules that have MICs > 1 mg/mL should not be considered as inhibitors.

TABLE 2-1 Selected Plants Whose Extracts and Essential Oils Have Anti-Methicillin-Resistant *Staphylococcus aureus* Activities

Plant	Family	Plant part	MIC	Reference
Alchornea cordifolia	Euphorbiaceae	Leaves	3.1 mg/mL	[12]
Persea americana	Lauraceae	Leaves	6.3 mg/mL	
Solanum verbascifolium	Solanaceae	Leaves	6.3 mg/mL	
Rosmarinus officinalis	Lamiaceae	Leaves	0.13–3.13 mg/mL	[13]
Thymus vulgaris	Lamiaceae	Aerial parts	18.5 µg/mL	[14]
Eucalyptus globulus	Myrtaceae	Aerial parts	85.6 µg/mL	
Chaerophyllum libanoticum	Apiaceae	Fruit	0.25 mg/mL	[15]
Zataria multiflora	Lamiaceae	Aerial parts	0.5–1 µL/mL (Eo)	[16]
Atuna racemosa	Chrysobalanaceae	Seeds	16–32 µg/mL	[17]
Cleistocalyx operculatus	Myrtaceae	Buds	4–16 µL/mL (Ext) 8 mg/mL (Eo)	[18]
Punica granatum	Lythraceae	Fruit, pericarp	250 mg/L	[19]
Tabebuia avellanedae	Bignoniaceae	Wood	125 − > 250 mg/L	
Dendrobenthamia capitata	Cornaceae	Aerial parts	1.25 ± 0.18 mg/mL	[20]
Elsholtzia rugulosa	Lamiaceae	Whole plant	1.43 ± 0.13 mg/mL	
Elsholtzia blanda	Lamiaceae	Aerial parts	1.32 ± 0.16 mg/mL	
Geranium strictipes	Geraniaceae	Root	1.34 ± 0.30 mg/mL	
Polygonum multiflorum	Polygonaceae	Root	1.34 ± 0.22 mg/mL	
Emblica officinalis	Euphorbiaceae	Seeds	Synergy with AMX	[21]
Nymphaea odorata	Nymphaeaceae	Stamen	Synergy with AMX	
Hypericum perforatum	Hypericaceae	Aerial parts	1 µg/mL	[22]
Ecballium elaterium	Cucurbitaceae	Fruit	0.19–1.56 mg/mL	[23]
Arthrocnemum indicum	Amaranthaceae	Buds	4–8 mg/mL	[24]
Salicornia brachiata	Amaranthaceae	Buds	4–8 mg/mL	
Suaeda monoica	Chenopodiaceae	Leaves	4–8 mg/mL	
Suaeda maritime	Chenopodiaceae	Leaves	4–8 mg/mL	
Sesuvium portulacastrum	Aizoaceae	Leaves	4–8 mg/mL	
Avicennia officinalis	Acanthaceae	Leaves	2–4 mg/mL	
Ceriops decandra	Rhizophoraceae	Leaves	1–2 mg/mL	
Aegiceras corniculatum	Myrsinaceae	Leaves	0.5–1 mg/mL	

TABLE 2-1 Selected Plants Whose Extracts and Essential Oils Have Anti-Methicillin-Resistant *Staphylococcus aureus* Activities (*cont'd*)

Plant	Family	Plant part	MIC	Reference
Excoecaria agallocha	Euphorbiaceae	Leaves	0.1−0.2 mg/mL	
Lumnitzera racemosa	Combretaceae	Leaves	0.5−1 mg/mL	
Acanthus ilicifolius	Acanthaceae	Leaves	2−4 mg/mL	
Olea europaea	Oleaceae	Leaves	0.8−12.5% v/v (Eo)	[25]
Chaerophyllum libanoticum	Apiaceae	Fruit	0.25 mg/mL (EO)	[26]
Boswellia spp.	Burseraceae	Bark	17.3−42.1 mg/mL (oleo-gum resin)	[27]
Melaleuca alternifolia	Myrtaceae	Leaves	0.04% v/v	[28]

AMX, amoxicillin, Eo, essential oil; Ext, extract; MIC, minimal inhibitory concentration.

Table 2-2 lists some molecules with demonstrated antimicrobial effect against MRSA. It can be noted that the different MICs reported are almost all in the range of micrograms per milliliter (μg/mL), but concentrations expressed in micromoles (μM) are also reported. Micromoles is considered to be the most appropriate unit for assessing molecules (because of size variations), but it is confusing to compare micromoles with concentrations expressed in mg/mL (or similar) units. To avoid confusion, it is recommended to report molecular concentrations in terms of μM. Rescaling can be then done by dividing the amount in μg/mL by the molecular weight of the substance and multiplying this result by 1,000 to obtain μM or μmol/L, which is the same.

Looking at Tables 2-1 and 2-2, we can see that the entries in the former can be more extensively compared. This is because the technologies necessary to extract and purify molecules are expensive and require specialized personnel. Although this can restrict the development of new antibiotics from natural plant sources, it is the logical way to develop most studies.

Regarding the type of molecules that can be studied, these belong to different classes but are mainly flavonoids and terpenoids.

SYNERGY BETWEEN NATURAL PRODUCTS AND ANTIBIOTICS USED FOR THE TREATMENT OF METHICILLIN-RESISTANT *STAPHYLOCOCCUS AUREUS*

Synergy is the effect of the combined action of two or more substances (in this case antimicrobials), characterized by having a greater effect than that resulting from the sum of the effects of the individual substances [49].

The synergistic effect occurs when different constituents of an extract interact to increase a specific antibacterial effect or when they act on different targets, thus increasing the possibility of affecting several vital processes in bacterial reproduction or metabolism. This effect is most striking when an antibiotic is combined with an agent that antagonizes bacterial resistance mechanisms (as in the case of β-lactamase) [50].

Herbal medicine has the particular feature of providing consumers with synergistic effects of metabolites contained within extracts and may even generate effects in different systems, thus contributing to patient healing by treating several different sets of symptoms. This diversity is also observed in the mechanisms of action

TABLE 2-2 Molecules Extracted from Plants that Have Anti-Methicillin-Resistant *Staphylococcus aureus* Activity

Plant/part	Family	Molecule	MIC	References
Rosmarinus officinalis/ aerial parts	Lamiaceae	Carnosic acid	32–64 µg/mL	[31]
		Carnosol	16 µg/mL	
		4′,7-dimethoxy-5-hydroxy-flavone	16–32 µg/mL	
		12-methoxy-trans-carnosic acid	16–64 µg/mL	
Hypericum japonicum/ aerial parts	Hypericaceae	taxifolin-7-O-α-L-rhamnopyranoside	8–64 µg/mL	[32]
		aromadendrin-7- O-α-L-rhamnopyranoside	64–128 µg/mL	
		quercetin-7-O-α-L-rhamnopyranoside	> 2048 µg/mL	
Inula hupehensis/root	Asteraceae	8-hydroxy-9,10-diisobutyloxythymol	62.3 µg/mL	[33]
Calophyllum thwaitesii/NR	Clusiaceae	calozeyloxanthone	4.1–8.1 µg/mL	[34]
Calophyllum moonii/NR		6-deoxy-γ-mangostin	0.25 mg/mL	
Punica granatum/pericarp	Lythraceae	α-lapachone	62.5 mg/L	[35]
		α-xyloidone	125–250 mg/L	
Tabebuia avellanedae/wood	Bignoniaceae	α -nor-lapachone	15.6–31.2 mg/L	
		α -nor-hydroxylapachone	15.6–62.5 mg/L	
Sophora flavescens/roots	Fabaceae	Sophoraflavanone G	2–4 µg/mL	[36]
		Kuraridin	8–16 µg/mL	
Dendrobenthamia capitata/ aerial parts	Cornaceae	Betulinic acid	62.5–125 mg/mL	[37]
Angelica dahurica/roots	Apiaceae	C17-polyacetylene falcarindiol	8–32 µg/mL	[38]
Hypericum perforatum L./ aerial parts	Hypericaceae	Phloroglucin	1 µg/mL	[39]
Momordica balsamina/ aerial parts	Cucurbitaceae	Balsaminol A	50 µM	[40]
		Balsaminol B	25 µM	
		Balsaminagenin F	100 µM	
		Balsaminoside A	50 µM	
		Karavilagenin C	200 µM	
		7β-methoxycucurbita-5, 24-diene-3β,23(R)-diol	25 µM	
Psoroma species/NR	Pannariaceae	Pannarin	8 µg/mL (MIC90)	[41]
Scutellaria baicalensis/NR	Lamiaceae	Baicalein	64–256 µg/mL	[42]

TABLE 2-2   Molecules Extracted from Plants that Have Anti-Methicillin-Resistant *Staphylococcus aureus* Activity (*cont'd*)

Plant/part	Family	Molecule	MIC	References
Podocarpus totara/leaves	Podocarpaceae	Totarol	4 µg/mL	[43]
Artemisia gilvescens/leaves	Asteraceae	Secoguaianolide sesquiterpene stereoisomer	1.95 µg/mL	[44]
Ulmus davidiana var. japonica/NR	Ulmaceae	Mansonone F	0.39−3.13 µg/mL	[45]
Sophora alopecuroides/NR	Fabaceae	Alopecurones A−C	3.13−6.25 µg/mL	[46]
Garcinia mangostana/pericarp	Clusiaceae	Rubraxanthone	0.31−1.25 µg/mL	[47]
Myrtus communis/leaves	Myrtaceae	Myrtucommulone A	0.5−2 µg/mL	[48]

MIC, minimal inhibitory concentration; NR, not reported.

that various phytochemicals have on the multi-drug-resistant bacteria. Among them are:

1. Modification of the receptor or active site (enhancing affinity to the antibiotic);
2. Enzymatic degradation and modification of bacterial enzymes that degrade antibiotics;
3. Increased permeability of the outer membrane; and
4. Inhibition of the efflux pump systems [51].

Several studies have been carried out to combat multidrug resistance in MRSA and have demonstrated the synergism between some naturally occurring molecules and chemotherapeutics to which the bacteria have become resistant (Table 2-3).

As shown in this table, there are more individual molecules than extracts studied. This may be explained by the large number of possible interactions that each of the molecules present in the extract has (some of which may exhibit antagonistic effects that mask a possible synergistic effect).

Synergism has been reported in combination with various antibiotics to which MRSA has resistance, suggesting that these antibiotics combined with the studied molecules could be used in the treatment of infections caused by this pathogen.

The strength of the studies mentioned here varies according to whether the whole extract or active molecule was tested, the number of combinations with different antibiotics, the strain tested (reference or clinical isolate), and the use of a positive control, among other features. Regarding the use of a positive control, it is important to note that these are not always available for all strains of MRSA due to the mechanism of resistance varying for each antibiotic. A possible control that could be used for molecules with a potential synergistic effect is reserpine; this alkaloid acts on resistant bacteria by inhibiting efflux pumps [55] and its use should be considered when testing selected antibiotics and MRSA strains that specifically exhibit this type of resistance.

OTHER NATURAL SOURCES WITH EFFECTS ON ANTI-METHICILLIN-RESISTANT *STAPHYLOCOCCUS AUREUS*

Although plant extracts and their active molecules are widely studied by research groups around the world, other groups study alternative methods to overcome bacteria resistant to multiple antibiotics. Nature has provided a variety of compounds with potential antimicrobial activity and these sources are currently being explored with good results. The following

TABLE 2-3 Molecules and Extracts that have Synergy With Antibiotics Against Methicillin-Resistant *Staphylococcus Aureus*

Plant	Part	Type of phytochemical	Antibiotic	Reference
Rosmarinus officinalis	Leaves	Ethanol extract	Cefuroxime	[13]
	Aerial parts	Carnosic acid	Erythromycin, tetracycline	[31]
		Carnosol	Tetracycline	
Hypericum japonicum	Aerial parts	taxifolin-7-O-α-L-rhamnopyranoside	Ampicillin, ceftazidime, levofloxacin	[32]
Zataria multiflora	Aerial parts	Essential oil	Vancomycin	[16]
Sophora flavescens	Roots	Sophoraflavanone G	Ciprofloxacin, erythromycin, gentamicin, fusidic acid	[36]
		Kuraridin	Ciprofloxacin, erythromycin, gentamicin, kanamycin, fusidic acid, oxacillin	
Artemisia herba-alba	Leaves	Methanol extract	Chloramphenicol, erythromycin, gentamicin, Penicillin G	[52]
Achillea santolina	Leaves and Flowers			
Emblica officinalis	Seed	Ethanol extract	Amoxicillin	[21]
Nymphaea odorata	Stamen			
Scutellaria baicalensis	NR	Baicalein	Ciprofloxacin, oxacillin	[42]
Ecballium elaterium	Fruit	Ethanol extract	Penicillin G	[23]
Psoroma spp.	NR	Pannarin	Gentamicin	[41]
Podocarpus totara	Leaves	Totarol	Methicillin	[43]
Stephania tetrandra	NR	Tetrandrine	Ampicillin, azithromycin, cefazolin, levofloxacin	[53]
		Demethyltetrandrine		
Turnera ulmifolia	Leaves	Ethanol extract	Gentamicin, kanamycin	[54]
Sophora alopecuroides	NR	Alopecurones A–C	Erythromycin, gentamicin, methicillin	[46]
Garcinia mangostana	Pericarp	Rubraxanthone	Vancomycin	[47]

NR, no report.

sections provide a brief description of some natural alternatives to plant products that can be used against MRSA.

Small Peptides

It is known that some peptides have antimicrobial effects; these are distributed in many living organisms throughout the planet. One peptide that has given good results both *in vitro* and *in vivo* is the omiganan pentahydrochloride (MBI-226), a cationic peptide composed of 12 amino acids that interacts with the cytoplasmic membrane of Gram-positive and Gram-negative bacteria, causing changes in its polarity with subsequent destruction of the bacteria [56]. This peptide is effective against MRSA at an MIC of ≤ 64 mg/mL [57].

Recently, some studies using peptides generated *de novo* have given good results; for example, in 2010 Lee and colleagues reported that a peptide of 11 residues had an MIC of 5 µg/mL against MRSA, which makes it more effective than omiganan *in vitro* [58].

Molecules Derived from Animals

Some chemically modified molecules produced by animals are utilized as antimicrobials; such is the case for chitosan (chitin polysaccharide derived by *N*-deacetylation), which has a synergistic effect with β-lactams (ampicillin, oxacillin, and penicillin). The anti-MRSA activities of two deacetylated forms (aminoethyl-chitosan 90 and 50) have been studied. Results of this study showed synergy between aminoethyl-chitosan 50 and the three antibiotics mentioned above; these synergistic combinations dramatically decreased the MICs of the antibiotics [59].

Bacteriocins

Bacteriocins are antimicrobial compounds derived from bacteria that are generated as a competitive mechanism (antagonism) against other bacteria. An example of this type of bacterium is *Pseudoalteromonas phenolica* O-BC30T, which produces a bacteriocin chemically defined as 2,2′,3-tribromobiphenyl-4,4′-dicarboxylic acid that inhibits clinical isolates of MRSA with MICs in the range of 1−4 µg/mL [60].

Lactobacilli are also known for their ability to produce such substances; *L. acidophilus* and *L. casei* were reported to generate lactic acid derivatives that act as bacteriocins, thus inhibiting the growth of MRSA by direct antagonism [61].

Nucleotides

There is such a diversity of molecules with potential antimicrobial activity that even the *letters of the genetic code* (i.e., nucleotides) have been shown to possess antimicrobial functions. A study in mice with burns infected with MRSA demonstrated that interleukin-10 (Il-10) antisense oligodeoxynucleotides have the capacity to exert a protective role against this microorganism *in situ* by acting as a mediator of abscess formation, which limits the infection and makes it impossible for the bacteria to migrate to the blood and hence to produce sepsis [62].

Nanoparticles

Since nanotechnology was established as a field of study with various applications in science and technology, its use has been increasing dramatically and some years ago nanoparticles began to be studied as potential antimicrobial agents. Research into this application has been increasing and it now has great potential to become a viable alternative to traditional antimicrobials. For example, studies developed by Nanda and colleagues [63] tested the activity of silver (Ag) nanoparticles against MRSA, with favorable results.

Another report described the use of polyacrylate nanoparticles covalently bound to penicillin, which prevent β-lactamases from inactivating penicillin, thus maintaining the activity of this antibiotic [64].

As described in previous sections, research into different therapeutic alternatives to traditional antibiotics is being carried out using a range of different sources and technologies, which comprise an important advance in the fight against multidrug resistance exhibited by some bacterial pathogens such as MRSA.

CONCLUSIONS

As has been said throughout this chapter, multiple resistance to antibiotics is a growing public health problem; more specifically, MRSA represents a high risk of infection in both hospital and community environments.

Since the 1980s, the production of new antibiotics has been declining such that 25 years later it has been reduced by two-thirds [65], suggesting that we are currently underestimating the ability of bacteria to generate mechanisms of resistance. The versatility of these microorganisms puts us at a disadvantage; thus, research into new sources of antibiotics should be a priority worldwide.

Following great efforts, different lines of research are now consolidating to try to solve this problem, and it is evident that several plants contain compounds that either individually or combined with known antibiotics are effective against multidrug-resistant bacteria like MRSA.

There is a need to identify alternative therapies against bacterial resistance and it is evident that compounds with potential antimicrobial activities exist in Nature. Evidence for this is provided by the large number of phytochemicals (extracts and EOs) and isolated molecules with antimicrobial activity that act in many different ways to combat MRSA and other multidrug-resistant bacteria.

The diversity of globally distributed flora provides the investigation of natural extracts with an almost infinite range of possibilities. Nature will surely give us the necessary substrates, along with suitable tools and knowledge, to contribute to the fight against pathogens with multiple antibiotic resistance.

References

[1] Rammelkamp CH, Maxon T. Resistance of Staphylococcus aureus to the action of penicillin. Proc Soc Exp Biol Med 1942;51:4.

[2] Barber M. Methicillin-resistant staphylococci. J Clin Pathol 1961;14:385–93.

[3] Beck WD, Berger-Bächi B, Kayser FH. Additional DNA in methicillin-resistant Staphylococcus aureus and molecular cloning of mec-specific DNA. J Bacteriol 1986;165:373–8.

[4] Brown DFJ, Reynolds PE. Intrinsic resistance to β-lactam antibiotics in Staphylococcus aureus. FEBS Letters 1980;122:275–8.

[5] Graninger W, Leitha T, Breyer S, Francesconi M, Lenz K, Georgopoulos A. Methicillin- and gentamicin-resistant Staphylococcus aureus: susceptibility to fosfomycin, cefamandole, N-formimidoyl-thienamycin, clindamycin, fusidic acid and vancomycin. Drugs Exp Clin Res 1985;11:23–7.

[6] Hiramatsu K, Aritaka N, Hanaki H, Kawasaki S, Hosoda Y, Hori S, et al. Disseminations in Japanese hospitals of strains of Staphylococcus aureus heterogenously resistant to vancomycin. Lancet 1997;350: 1670–3.

[7] Centers for Disease Control. Staphylococcus aureus resistant to vancomycin, United States. Morb Mortal Wkly Rep 2002;51:565–7.

[8] Goulda IM, David MZ, Esposito S, Garau J, Linae G, Mazzei T, et al. New insights into methicillin-resistant Staphylococcus aureus (MRSA) pathogenesis, treatment and resistance. Int J Antimicrob Agents 2012;39: 96–104.

[9] Cosa P, Vlietinck AJ, Vanden Berghe D, Maesa L. Anti-infective potential of natural products: How to develop a stronger in vitro proof-of-concept. J Ethnopharmacol 2006;106:290–302.

[10] Ríos JL, Recio MC. Medicinal plants and antimicrobial activity. J Ethnopharmacol 2005;100:80–4.

[11] Rios JL, Recio MC, Villar A. Screening methods for natural products with antimicrobial activity: A review of the literature. J Ethnopharmacol 1988;23:127–49.

[12] Pesewu GA, Cutler RR, Humber DP. Antibacterial activity of plants used in traditional medicines of Ghana with particular reference to MRSA. J Ethnopharmacol 2008;116:102–11.

[13] Jarrar N, Abu-Hijleh A, Adwan K. Antibacterial activity of Rosmarinus officinalis L. alone and in combination with cefuroxime against methicillin-resistant Staphylococcus aureus. Asian Pac J Trop Med 2010;3:121–3.

[14] Tohidpour A, Sattari M, Omidbaigi R, Yadegar A, Nazemi J. Antibacterial effect of essential oils from two medicinal plants against Methicillin-resistant Staphylococcus aureus (MRSA). Phytomedicine 2010; 17:142–5.

[15] Demirci B, Kosar M, Demirci F, Dinc M, Baser KHC. Antimicrobial and antioxidant activities of the essential oil of Chaerophyllum libanoticum Boiss. et Kotschy. Food Chem. 2007;105:1512–7.

[16] Mahboubi M, Ghazian Bidgoli F. Antistaphylococcal activity of Zataria multiflora essential oil and its synergy with vancomycin. Phytomedicine 2010;17: 548–50.

[17] Buenz E, Bauer BA, Schnepple DJ, Wahner-Roedler DL, Vandell AG, Howe CL. A randomized Phase I study of Atuna racemosa: A potential new

anti-MRSA natural product extract. J Ethnopharmacol 2007;114:371−6.

[18] Dung NT, Kim JM, Kang SC. Chemical composition, antimicrobial and antioxidant activities of the essential oil and the ethanol extract of Cleistocalyx operculatus (Roxb.) Merr and Perry buds. Food Chem Toxicol 2008;46:3632−9.

[19] Machado TB, Pinto AV, M.C.F.R., P., Leal ICR, Silva MG, et al. In vitro activity of Brazilian medicinal plants, naturally occurring naphthoquinones and their analogues, against methicillin-resistant Staphylococcus aureus. Int J Antimicrob Agents 2003; 21:279−84.

[20] Zuo GY, Wang GC, Zhao YB, Xua GL, Haob XY, Hanc J, et al. Screening of Chinese medicinal plants for inhibition against clinical isolates of methicillin-resistant Staphylococcus aureus (MRSA). J Ethnopharmacol 2008;120:287−90.

[21] Mandal S, DebMandal M, Pal NK, Saha K. Synergistic anti-Staphylococcus aureus activity of amoxicillin in combination with Emblica officinalis and Nymphae odorata extracts. Asian Pac. J Trop Med 2010;3:711−4.

[22] Reichling J, Weseler A, Saller R. A current review of the antimicrobial activity of Hypericum perforatum L. Pharmacopsychiatry 2001;34(S1):116−8.

[23] Adwan G, Salameh Y, Adwan K. Effect of ethanolic extract of Ecballium elaterium against Staphylococcus aureus and Candida albicans. Asian Pac J Trop Biomed 2011;1:456−60.

[24] Chandrasekaran M, Kannathasan K, Venkatesalu V, Prabhakar K. Antibacterial activity of some salt marsh halophytes and mangrove plants against methicillin resistant Staphylococcus aureus. World J Microbiol Biotechnol 2009;25:155−60.

[25] Sudjana AN, D'Orazio C, Ryan V, Rasoold N, Ng J, Islam N, et al. Antimicrobial activity of commercial Olea europaea (olive) leaf extract. Int J Antimicrob Agents 2009;33:461−3.

[26] Demirci B, Kosar M, Demirci F, Dinc M, Baser KHC. Antimicrobial and antioxidant activities of the essential oil of Chaerophyllum libanoticum Boiss. et Kotschy. Food Chem 2007;105:1512−7.

[27] Hasson SS, Al-Balushi MS, Sallam TA, Idris MA, Habbal O, Al-Jabri AA. In vitro antibacterial activity of three medicinal plants-Boswellia (Luban) species. Asian Pac J Trop Biomed 2011;1:178−82.

[28] Mann CM, Markham JL. A new method for determining the minimum inhibitory concentration of essential oils. J Appl Microbiol 1998;84:538−44.

[29] Dryden MS, Dailly S, Crouch M. A randomized, controlled trial of tea tree topical preparations versus a standard topical regimen for the clearance of MRSA colonization. J Hosp Infec 2004;56:283−6.

[30] Houghton P, Mukherjee PK. Evaluation of herbal medicinal products. London: Pharmaceutical Press; 2009.

[31] Oluwatuyi M, Kaatz G, Gibbons S. Antibacterial and resistance modifying activity of Rosmarinus officinalis. Phytochemistry 2004;65:3249−54.

[32] An J, Zuo GY, Hao XY, Wang GC, Li ZS. Antibacterial and synergy of a flavanonol rhamnoside with antibiotics against clinical isolates of methicillin-resistant Staphylococcus aureus (MRSA). Phytomedicine 2011; 18:990−8.

[33] An J, Liu YLQ, Gao K. Antimicrobial activities of some thymol derivatives from the roots of Inula hupehensis. Food Chem. 2010;120:512−6.

[34] Dharmaratne HRW, Wijesinghe WMNM, Thevanasem V. Antimicrobial activity of xanthones from Calophyllum species, against methicillin-resistant Staphylococcus aureus (MRSA). J Ethnopharmacol 1999;66:339−42.

[35] Machado TB, Pinto AV, Pinto MCF, Leal ICR, Silva MG, Amaral ACF, et al. In vitro activity of Brazilian medicinal plants, naturally occurring naphthoquinones and their analogues, against methicillin-resistant Staphylococcus aureus. Int J Antimicrob Agents 2003;21:279−84.

[36] Chan B, Yu H, Wong C, Lui S, Jolivalt C, Ganem-Elbaz C, et al. Quick identification of kuraridin, a noncytotoxic anti-MRSA (methicillin-resistant Staphylococcus aureus) agent from Sophora flavescens using high-speed counter-current chromatography. J Chromatogr B Analyt Technol Biomed Life Sci 2012;880:157−62.

[37] Zuo GY, Wang GC, Zhao YB, Xu GL, Hao XY, Han J, et al. Screening of Chinese medicinal plants for inhibition against clinical isolates of methicillin-resistant Staphylococcus aureus (MRSA). J Ethnopharmacol 2008;120:287−90.

[38] Lechner D, Stavria M, Oluwatuyia M, Pereda-Mirandab R, Gibbons S. The anti-staphylococcal activity of Angelica dahurica (Bai Zhi). Phytochemistry 2004;65:331−5.

[39] Reichling, Weseler A, R S. A current review of the antimicrobial activity of Hypericum perforatum L. Pharmacopsychiatry 2001;34:S116−8.

[40] Ramalhete C, Spengler G, Martins A, Martins M, Viveiros M, Mulhovoc S, et al. Inhibition of efflux pumps in meticillin-resistant Staphylococcus aureus and Enterococcus faecalis resistant strains by triterpenoids from Momordica balsamina. Int J Antimicrob Agents 2011;37:70−4.

[41] Celenza G, Segatore B, Setacci D, Bellio P, Brisdelli F, Piovano M, et al. In vitro antimicrobial activity of pannarin alone and in combination with antibiotics

against methicillin-resistant Staphylococcus aureus clinical isolates. Phytomedicine 2012;19:596−602.

[42] Chan BCL, Ip M, Lau CBS, Lui SL, Jolivalt C, Ganem-Elbaz C, et al. Synergistic effects of baicalein with ciprofloxacin against NorA over-expressed methicillin-resistant Staphylococcus aureus (MRSA) and inhibition of MRSA pyruvate kinase. J Ethnopharmacol 2011;137:767−73.

[43] Nicolson K, Evans G, O'Toole PW. Potentiation of methicillin activity against methicillin-resistant Staphylococcus aureus by diterpenes. FEMS Microbiol Lett 1999;179:233−9.

[44] Kawazoe K, Tsubouchi Y, Abdullah N, Takaishi Y, Shibata H, Higuti T, et al. Sesquiterpenoids from Artemisia gilvescens and an Anti-MRSA Compound. J Nat Prod 2003;66:538−9.

[45] Shin DY, Kim HS, Min KH, Hyun SS, Kim SA, Huh H, et al. Isolation of a potent anti-MRSA sesquiterpenoid quinone from Ulmus davidiana var. japonica. Chem Pharm Bull 2000;48:1805−6.

[46] Sato M, Tsuchiya H, Miyazaki T, Ohyama M, Tanaka T, Iinuma M. Antibacterial activity of flavanostilbenes against methicillin-resistant Staphylococcus aureus. Lett Appl Microbiol 1995;21:219−22.

[47] Iinuma M, Tosa H, Tanaka T, Asai F, Kobayashi Y, Shimano R, et al. Antibacterial activity of xanthones from guttiferaeous plants against methicillin-resistant Staphylococcus aureus. J Pharm Pharmacol 1996;48:861−5.

[48] Appendino G, Bianchi F, Minassi A, Sterner O, Ballero M, Gibbons S. Oligomeric acylphloroglucinols from myrtle (Myrtus communis). J Nat Prod 2002;65:334−8.

[49] Acar JF. Antibiotic sinergy and antagonism. Med Clin North Am 2000;84:1391−406.

[50] Wagner H, Ulrich-Merzenich G. Synergy research: Approaching a new generation of phytopharmaceuticals. Phytomedicine 2000;16:97−110.

[51] Hemaiswarya S, Kruthiventi AK, Doble M. Synergism between natural products and antibiotics against infectious diseases. Phytomedicine 2008;15:639−52.

[52] Darwish RM, Aburjai T, Al-Khalil S, Mahafzah A. Screening of antibiotic resistant inhibitors from local plant materials against two different strains of Staphylococcus aureus. J Ethnopharmacol 2002;79:359−64.

[53] Zuo GY, Li Y, Wang T, Han J, Wang GC, Zhang YL, et al. Synergistic antibacterial and antibiotic effects of bisbenzylisoquinoline alkaloids on clinical isolates of methicillin-resistant Staphylococcus aureus (MRSA). Molecules 2011;16:9819−26.

[54] Coutinho D, Costa G, Lima E, Falcão-Silva V, Siqueira J. Herbal therapy associated with antibiotic therapy: potentiation of the antibiotic activity against methicillin-resistant Staphylococcus aureus by Turnera ulmifolia L. BMC Complement Altern Med 2009;9:13.

[55] Schmitz FJ, Fluit AC, Lückefahr M, Engler B, Hofmann B, Verhoef J, et al. The effect of reserpine, an inhibitor of multidrug efflux pumps, on the in-vitro activities of ciprofloxacin, sparfloxacin and moxifloxacin against clinical isolates of Staphylococcus aureus. J Antimicrob Chemother 1998; 42:807−10.

[56] Isaacson RE. MBI-226. Micrologix/Fujisawa. Curr Opin Investig Drugs 2003;4:999−1003.

[57] Fritsche TR, Rhomberg PR, Sader HS, Jones RN. In vitro activity of omiganan pentahydrochloride tested against vancomycin-tolerant, intermediate, and resistant Staphylococcus aureus. Diagn Microbiol Infect Dis 2008;60:399−403.

[58] Lee SH, Kim S, Lee Y, Song M, Kim I, Won H. De novo generation of short antimicrobial peptides with simple amino acid composition. Regul Peptides 2011; 166:36−44.

[59] Dae-Sung L, Young-Mog K, Myung-Suk L, Chang-Bum A, Won-Kyo J, Jae-Young J. Synergistic effects between aminoethyl-chitosans and b-lactams against methicillin-resistant Staphylococcus aureus (MRSA). Bioorg Medicinal Chem Lett 2010;20:975−8.

[60] Isnansetyo A, Kamei Y. Anti-methicillin-resistant Staphylococcus aureus (MRSA) activity of MC21-B, an antibacterial compound produced by the marine bacterium Pseudoalteromonas phenolica O-BC30T. Int J Antimicrob Agents 2009;34:131−5.

[61] Karska-Wysocki B, Bazo M, Smoragiewicz W. Antibacterial activity of Lactobacillus acidophilus and Lactobacillus casei against methicillin-resistant Staphylococcus aureus (MRSA). Microbiol Res 2010;165: 674−86.

[62] Asai A, Kogiso M, Kobayashi M, Herndon DN, Suzuki F. Effect of IL-10 antisense gene therapy in severely burned mice intradermally infected with MRSA. Immunobiology 2012;217:711−8.

[63] Nanda A, Saravanan M, Phil M. Biosynthesis of silver nanoparticles from Staphylococcus aureus and its antimicrobial activity against MRSA and MRSE. Nanomedicine 2009;5:452−6.

[64] Turos E, Reddy GSK, Greenhalgh K, Ramaraju P, Abeylath SC, Jang S, et al. Penicillin-bound polyacrylate nanoparticles: Restoring the activity of b-lactam antibiotics against MRSA. Bioorg Medicinal Chem Lett 2007;17:3468−72.

[65] Taubes G. The bacteria fight back. Science 2008;321: 356−61.

Bioactivity of Plant Constituents against Vancomycin-Resistant Enterococci

Victor Kuete

Department of Biochemistry, Faculty of Science, University of Dschang,
P.O. Box 67 Dschang, Cameroon

INTRODUCTION

Enterococci are a common cause of nosocomial infections, particularly affecting the urinary tract, wounds, and soft tissues [1]. The appearance of enterococcal resistance to glycopeptide agents in 1988 made treatment of such infections much more difficult [2–4]. Because of their remarkable ability to acquire antibiotic resistance, enterococci have become increasingly recognized as important nosocomial pathogens playing a causal role in 12% of all hospital-acquired infections in the United States [5]. Resistance to many antibiotics including β-lactams, aminoglycosides, lincosamides (e.g., clindamycin), and trimethoprim-sulfamethoxazole has been reported for enterococci [6,7]. Vancomycin is a glycopeptide antibiotic used in the prophylaxis and treatment of infections caused by Gram-positive bacteria [6]. Since 1988, many vancomycin-resistant enterococci

(VRE) involving *Enterococcus faecalis* and *Enterococcus faecium* strains have been reported in the USA and Europe [5,8,9] and by 1991 large numbers of them were also found to be resistant to penicillins and aminoglycosides [10]. Vancomycin-resistant *E. faecium* was associated was estimated to be the cause of 4% of healthcare-associated infections reported in the USA from January 2006 to October 2007 [11]. Genes associated with vancomycin resistance exhibited by enterococci can be divided into six types: Van-A (resistant to vancomycin and teicoplanin), Van-B (resistant to vancomycin but sensitive to teicoplanin), Van-C (partly resistant to vancomycin and sensitive to teicoplanin), Van-D, Van-E, and Van-F (12).

The development of vancomycin resistance has made these organisms even more difficult, and in many cases impossible, to treat. New antibiotics, including daptomycin, linezolid, oritavancin, quinupristin, and tigecycline, have

Fighting Multidrug Resistance with Herbal Extracts,
Essential Oils and Their Components
http://dx.doi.org/10.1016/B978-0-12-398539-2.00003-3

been developed against VRE. However, resistance to linezolid has already been detected [13]. Therefore, the search for new antimicrobial compounds active against VRE is of great relevance. Further development of new anti-VRE agents with fewer side effects is also needed.

MEDICINAL PLANT METABOLITES USED IN THE FIGHT AGAINST VANCOMYCIN-RESISTANT ENTEROCOCCI

Medicinal plants that have been used for a long time may be good sources of safe antibacterial agents. The plant defense system against pathogenic microorganisms comprises a panel of bioactive constituents [14,15]. For novel compounds, the stringent endpoint criteria for antimicrobial activity have been defined as following: significant activity [minimal inhibitory concentration (MIC) of $< 10\,\mu g/mL$]; moderate activity $(10 < MIC \leq 100\,\mu g/mL)$; and low or negligible activity (MIC of $> 100\,\mu g/mL$) [14,15]. These criteria will be used in this chapter to evaluate the activity of natural compounds reported against VRE. There are scientific reports available on natural compounds used against VRE and this chapter will describe the current status of the search for plant products active against resistant bacteria.

Natural Terpenoids Active against Vancomycin-Resistant Enterococci

Terpenoids (or isoprenoids) are a large and diverse class of naturally occurring organic chemical compounds found in all classes of living things that can be assembled and modified in thousands of ways; they form the largest group of natural products [16]. Most are multicyclic structures that differ from one another not only in their functional groups but also in their basic carbon skeletons. Owing to their

aromatic properties, plant terpenoids are used extensively. They play a pivotal role in oriental herbal medicines and are currently under investigation for their antimicrobial activities. Terpenoids contribute to the scent of the cinnamon, clove, eucalyptus, and ginger, the yellow color of sunflowers, and the red color of tomatoes [16]. Some common terpenoids are camphor, citral, menthol, and salvinorin A found in the plant *Salvia divinorum*, and the cannabinoids formed in *Cannabis* spp.

Some terpenoids can significantly inhibit the growth of VRE (Fig. 3-1). Abietane diterpenes isolated from *Plectranthus grandidentatus* and *Plectranthus hereroensis* acetone extracts were found to be active against VRE [17]. The reported MIC values are $15.63\,\mu g/mL$ for 7α-acetoxy-6β-hydroxyroyleanone (compound 1, Fig. 3-1) and horminone (compound 2, Fig. 3-1) [17], and $31.25\,\mu g/mL$ for coleon U (compound 3, Fig. 3-1), 16-acetoxy-7α,12-dihydroxy-8,12-abietadiene-11,14-dione (compound 4, Fig. 3-1), and 3α-acetoxy-3α,7β,12-trihydroxy-17(15−16),18(4−3)-bisabeo-abieta-4[19],8,12,16-tetraene-11,14-dione (compound 5, Fig. 3-1). MIC values of $62.50\,\mu g/mL$ have also been reported for 6β,7α-dihydroxyroyleanone (compound 6, Fig. 3-1) and 7α,12-dihydroxy-17(15−16)-abeoabieta-8,12,16-triene-11,14-dione (compound 7, Fig. 3-1) [17]. Under similar experimental conditions, these terpenoids were more active than vancomycin (MIC of $250\,\mu g/mL$) [17]. Poor activities against vancomycin-resistant *E. faecalis* and *E. faecium* (MICs of $1-4\,mg/mL$) were also obtained for aromadendrene and globulol isolated from the essential oil of *Eucalyptus globulus* fruit [18]. Another abietane diterpenoid is parvifloron D (compound 8, Fig. 3-1), an antibacterial metabolite from *Plectranthus ecklonii* [19], which exhibits significant activity against low-level vancomycin-resistant *E. faecalis* (MIC of $7.81\,\mu g/mL$) [20]. A triterpenoid obtained from the aerial parts of the Argentinean plant *Caiophora coronata*, 1β,3β-dihydroxyurs-12-en-27-oic acid (compound

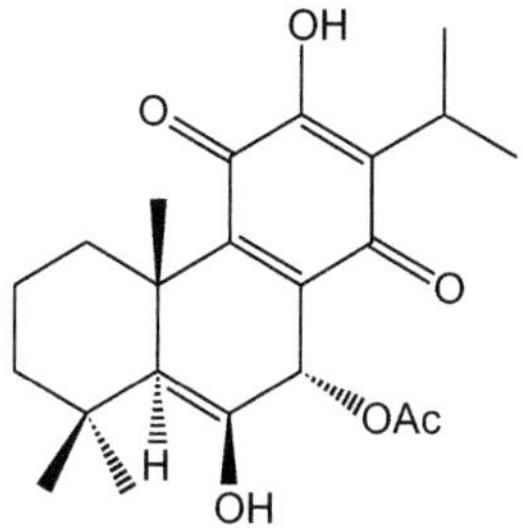

FIGURE 3-1 Plant terpenoids active against vancomycin-resistant enterococci.

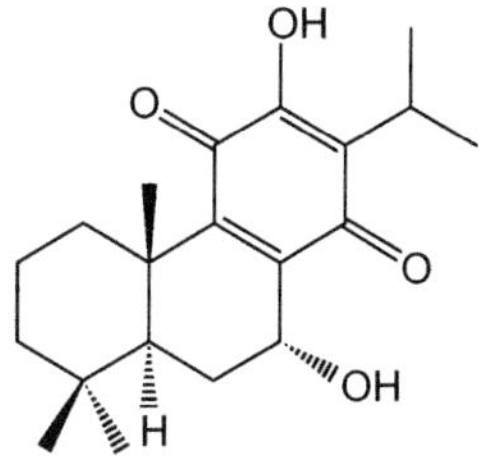

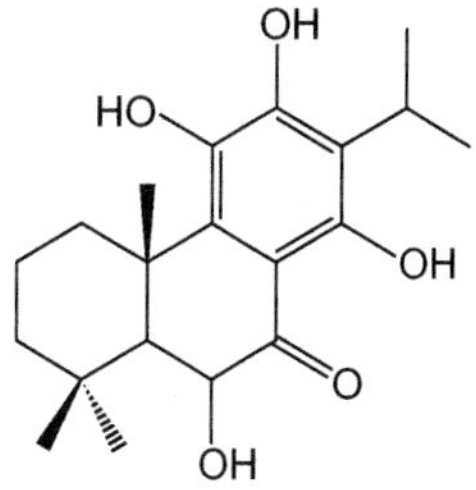

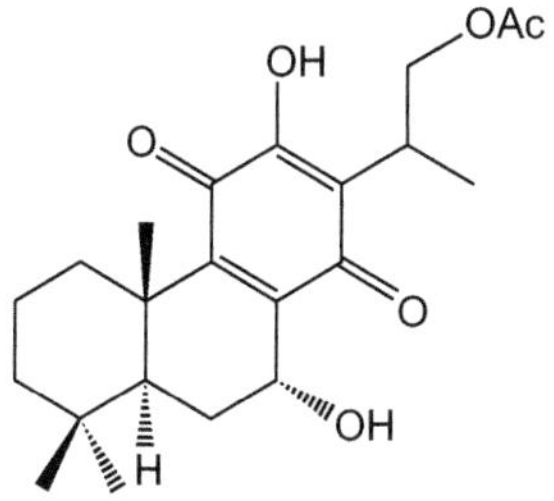

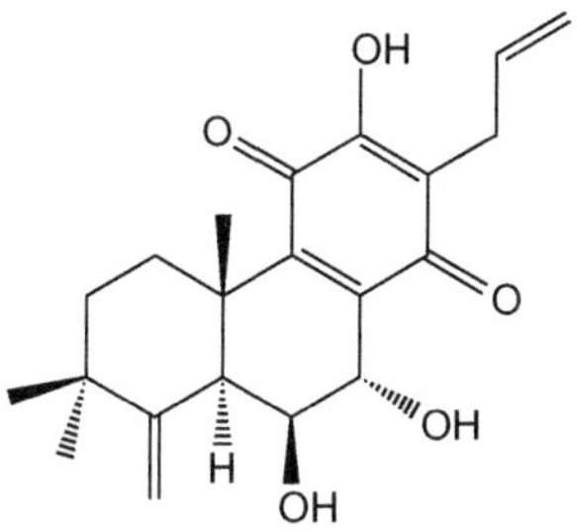

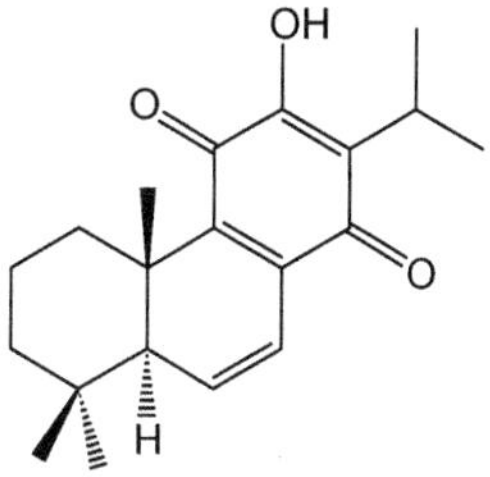

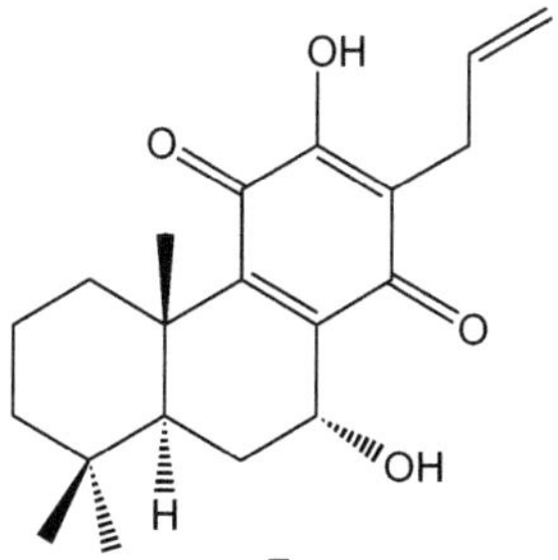

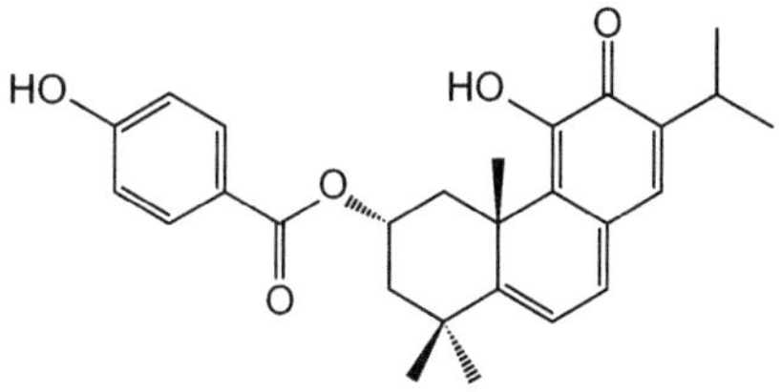

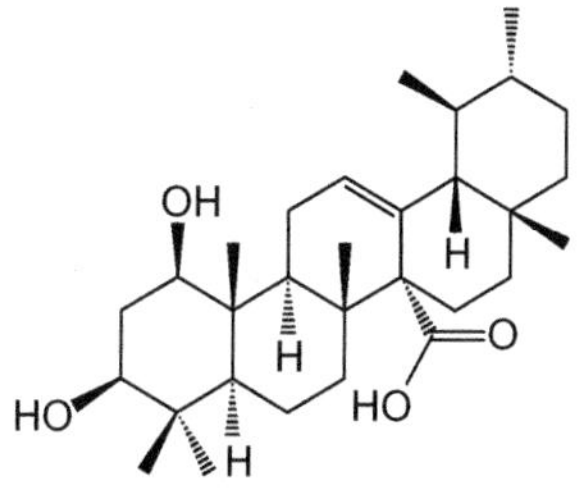

9, Fig. 3-1), significantly inhibits the growth of vancomycin-resistant *E. faecalis* (MIC of 4 μg/mL) [21]. Further, two taxane diterpenoids, demethylfruticuline A and fruticuline A, which are components of the exudate produced by the aerial parts of *Salvia corrugata*, are moderately active against many strains of vancomycin-resistant *E. faecalis* (MIC of 64 μg/mL) [22].

Natural Phenolic Compounds Active against Vancomycin-Resistant Enterococci

Plant phenolics are small molecules containing one or more phenolic groups and can be classified as simple phenols (monophenols); which contain only one phenolic group, or di-

(bi), tri- and oligophenols, which contain two, three, or many phenolic groups, respectively. They are the most widely distributed class of plant secondary metabolites, with several thousand different compounds identified so far [23]. As they are also present in food, they may have an impact on health. Most are known to have interesting biologic properties [15] and some have been shown to be active against VRE.

Some simple phenolics, phenolic acids, and phenylpropanoids reported to exhibit bioactivity against vancomycin-resistant *E. faecium* include brazilin and methyl gallate (MICs of 16 μg/mL), caffeic acid, gallic acid, and phloroglucinol (MICs of 32 μg/mL), and 4-hydroxycinnamic acid (MIC of 64 μg/mL) [24].

The coumarin, 5-methoxypsoralen, has moderate activity against vancomycin-resistant *E. faecium* [24]. In addition, several coumarins isolated from *Aegle marmelos*, such as 8-hydroxysmyrindiol, isogosferol, scoparone, xanthotoxin, and xanthotoxol, can inhibit the growth of vancomycin-resistant *E. faecalis in vitro* [25]. However, these compounds have rather low activities (MICs of 300 μg/mL) and are less active than vancomycin [25].

Several xanthones and neolignans significantly inhibit the growth of VRE *in vitro* (Fig. 3-2). α-Mangostin (compound 10, Fig. 3-2), a xanthone isolated from the stem bark of *Garcinia mangostana*, was found to be significantly active against various strains of vancomycin-resistant *E. faecalis* and *E. faecium* (MICs of 6.25 μg/mL) [26]. β-Mangostin also exerts moderate inhibitory growth effects on VRE strains [26]. Many other xanthones isolated from the roots of the Chinese *Cudrania cochinchinensis*, including gerontoxanthone H (or cudraxanthone H; compound 11, Fig. 3-2), gerontoxanthone I (compound 12, Fig. 3-2), alvaxanthone (compound 13, Fig. 3-2), isoalvaxanthone (compound 14, Fig. 3-2), and 1,3,7-trihydroxy-2-prenylxanthone (compound 15, Fig. 3-2), exhibit significant inhibitory activities against vancomycin-resistant *E. faecalis*, *E. faecium*, and *E. gallinarum*, with MIC values varying between 1.56 and 6.25 μg/mL against most strains tested [27]. Other xanthones isolated from this plant, including gerontoxanthone G (compound 16, Fig. 3-2), toxyloxanthone C (compound 17, Fig. 3-2), and cudraxanthone S (compound 18, Fig. 3-2), also exhibit good activities, with MICs of 6.25–25 μg/mL [27].

Neolignans form a class of phenylpropanoid dimers in which the phenylpropane units are linked head-to-tail instead of tail-to-tail, as found in lignans. Some neolignans isolated from the methanol extract of *Magnolia officinalis*, such as piperitylmagnolol (compound 19, Fig. 3-2) (MIC of 6.25 μg/mL), magnolol (MIC of 25 μg/mL), and honokiol (MIC of 25 μg/mL), were found to be active against VRE [28]. Under similar experimental conditions, piperitylmagnolol was found to be more active than vancomycin (MIC of > 100 μg/mL) and oxacillin (MIC of > 100 μg/mL) and as active as chloramphenicol (MIC of 6.25 μg/mL), highlighting its potential as an effective antibacterial against VRE [28].

One of the largest classes of naturally occurring polyphenolic compounds are the flavonoids. Flavonoids are polyphenolic compounds that occur commonly in nature and can be differentiated according to their chemical structures, into anthocyanins, catechins, chalcones, flavanones, flavones, flavonols, and isoflavones [29], and can be divided into three main classes

(1) Flavonoids, derived from 2-phenylchromen-4-one (2-phenyl-1,4-benzopyrone) structure (e.g., quercetin and rutin);

(2) Isoflavonoids, derived from 3-phenylchromen-4-one (3-phenyl-1,4-benzopyrone) structure; and

(3) Neoflavonoids, derived from 4-phenylcoumarine (4-phenyl-1,2-benzopyrone).

There are approximately 4000 flavonoids in nature, many of which are present in fruits, vegetables, and beverages (tea, coffee, beer, wine, and fruit drinks) [29]. Recently, there has been increasing interest in flavonoids because

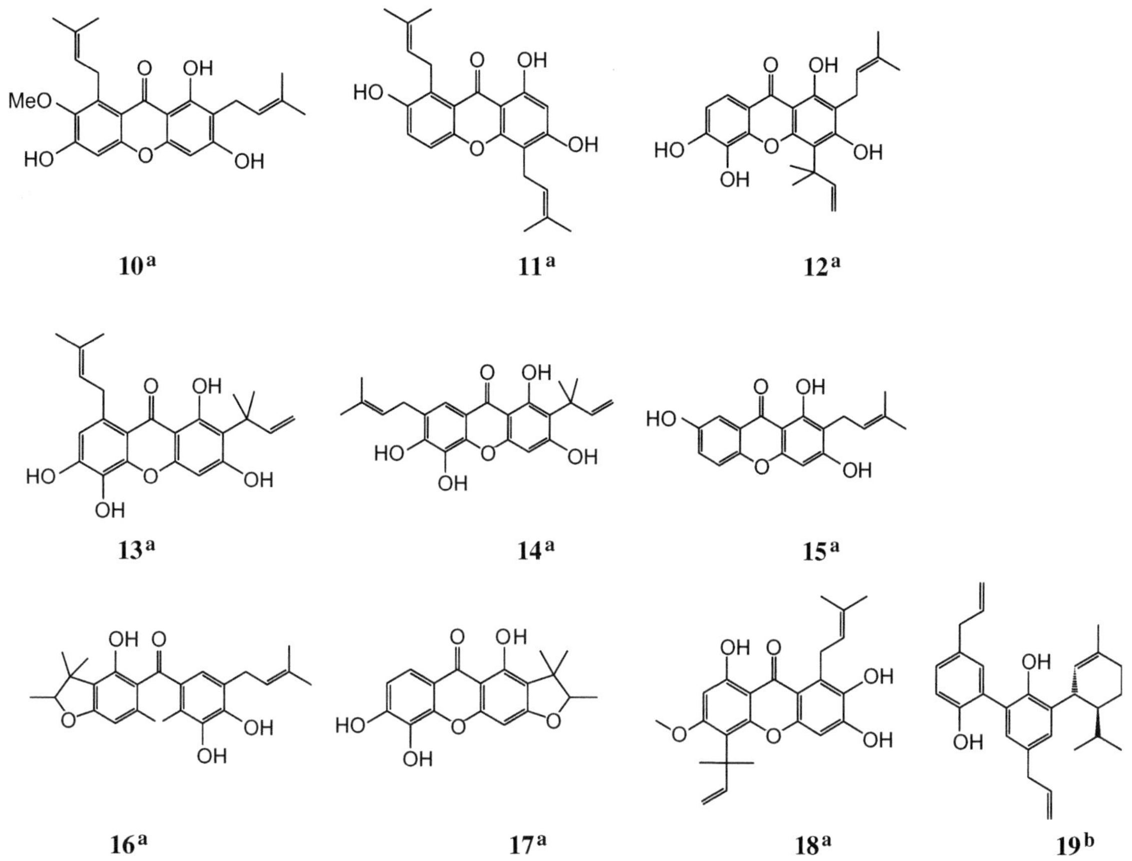

FIGURE 3-2 **Plants xanthones and neolignans with significant activity against vancomycin-resistant enterococci.** Significant activity is defined as an MIC of $< 10\,\mu g/mL$. [a], xanthones; [b], neolignans.

of their potential benefits to humans. These flavonoids possess antiallergic, anti-inflammatory, antioxidant, antiplatelet, antitumor, and antiviral activities [29]. Several flavonoids are reported to exhibit activities against VRE, some of which are significant (Fig. 3-3). The isoflavonoids erybraedin A (compound 20, Fig. 3-3; MIC of $1.53-3.13\,\mu g/mL$) and eryzerins A (MIC of $12.5-25\,\mu g/mL$), C (compound 21, Fig. 3-3; MIC of $6.25\,\mu g/mL$), D (MIC of $12.5\,\mu g/mL$), and E (MIC of $12.5\,\mu g/mL$) isolated from *Erythrina zeyheri* were found to be active against various strains of VRE [30]. Of these isoflavonoids, erybraedin A and eryzerin C were found to be more active than vancomycin (MICs of $12.5-100\,\mu g/mL$) under similar experimental conditions [30]. In addition to the growth inhibitory potency of erybraedin A and eryzerin C, these compounds were reported to show synergistic effects with vancomycin against VRE [30]. Kurarinone (compound 22, Fig. 3-3) (MIC of $2\,\mu g/mL$), a flavonoid isolated from *Sophora flavescens*, also exhibited greater activity against VRE than ampicillin (MIC of $250\,\mu g/mL$) or vancomycin (MIC of $150\,\mu g/mL$) [31].

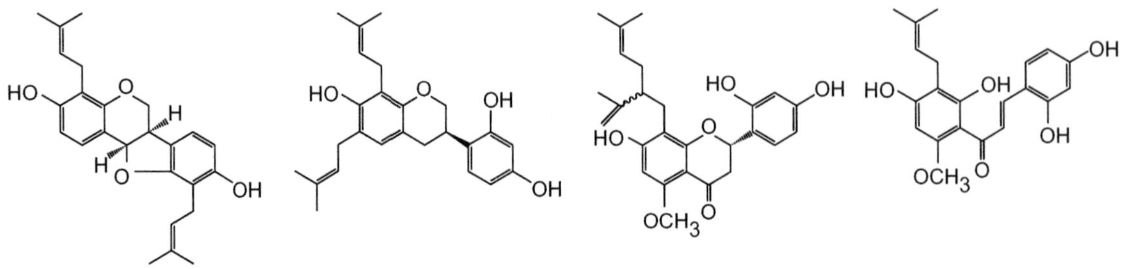

FIGURE 3-3 **Plant flavonoids with significant activity against vancomycin-resistant enterococci.** Significant activity is defined as an MIC of < 10 μg/mL.

When combinations of vancomycin and a panel of flavonoids were tested against VRE, the addition of 12.5 μg/mL of 3,5,7-trihydroxyflavone (galangin) or 6.2 μg/mL of 3,7-dihydroxyflavone was found to affect resistant strains of *E. faecium* [32] and *E. faecalis* so severely that the MICs for vancomycin were reduced to values characteristic of vanco-mycin-sensitive strains (from > 250 μg/mL to < 4 μg/mL) [32]. The compound 7,9,2',4'-tetrahydroxy-8-isopentenyl-5-methoxychalcone (compound 23, Fig. 3-3), isolated from the roots of *Sophora flavescens*, was found to be active against VRE, either alone or in combination with ampicillin or gentamicin. Moreover, the MIC values of 7,9,2',4'-tetrahydroxy-8-isopen-tenyl-5-methoxychalcone ranged from 7.8 to 15 μg/mL and in combination with ampicillin and gentamicin yielded a fractional inhibitory concentration ranging from 0.188 to 0.375 μg/mL, which indicates a synergistic effect [33].

Natural Alkaloids Active against Vancomycin-Resistant Enterococci

Alkaloids are a group of naturally occurring chemical compounds that contain mostly basic nitrogen atoms. Some related compounds with neutral or even weakly acidic properties also belong to this group [34]. In addition, some synthetic compounds with similar structures are also classified as alkaloids [35]. Alkaloid-containing plants have been used by humans since ancient times for therapeutic and recrea-tional purposes [35].

Although linezolid, a synthetic alkaloid, is one of the most effective drugs against VRE, there are still very few reports on the antibacte-rial activity of plant alkaloids against VRE. Line-zolid-resistant *E. faecium* isolates harbor a G2576T mutation in the *23S rRNA* gene and share the same allelic profile, which clusters in the C1 multilocus sequence typing epidemic lineage [36]. Deoxypseudophrynaminol, a natu-rally occurring alkaloid was also found to be a potent antibacterial agent against VRE, although the reported MIC values were rather high (20−40 μg/mL) [37]. Moreover, 12-methoxy-4-methylvoachalotine, an alkaloid iso-lated from the ethanol extract of the root bark of *Tabernaemontana catharinensis* showed moderate activity against VRE, with an MIC value of 0.08 to 0.31 mg/mL [38]. Interestingly, reserpine, another natural alkaloid isolated in 1952 from the dried root of *Rauwolfia serpentina*, was found to be a good inhibitor of multidrug-resistant efflux pumps and potentiates the activity of conventional antibiotics against VRE [39].

CONCLUSION

From the foregoing discussion, it is clear that the numbers of plant secondary metabolites active against VRE are minimal. Most of those discovered are either terpenoids or phenolics. One of the most likely explanations is insufficient screening of natural compounds. In fact, although synthesized alkaloids, such as linezolid (an antibiotic) or natural reserpine (an efflux pumps inhibitor), have been identified as good tools to fight VRE, researchers do not always seem very keen to investigate plant alkaloids that are active against VRE. However, we have identified some terpenoids and phenolic compounds that may lead to new drugs for clinical use after further study. It is probable that comprehensive screening of natural products and clinical trials of some of the molecules we have described could yield interesting results. Moreover, this chapter is the first scientific review to list the secondary metabolites of plants exhibiting activities on VRE. Finally, the present report provides a reliable database for future scientific studies.

References

[1] Hall LMC. Recent advances in understanding the epidemiology of enterococci. Rev Med Microbiol 1993;4:192−7.

[2] Leclerq R, Perlot E, Duval J, Courvalin P. Plasmid-mediated resistance to vancomycin and teicoplanin in Enterococcus faecium. N Engl J Med 1998;319:157−61.

[3] Uttley AHC, Collins CH, Naidoo J, George RC. Vancomycin-resistant enterococci. Lancet 1998;1:57−8.

[4] Nelson RRS. Selective isolation of vancomycin-resistant enterococci. J Hosp Infect 1998;39:13−8.

[5] Noskin GA. Vancomycin-resistant enterococci: Clinical, microbiologic, and epidemiologic features. J Lab Clin Med 1997;130:14−9.

[6] Moellering RC. Emergence of Enterococcus as a significant pathogen. Clin Infect Dis 1992;14:1173−8.

[7] Murray BE. The life and the times of the Enterococcus. Clin Microbiol Rev 1990;3:46−65.

[8] Rubin LG, Tucci V, Cercenado E, Eliopoulos G, Isenberg HD. Vancomycin-resistant Enterococcus faecium in hospitalized children. Infect Con Hosp Epidemiol 1992;13:700−5.

[9] Karanfil LV, Murphy M, Josephson A, Gaynes R, Mandel L, Hill BC, et al. A cluster of vancomycin-resistant Enterococcus faecium in an intensive care unit. Infect Con Hosp Epidemiol 1992;13:195−200.

[10] Wade J, Rolando N, Casewell M. Resistance of Enterococcus faecium to vancomycin and teicoplanin. Lancet 1991;337:1616.

[11] Hidron AI, Edwards JR, Patel J, Horan TC, Sievert DM, Pollock DA, et al. NHSN annual update: antimicrobial-resistant pathogens associated with healthcare-associated infections: annual summary of data reported to the National Healthcare Safety Network at the Centers for Disease Control and Prevention, 2006−2007. Infect Con Hosp Epidemiol 2008;29:996−1011.

[12] Pootoolal J, Neu J, Wright GD. Glycopeptide antibiotic resistance. Ann Rev Pharmacol Toxicol 2002;42:381−408.

[13] Mutnick AH, Enne V, Jones RN. Linezolid resistance since 2001: SENTRY Antimicrobial Surveillance Program. Ann Pharmacother 2003;37:769−74.

[14] Kuete V. Potential of Cameroonian plants and derived products against microbial infections: a review. Planta Med 2010;76:1479−91.

[15] Kuete V, Efferth T. Cameroonian medicinal plants: pharmacology and derived natural products. Front Pharmacol 2010;1:123.

[16] Specter M. A Life of Its Own. The New Yorker. http://www.newyorker.com/reporting/2009/09/28/090928fa_fact_specter?currentPage=all 2009 [accessed on 28.04.12].

[17] Gaspar-Marques C, Rijo P, Simões MF, Duarte MA, Rodriguez B. Abietanes from Plectranthus grandidentatus and P. hereroensis against methicillin- and vancomycin-resistant bacteria. Phytomedicine 2006;13:267−71.

[18] Mulyaningsih S, Sporer F, Zimmermann S, Reichling J, Wink M. Synergistic properties of the terpenoids aromadendrene and 1,8-cineole from the essential oil of Eucalyptus globulus against antibiotic-susceptible and antibiotic-resistant pathogens. Phytomedicine 2010;17:1061−6.

[19] Nyila MA, Leonard CM, Hussein AA, Lall N. Bioactivities of Plectranthus ecklonii constituents. Nat Prod Commun 2009;4:1177−80.

[20] Simões MF, Rijo P, Duarte A, Matias D, Rodriguez B. An easy and stereoselective rearrangement of an abietane diterpenoid into a bioactive microstegiol derivative. Phytochem Lett 2010;3:234−7.

[21] Khera S, Woldemichael GM, Singh MP, Suarez E, Timmermann BN. A novel antibacterial iridoid and triterpene from Caiophora coronata. J Nat Prod 2003;66:1628−31.

[22] Schito AM, Piatti G, Stauder M, Bisio A, Giacomelli E, Romussi G, et al. Effects of demethylfruticuline A and fruticuline A from Salvia corrugata Vahl. on biofilm production in vitro by multiresistant strains of Staphylococcus aureus, Staphylococcus epidermidis and Enterococcus faecalis. Int J Antimicrob Agent 2011;37:129−34.

[23] Hättenschwiler S, Vitousek PM. The role of poly-phenols in terrestrial ecosystem nutrient cycling. Trends Ecol Evol 2000;15:238−43.

[24] Rivero-Cruz JF. Antimicrobial compounds isolated from Haematoxylon brasiletto. J Ethnopharmacol 2008;119:99−103.

[25] Chakthong S, Weaaryee P, Puangphet P, Mahabusarakam W, Plodpai P, Voravuthikunchai SP, et al. Alkaloid and coumarins from the green fruits of Aegle marmelos. Phytochemistry 2012;75:108−13.

[26] Sakagami Y, Iinuma M, Piyasena KG, Dharmaratne HR. Antibacterial activity of alpha-mangostin against van-comycin resistant Enterococci (VRE) and synergism with antibiotics. Phytomedicine 2005;12:203−8.

[27] Fukai T, Oku Y, Hou AJ, Yonekawa M, Terada S. Anti-microbial activity of isoprenoid-substituted xanthones from Cudrania cochinchinensis against vancomycin-resistant enterococci. Phytomedicine 2005;12:510−3.

[28] Syu WJ, Shen CC, Lu JJ, Lee GH, Sun CM. Antimi-crobial and Cytotoxic Activities of Neolignans from Magnolia officinalis. Chem Biodivers 2004;1:530−7.

[29] Buhler DR, Miranda C. Antioxidant Activities of Fla-vonoids, http://lpi.oregonstate.edu/f-w00/flavonoid. html 2000 [accessed on 02.05.12].

[30] Sato M, Tanaka H, Oh-Uchi T, Fukai T, Etoh H, Yamaguchi R. Antibacterial activity of phytochemicals isolated from Erythrina zeyheri against vancomycin-resistant enterococci and their combinations with vancomycin. Phytother Res 2004;18:906−10.

[31] Chen L, Cheng X, Shi W, Lu Q, Go VL, Heber D, et al. Inhibition of growth of Streptococcus mutans, methicillin-resistant Staphylococcus aureus, and vancomycin-resistant enterococci by kurarinone,

a bioactive flavonoid isolated from Sophora fla-vescens. J Clin Microbiol 2005;43:3574−5.

[32] Liu IX, Durham DG, Richards ME. Vancomycin resistance reversal in enterococci by flavonoids. J Pharm Pharm 2001;53:129−32.

[33] Lee G-S, Kim E-S, Cho S-I, Kim Jung-Hoon, Choi G, Ju Y-S, et al. Antibacterial and synergistic activity of prenylated chalcone isolated from the roots of Soph-ora flavescens. J Korean Soc Appl Biol Chem 2010;53:290−6.

[34] Manske RHF. The Alkaloids. Chemistry and Physi-ology, Vol. VIII. New York: Academic Press; 1965. p. 673.

[35] Lewis AR. Lewis' dictionary of toxicology. Boca Raton, FL: CRC Press; 1998. 51.

[36] Bonora MG, Solbiati M, Stepan E, Zorzi A, Luzzani A, Catania MR, et al. Emergence of linezolid resistance in the vancomycin-resistant Enterococcus faecium mul-tilocus sequence typing C1 epidemic lineage. J Clin Microbiol 2006;44:1153−5.

[37] Hodgson JW, Mitchell MO, Thomas ML, Waters KF. Deoxypseudophrynaminol: a novel antibacterial alkaloid. Bioorg Med Chem Lett 1995;5:2527−8.

[38] Medeiros MR, Prado LA, Fernandes VC, Figueiredo SS, Coppede J, Martins J, et al. Antimi-crobial activities of indole alkaloids from Taber-naemontana catharinensis. Nat Prod Commun 2011;6:193−6.

[39] Mullin S, Mani N, Grossman TH. Inhibition of antibiotic efflux in bacteria by the novel multidrug resistance inhibitors biricodar (VX-710) and timcodar (VX-853). Antimicrob Agents Chemother 2004;48:4171−6.

Natural Products as Alternative Treatments for *Candida* Species Resistant to Conventional Chemotherapeutics

Maria de las Mercedes Oliva, Mauro Nicolás Gallucci,
Maria Evangelina Carezzano, Mirta Susana Demo

Universidad Nacional de Rio Cuarto, Córdoba, Argentina

INTRODUCTION

Candida spp. are among the most common causes of fungal diseases in humans and candidiasis represents one of the most frequent infections of the skin. *Candida* spp. are often referred to as opportunistic pathogens because they are normally found as skin commensals, but can rapidly colonize intertriginous areas and damaged skin. These yeasts are spread by direct contact or a change in the skin's natural protective mechanism, which allows existing organisms to proliferate. There are many predisposing factors for the development of candidiasis, some of which are specific to a location (e.g., the mouth or vagina): these include anemia, diabetes, impaired immunity, long-term broad-spectrum antibiotic therapy, debilitating illness, menopause, pregnancy, vitamin C deficiency, and the use of contraceptive pills, immunosuppressive medication, and

spermicidal creams [1,2]. Oral candida infections are the most common fungal infections in human immunodeficiency virus (HIV)-positive patients, while immunosuppressive therapies, catheter use, and endocrinopathies are responsible for systemic yeast infections, which can become life threatening, especially if they are caused by a drug-resistant strain [1,3–6]. Candida infections involve complex interactions between a wide range of host factors and yeast virulence determinants that may be differentially expressed depending on the prevailing environmental conditions [7]. The genus *Candida* comprises more than 200 species, but only a dozen or so of these have been associated with human infection, with *Candida albicans* being the most important cause of disease. However, other species such as *Candida dubliniensis*, *Candida glabrata*, *Candida guilliermondii*, *Candida krusei*, *Candida parapsilosis*, and *Candida tropicalis* are also increasingly recognized as

Fighting Multidrug Resistance with Herbal Extracts,
Essential Oils and Their Components
http://dx.doi.org/10.1016/B978-0-12-398539-2.00004-5

significant human pathogens [1,7,8]. *C. albicans* is the most prevalent pathogenic microorganism in oral candidiasis, but *C. krusei*, *C. guilliermondii*, *C. parapsilosis*, and *C. tropicalis* are also present in the disease course. These yeast species represent more than 80% of the clinical isolates in this pathogenia and are responsible for many cutaneous fungal infections [9,10]. In vulvovaginal candidiasis, *C. albicans* affect 85–95% of women, while *C. glabrata* affects 10–20%, and *C. tropicalis* and *C. krusei* are also associated with these infections. In addition, onychomycoses caused by *Candida* infection have been associated with *C. parapsilosis* [11]. The next section describes some characteristics of the most important yeast species that cause candidiasis.

Candida albicans

Candida albicans is a ubiquitous polymorphic species and the most common opportunistic pathogen, mainly in people with impaired immune system (i.e., cancer, transplant, or HIV patients). It can grow as yeast cells, pseudohyphae, and hyphae; this dimorphism is required for full virulence. Oral candidiasis occurs as a result of the overgrowth of cells of the resident microflora. In contrast, vaginal infections often occur in individuals without obvious immune defects. *C. albicans* cells can adhere to and colonize certain human tissues and can adhere to prostheses, leading to the formation of biofilms, which further facilitates adhesion, infection, and resistance to antifungals. The production of a range of extracellular hydrolases has been implicated in *C. albicans* pathogenicity [4,7,12–15].

Candida dubliniensis

Candida dubliniensis is an opportunistic pathogen, particularly in HIV-positive individuals. This species has the ability to colonize deep periodontal pockets. It was found to be responsible for some cases of vulvovaginal candidiasis and

for increasing numbers of candidemia cases resulting in the death of the patient. It shares morphologic and physiologic features with *C. albicans*. *C. dubliniensis* cells adhere to specific host proteins, e.g., mucin, as strongly as do those of *C. albicans*, and they express a large number of surface adhesins, very few of which have actually been characterized. To distinguish the most medically important species, identification criteria use phenotypic characteristics and rapid nucleic acid-based methods (e.g., the polymerase chain reaction) [7,15,16].

Candida krusei

Candida krusei is responsible for multiple drug-resistant opportunistic fungal infections, predominantly in immunocompromised patients. It is the fourth most common non-*albicans* species implicated in invasive infections and has emerged as a particularly important pathogen in patients with hematologic malignancies. Factors predisposing to infection are prior administration of fluconazole, neutropenia, and gastrointestinal mucosal barrier breakdown due to cytotoxic chemotherapy or radiation [6,17].

Candida tropicalis

This species is a commensal organism of the gastrointestinal tract that, under appropriate conditions, can invade deeply into the gastrointestinal mucosa and cause disseminated infection. The major predisposing factors occur in patients with acute leukemia, bone marrow transplantation, gastrointestinal mucosal damage caused by chemotherapy, and neutropenia [17].

Candida glabrata

Candida glabrata is also a commensal organism. Predisposing factors for candidemia include surgery (especially involving the abdominal cavity), the presence of intravenous catheters, and prior administration of fluconazole and amphotericin B. Features suggestive of *C. glabrata* candidemia are high fever and hypotension, mimicking endotoxic shock from bacteria. A large repertoire of adhesins, genome plasticity, phenotypic switching, and the remarkable ability to persist and survive inside host immune cells further contributes to the pathogenicity of this species [17,18].

Candida parapsilosis

This *Candida* species can be found on the hands of healthcare workers, which naturally increases the risk of acquiring this infection within hospitals. It induces severe complications and carries a high mortality rate. Uniquely among all *Candida* species, it has been shown to form loose biofilm clusters on prosthetic valves. It is also the leading cause of candidemia in neonatal intensive care units [17,19].

Candida lusitaniae and Candida guilliermondii

These species are rare causes of invasive infections that are increasingly encountered among severely immunosuppressed patients. Both species exist in the flora of the gastrointestinal tract and candidemic patients are generally infected with an endogenous strain. Exogenous transmission of *C. lusitaniae* has also been reported in several outbreaks. The hallmark of these organisms is their ability to develop resistance to amphotericin B [17].

It has been difficult to develop effective therapies for serious *Candida* infections due to the limited number of available antifungal agents. Moreover, complete eradication of the infection, especially in patients with advanced HIV infection and AIDS, is extremely difficult to accomplish. Some of the drugs used in the treatment of *Candida* diseases, such as amphotericin B, are very toxic and others, such as fluconazole, are limited by the high rate of spontaneous

mutations conferring resistance. Thus, the search for alternative antifungal compounds has been a major concern in recent years [5].

CONVENTIONAL ANTIFUNGAL DRUGS

Conventional treatments for fungal diseases are limited due to the fact that fungi are eukaryote organisms, thus making it difficult to develop a drug that is selectively toxic to the fungal cell and not to the host. Drugs for the treatment of candidiasis include a variety of imidazoles and triazoles. Historically, azole drugs have been widely used as first-line treatments; in particular, fluconazole is commonly used to treat candidiasis. Azoles disrupt the biosynthesis of ergosterol, a fungal-specific sterol of cellular membranes [13,20]. The triazole antifungal drugs inhibit fungal cytochrome P450-dependent lanosterol 14-α-demethylase, which is essential for the conversion of lanosterol to ergosterol in fungal cell membranes [1,9]. The polyene drugs (amphotericin B deoxycholate and its lipid-associated formulations) are other antimycotic treatment options and act by inserting into the fungal membrane in close association with ergosterol. The subsequent formation of porin channels leads to the loss of transmembrane potential and impaired cellular function [21]. New classes of antifungal drugs (echinocandins, nikkomycins, pneumocandins, and sordarin derivatives) are now being used. The echinocandins (anidulafungin, caspofungin, and micafungin) kill fungal cells by destabilizing their cell walls through inhibiting synthesis of β-1,3-glucans. Sordarin blocks fungal protein synthesis by inhibiting elongation factor-2. The nikkomycins competitively inhibit chitin synthase, thus blocking the formation of chitin, a component of the fungal cell wall. Flucytosine is a base pyrimidine analog that inhibits cellular DNA and RNA synthesis. Cispentacin, a cyclic β-amino acid originally isolated from *Bacillus cereus*, is active against *Candida* spp. Its antifungal derivative, PLD118, is actively accumulated in sensitive fungi and inhibits intracellular iso-leucyl-tRNA synthetase, an enzyme vital for protein synthesis and cell growth. [1,9,22]. Nystatin is the drug of choice for topical treatment of *C. albicans* infections within the oral cavity; however, microbial resistance is a growing problem. For the systemic treatment of patients at a high risk of developing fungal infections, the use of fluconazole is now recommended. [10] It is important to note, however, that virtually all *Candida* isolates have the potential to develop resistance to antifungal agents *in vitro*, especially to the azole agents. The increased use of conventional antifungal agents may select for non-*albicans* spp. that exhibit decreased suscepti-bility to these agents [17].

Resistance to Conventional Antifungals

Microbiologic resistance occurs when the growth of a pathogen is inhibited by an antimi-crobial agent at a concentration higher than the range required for wild-type strains. Clinical resistance is defined as the failure to eradicate a fungal infection despite the administration of an antifungal agent with *in vitro* activity against the organism. Such failures can be attributed to a combination of factors that may be related to the host, the antifungal agent, or the pathogen. Microbiologic resistance can be primary (intrinsic) or secondary (acquired). Primary resistance is found naturally in certain fungi without prior exposure to the drug; this empha-sizes the importance of identifying fungal species from clinical specimens. Secondary resistance develops among previously suscep-tible strains after drug exposure. Mechanisms of antifungal resistance, primary or secondary, are related to intrinsic or acquired characteris-tics of the fungal pathogen that either interfere with the antifungal mechanism of the respective drug/drug class or lower target drug levels [9,13,21,23–25].

The widespread application of triazole anti-fungal drugs promotes colonization by less susceptible species like *C. glabrata* or *C. krusei*, but also aids the selection of resistant subpopu-lations of normally susceptible organisms like *C. albicans*. The use of new echinocandin drugs (anidulafungin, caspofungin, and micafungin) is expanding because they are effective against a wide range of *Candida* spp., including azole-resistant species [9]. Prolonged and prophy-lactic treatment with azoles and echinocandins often results in the emergence of clinically resis-tant *C. glabrata* and *C. parapsilosis* strains. This limits the choice of possible therapeutic agents, often necessitating surgical intervention owing to the inability to eliminate the biofilm present on cardiac valves [18,19]. Another aspect of yeast resistance is that biofilm formation may correlate with the development of resistance in *C. albicans*. Actually, these yeasts exist predomi-nantly within biofilms; this is likely to be the cause of recalcitrant *Candida* persistence on inert, inserted surfaces or superficial mucosae. It was observed that adherent *C. albicans* cells are more resistant to antimycotics than are stationary-phase cells, and that resistance increases further with the formation of biofilms [3,13,26].

In recent years, the number of fungal infec-tions worldwide by *Candida* species has increased considerably and resistance to tradi-tional antifungal therapies is also rising. The current therapeutic options appear to be highly toxic and there are a lot of drug interactions. The lack of disponibility of conventional antifungals has encouraged the search for new alternatives among natural products [4]. High levels of anti-fungal drug resistance have been reported in *Candida* sp.; these exhibit primary resistance patterns toward amphotericin B, clotrimazole, fluconazole, itraconazole, and nystatin. Recent studies indicate *C. albicans* resistance to azoles and hepatotoxicity and nephrotoxicity linked to polyene use, particularly amphotericin B. Thus, there is a necessity to develop new agents to be used as a supplementary strategy for anti-mycotic chemotherapy. Natural compounds, mainly obtained from higher plants, are a potential source of antimycotic agents, either in their nascent forms or as template structures for more effective derivatives [8,14,22].

ANTIFUNGAL ACTIVITY OF NATURAL PRODUCTS OBTAINED FROM PLANTS

In recent years, medicinal plant research has attracted a lot of attention worldwide. Accumulated evidence has demonstrated the potential of medicinal plants used in various traditional, complementary, and alternative systems for the treatment of human diseases [27]. Medicinal plants have been used since antiquity and they constitute an important source of new drugs. The native people of America have a vast heritage of knowledge and the use of medicinal plants that may surprise the scientific community [28]. A medicinal plant is defined as a plant that is useful in therapeutics because it contains active compounds. These substances are mainly secondary metabolites, such as alkaloids, essential oils (EOs), flavonoids, glycosides, resins, saponins, tannins, and terpenoids, reported to play a natural role as plant defense mechanisms against predation by microorganisms, insects, and herbivores. In this way, terpenoids are responsible for plant odors and flavors, while quinolones and tannins are responsible for plant pigments. The amount of these compounds in a plant depends on environmental factors (e.g., moisture, soil, and temperature) related to where the plant has grown [27–30]. There are several steps involved in the primary bioassays used to detect biologic activity in natural products. First, the rapid screening of large numbers of products or extracts should be performed. The chosen assays should be simple, easy to implement, and produce results quickly.

A compound or extract identified as having specific activity at a nontoxic dose is called a *hit*. *Lead* status may be given to an extract or compound following its further evaluation in secondary or specialized *in vitro* bioassays and animal models. The last step includes a study of the kinetic and toxicologic properties of the extract or compound and defines the *proof-of concept* and *development candidate* categories [31].

In developing countries, medicinal plants have been long used as alternative treatments for many health problems. Many plant extracts and EOs have been shown to exert biologic activity both *in vitro* and *in vivo*, including anti-cancer and anti-infective activities. This has justified research into traditional medicines, focusing on characterization of the antimicrobial activities of plants derivatives. Brazil, Cuba, India, Jordan, and Mexico are examples of countries that contain a diverse flora with antimicrobial uses. More than 200 plants species have been studied since the year 2000 in the search for natural products active against *Candida* spp. [32]. Thus, there are antifungal agents usually used as chemotherapeutic agents in clinics, such as aureobasidins, echinocandins/ pneumocandins, polyenes, and sordarins, which originated in natural products. The research into natural molecules and products is primarily focused on plants, since they can be obtained more easily and selected on the basis of their ethnopharmacologic use. When a natural antifungal product is obtained, it is better if it is shown to have fungicidal, rather than fungistatic, properties. This is particularly important for immunocompromised patients, such as those with HIV, because the prophylactic use of fungistatic drugs has been associated with an increased frequency of innate and acquired drug resistance in clinical isolates. Tangarife-Castaño et al. suggested a classification system for antifungal activity in plant derivatives based on the minimum inhibitory concentration (MIC): strong inhibitors (MIC of

≤ 0.5 mg/mL); moderate inhibitors (MIC of 0.6–1.5 mg/mL); and weak inhibitors (MIC of > 1.6 mg/mL) [4].

The antimycotic activity of natural products obtained from plants has been extensively described. The next section reviews recent studies into the antifungal activities of plant extracts, EOs, and terpenes.

Antimycotic Activity of Plant Extracts

An ethanol extract of the leaves of *Schinus terebinthifolius* (Anacardiaceae) has been shown to be active against *C. albicans*. This species grows in South America; it is important in traditional medicine because of its bark resins and it is used to treat diarrhea, ulcers, uterine and urinary pain, venereal diseases, and wounds [33]. Studies into the antifungal activity of ethanol extracts of 10 Argentinean plants used in native medicine against yeast, microfungi, and basidiomycetes have been reported. Extracts of *Larrea divaricata*, *Larrea cuneifolia* (Zygophyllaceae) and *Zuccagnia punctata* (Leguminosae) were found to display remarkable activity against the majority of the tested fungi. *Prosopanche americana* (Hydnoraceae) also inhibits yeast growth. Popular uses of *L. divaricata* and *L. cuneifolia* are as antirheumatic, antiinflammatory, and emmenagogue [34]. In tests, ethanol extracts of the leaves of *Lippia triphylla* Kunth and *Aloysia triphylla* L'Herit. Briton (Verbenaceae) showed activity against *C. albicans*. The results of this investigation supported the traditional use of *A. triphylla* leaves in the treatments of infectious diseases caused by *C. albicans* and some pathogenic bacteria and provided preliminary scientific validation for the use of *L. triphylla* leaves as an antibacterial and antifungal medicinal [35]. Another study demonstrated antifungal activity in *Lippia origanoides* Kunth (Verbenaceae) and *Morinda royoc* L (Rubiaceae) ethanol extracts against *Aspergillus flavus*, *Aspergillus fumigatus*, *C. krusei*, and *C. parapsilosis* [4].

Acetone, chloroform, ethanol (95%), and ether extracts of *Cassia alata* (Caesalpinaceae) leaves also showed significant antifungal activity *in vitro* against various fungi, namely *Aspergillus niger*, *C. albicans*, *C. tropiathis*, *Rhododendron japonicum*, and *Rhodotorula glutinis* [27].

A study of methanol and ethanol extract of the Algerian plants *Allium cepa*, *Allium schoenoprasum* (Amaryllidaceae), *Apium graveolens* (Apiaceae), *Cichorium intybus* (Asteraceae), *Citrus paradise* (Rutaceae), *Mentha longifolia* (Lamiaceae), *Vaccinium macrocarpon* (Ericaceae), and *Vicia faba* (Fabaceae), revealed activity against *C. albicans*, and *Candida maltose*. It is interesting to note that the yeasts showed more sensitivity than the Gram-positive bacteria to the extracts and that the methanol extracts were more active than the ethanol ones. [36]

Experiments into the antifungal activity of five different extracts (aqueous, chloroform, ethanol, ether, and methanol) of *Cassia fistula* L. (Caesalpiniaceae) were performed on *C. albicans*, *C. krusei*, *C. parapsilosis*, and *C. tropicalis* species. All extracts showed promising antifungal activity against all *Candida* spp. tested: the maximum activity against yeast was observed in methanol extract followed by the ethanol and aqueous extracts [8].

The antifungal activity of tinctures obtained from *Anacardium occidentale* (Anacardiaceae), *Malva sylvestris* (Malvaceae), and *Salvia officinalis* (Lamiaceae) was tested against *C. albicans*, *C. tropicalis*, and *C. krusei*. The *M. sylvestris* tincture was active against all tested strains at lower concentrations, while *A. occidentale* inhibited *C. albicans*, and *C. krusei* and *S. officinalis* only showed antifungal activity toward *C. krusei* [10].

Traditionally, the EOs of aromatic plants species have been used for several purposes. Evaluations of the biologic activities of these EOs have revealed that some exhibit antibacterial, antifungal, and insecticidal properties [11]. Various studies have demonstrated that EOs are well tolerated by humans and hold therapeutic value for the treatment of acne, dandruff, head lice, and recurrent herpes labialis. The demonstration of antifungal properties of EOs has broad importance because they offer significant potential as a topical or intraoral therapy for *Candida* (and possibly for other fungi/yeasts). Oropharyngeal candidiasis is the most common opportunistic infection among patients infected with HIV. Treatment of yeast infections with conventional drugs is not always efficient because of problems such as a lack of efficacy, emergence of resistance, adverse events, high costs, and the need for intravenous administration. EOs may have therapeutic value for both dermatophytic and oral/mucosal infection in both immunocompetent and immunocompromised patients [6].

There are several methods for obtaining plant volatile oils, but they generally use distillation methods, usually steam or hydrodistillation. EOs obtained from nonwoody plant material are mainly composed of variable mixtures of terpenoids, which are the chemicals responsible for the medicinal, culinary, and fragrant uses of aromatic and medicinal plants. Terpenes are derived from the condensation of branched five-carbon isoprene units and are categorized according to the number of these units present in the carbon skeleton, specifically monoterpenes [C_{10}] and sesquiterpenes [C_{15}]. Diterpenes [C_{20}] and a variety of other compounds, including low molecular weight aliphatic hydrocarbons (linear, ramified, saturated, and unsaturated), acids, alcohols, aldehydes, acyclic esters, lactones, and, exceptionally, nitrogen- and sulfur-containing compounds, coumarins and homologs of phenylpropanoids may also be present [37].

Several studies worldwide support the antifungal activity of EOs obtained from aromatic plants. Some of these studies are reviewed in the following paragraphs.

Hammer and coworkers investigated the antimicrobial activity of extracts and EOs of several plants and reported that plant extracts are effective for the inhibition of bacteria and

Candida spp. [30]. This group has subsequently continued this line of research, focusing on the antimicrobial activity of *Melaleuca alternifolia* (tea tree) EOs and elucidating the mechanism of action [38–40].

The EO of *Baccharis notosergila* (Asteraceae) shows antifungal activity against *C. albicans*. It is a shrub mainly found in Argentina, Paraguay, and Uruguay, and is used in folk medicine as a diuretic and digestive [41]. Another study found that EOs from leaves and inflorescences of *Solidago chilensis* (Asteraceae) exhibit antifungal activity against five different strains of filamentous fungi and *C. albicans* [42]. The EOs of *Schinus polygamus* (Anacardiaceae) also show moderate activity against *C. albicans*. The leaves of this aromatic and medicinal shrub, a native of Patagonia (Argentina), are chewed by local people to clean their teeth and the juice is drunk by child as a diuretic [43]. In another study, the antimicrobial properties of EOs of 14 medicinal plants from Argentina were tested against different microorganisms. The strongest activity was observed in the oil of *A. triphylla* and *Psila spartoides* (Asteraceae), which inhibited *C. albicans* and all Gram-positive and Gram-negative bacteria tested [44].

Plants from Brazil have been used as natural medicines by the local population for the treatment of several tropical diseases, as well as fungal and bacterial infections. A study to determine the antimicrobial activity against Gram-positive, Gram-negative bacteria and the yeast *C. albicans* was performed using EOs obtained from *A. triphylla* and seven aromatic plants of the Lamiaceae family (*Ocimum basilicum*, *Ocimium gratissimum*, *Origanum applii*, *Origanum vulgare* (Lamiaceae), *Mentha piperita* (Lamiaceae), *Mentha spicata*, and *Thymus vulgaris*): only EOs from *A. triphylla* and *M. piperita* were able to control the growth of *C. albicans* [45]. Another study tested EOs from 35 plants commonly used in Brazilian folk medicine as anti-inflammatories and anti-infectives and observed anti-*Candida* activity in 13. Strong activity was observed in EOs of *Achillea millefo-lium*, *Mikania glomerata* (Asteraceae), and *Stachys byzantina* (Lamiaceae). Furthermore, moderate antifungal activity was observed in the EOs of *A. triphylla*, *Anthemis nobilis* (Asteraceae), *Cymbopogon martini* (Poaceae), *Cyperus articulatus*, *Cyperus rotundus* (Cyperaceae), *Lippia alba*, *Mentha arvensis*, and *M. piperita* [32]. The EO obtained from the herb *Santolina chamaecyparis-sus* (Asteraceae) showed significant antifungal activity against 13 strains of *C. albicans* both *in vitro* and *in vivo* in experimentally induced vaginal and systemic candidiasis in mice. This EO also showed activity against experimentally induced superficial cutaneous mycosis in guinea pigs, using the hair root invasion test [27].

Lemongrass oil, obtained from *Cymbopogon citratus* (Poaceae) and citral (the main component of the oil) exhibited activity against *C. albicans*, *C. albicans* CI-I (clinical isolate) and CI-II, *C. glabrata*, *C. krusei*, *C. parapsilosis*, and *C. tropicalis* strains [11].

The antifungal activity of *A. triphylla* EO was tested against strains resistant to fluconazole, including *C. albicans*, *C. dubliniensis*, *C. glabrata*, *C. guilliermondii*, *C. krusei*, *C. parapsilosis*, and *C. tropicalis*, and had a fungicidal effect against all of these yeast strains [46].

The EOs from *Piper bredemeyeri* Jacq (Piperaceae) and *L. origanoides* Kunth were found to be active against *C. albicans*, inhibiting germ tube formation. In addition, the number of blastoconidia/mL was reduced [4].

Further research using *C. tropicalis* strains showed that the EOs of *Cinnamomum zeylanicum* (Lauraceae), *Eugenia caryophyllata* (syn. *Syzygium aromaticum*; Myrtaceae), and *Origanum vulgare* strongly inhibits the growth of these yeasts [47].

Several reports attribute the antimicrobial activity of the EOs to the terpenoid and phenolic compounds present in them [37,48–50]. The major phenolic components of EOs, such as carvacrol, have been suggested to have potent antifungal activity [14]. EOs that are rich in components such as the aldehydes citral (an isomeric mix of neral and geranial), citronellal, cinnamaldehyde, cumaldehyde, and phenols (thymol, carvacrol, and eugenol) have demonstrated antifungal properties. Other components, such as the primary alcohols geraniol and citronellol, have excellent antifungal activity without the risk of irritation that some of the aforementioned components possess [2]. In a study into the antifungal activities of the terpenes 1,8-cineol, limonene, menthol, α-pinene, and thymol, all were shown to inhibit *C. albicans* [51]. Moreover, terpene alcohols such as linalool exhibit strong antimicrobial activity, especially pronounced on whole bacterial or fungal cells, while hydrocarbon derivatives possess weaker antifungal properties, as their low solubility in water limits diffusion through the phospholipid bilayer of cell membranes. Greater antifungal potential could be attributed to oxygenated terpenes [52].

Important pathogens are susceptible to a broad range of EOs. These hydrophilic compounds are inexpensive and available worldwide, which make them promising alternatives for the treatment of localized infections even with severe hospital-acquired strains [6]. Their ease of use and relative lack of toxicity indicates numerous possible formulations for topical application for dermatomycoses and superficial candidiasis [2]. The EOs have been tested for both *in vitro* and *in vivo* antimycotic activity and have much potential as antifungal agents that can be used either alone or in conjunction with standard antifungal therapies [53]. Natural compounds constitute a potential source of antimycotic agents either in their nascent forms or as template structures for more effective derivatives. What is more, synergy between EOs and conventional antimycotics has previously been investigated, with the aim of enabling lower doses of antifungal agents to be administered. Synergism was shown by an increased rate of fungal killing; a shorter therapy time, which can prevent the emergence of drug resistance; an enlarged spectrum of antifungal activity; and decreased drug toxicity. In a study combining itraconazole and the EO derived from *P. bredemeyeri* Jacq, a synergistic effect was obtained, although no interaction was detected between this oil and amphotericin B [4]. It was also reported that natural compounds such as benzoic, cinnamic, 2-hydroxycinnamic, 3-hydroxycinnamic, 4-hydroxycinnamic, and salicylic acids, thymol, 2,3-dihydroxybenzaldehyde, 2,5-dihydroxybenzaldehyde, 3,4,5-trimethoxy-benzaldehyde, vanillin, and veratraldehyde improve the effectiveness of selected azoles and polyene antifungal drugs (amphotericin B, fluconazole, and itraconazole) against *C. albicans*, *C. glabrata*, *C. krusei*, *C. lusitaniae*, *C. neoformans*, *C. parapsilosis*, and *C. tropicalis*. This synergistic effect was named *chemosensitization* and depends on the yeast strain and drugs employed [22].

EOs could be a possible therapy for *difficult to treat* hospital-acquired infections with multi-drug-resistant strains. However, the antimicrobial activity of EOs may not be suitable as a stand-alone option for treating serious microbial infections, mainly due to variation in the terpenic composition of the oils. EOs of the same type, but from different producers, can vary in their active component content, leading to remarkable differences in antimicrobial activity. Variation can also depend on the plant family; for example, one tea tree oil extract may not be equivalent to another tea tree oil extract. Differences between oils can also depend on the country of origin; for example, lemon oil from Nepal has stronger antifungal activity than lemon oil from Australia [6]. In conclusion, additional pharmaceutical studies focusing on the oil chemical composition and identification of the active ingredients are important.

The EOs of *Baccharis spartioides* (Asteraceae) collected from different regions of Argentina show variations in both chemical composition and biologic activity. Oils from *B. spartioides* collected from central Argentina exhibited stronger antifungal activity against *C. albicans* than equivalent oils from plants collected in southern and northwestern regions. The differences in activity could be linked to antagonism or synergism between compounds found in the oil [54]. Another study investigated the antimicrobial activity of *A. triphylla* EO collected from different regions of Argentina and Paraguay against bacteria and yeast. *C. albicans* was the most sensitive microorganisms, being inhibited by all the EOs; however, differences in the antifungal activity of each *A. triphylla* EO preparation were observed, which could be attributed to differences in the quantity and quality of the terpenes of each sample and possible interactions between these compounds [46].

Mechanism of Action

EOs appears to mainly act on the fungal cell membrane, by disrupting its structure and causing leakage and cell death; blocking membrane synthesis; and inhibiting spore germination, fungal proliferation, and cellular respiration. The high volatility and lipophilicity of the EOs helps them to penetrate the cell membrane and exert their biologic effects [11,53].

The antifungal activity of *O. gratissimum* EOs against *C. albicans*, *C. krusei*, *C. parapsilopsis*, and *C. tropicalis* was evaluated. This species, known in Brazil as *alfa vaca*, is used to treat diarrhea, fever, headache, respiratory tract infections, and ophthalmic and skin diseases. The EO was found to have antibacterial, antifungal, and antiviral activity. The data obtained in this study showed that the oil produced remarkable ultrastructural alterations in target cells. Morphologic changes included thickening and detachment of the fungal cell wall, increased numbers of cytoplasmic vesicles, lamellar proliferation in the cytoplasm in continuity with the cell wall, and disruption to cell division [5]. Similar results were observed for lemongrass EO on *C. albicans* [55].

The biological activity of tea tree oil has been widely studied. An *in vitro* study reported that tea tree oil and some of its components cause leakage of intracellular compounds and inhibit respiration in bacteria and yeasts, particularly in *Candida*, suggesting adverse effects on mitochondria by inhibition of mitochondrial ATPase. This study also showed that tea tree oil and/or it components increases yeast cell permeability and membrane fluidity, and inhibits medium acidification. The mechanism of action of tea tree oil terpenes was reported to involve changing the properties of the lipid bilayer; the hydrophobicity of different terpene compounds determines the position of each compound within the membrane. Insertion of tea tree oil terpenes disrupts lipid packing, thus inducing changes in bilayer fluidity that affect cell permeability [2,40].

After treating *C. albicans* with the minimum fungicidal concentration of *A. triphylla* EO, cell morphology and shape became distorted and notable structural disorganization occurred within the cytoplasm: large vacuoles could be seen by high-resolution light microscopy. In

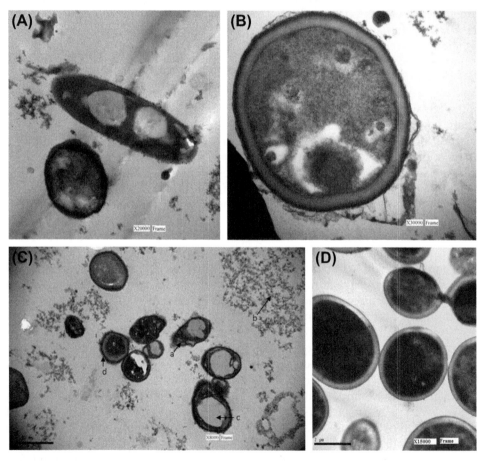

FIGURE 4.1 **Electron microscopy images of *Candida albicans* treated with the minimum fungicidal concentration of *Aloysia triphylla* essential oil.** (A) Elongation of the cell, with large vacuoles visible inside the cytoplasm. (B) Thickening of the cell wall and mitochondrial morphology changes. (C) Alterations to membrane integrity (a), loss of the cytoplasmic content (b), intracytoplasmic vacuoles (c), and thickening of the cell wall (d). (D) Normal *Candida albicans* cells (untreated control).

addition, the contents of some treated cells appeared depleted and amorphous. Loss of cellular material was also observed [56]. Electron microscopy observations of the effects of the same oil on *C. albicans* enabled the visualization of more detailed alterations (unpublished data; Fig. 4-1). These combined observations revealed cell elongation with large vacuoles inside the cytoplasm (Fig. 4-1A), alterations to budding, thickening of the cell wall, and changes to mitochondrial morphology (Fig. 4-1B). In addition, plasma membrane integrity was altered by the EO, leading to the loss of cytoplasmic contents (Fig. 4-1C). These changes were not seen in untreated controls (Fig. 4-1D). Thus, *A. triphylla* EO constitutes a promising natural product for the treatment of candidiasis, warranting further studies on its pharmacokinetics and toxicologic properties. These results represent an initial characterization of the anti-*Candida* activity of EOs from traditional Argentinian medicinal plants.

CONCLUSIONS AND FUTURE PROSPECTS

Microorganisms have the ability to develop resistance to chemotherapeutic agents and this seems to be a continuous process. Consequently, the life span of an antibiotic and its efficacy are limited. Natural products obtained from medicinal plants sources have become a very important alternative source of new substances for combating infections in human beings; these compounds may also have lower cost and reduced toxicity [27].

The data and studies described above confirm that medicinal plants can be a rich source of potential antifungal compounds. Several publications have documented the antimicrobial activity of plant extracts and EOs against *Candida* species. Plant secondary metabolites have potential applications in medical procedures and in the cosmetic and pharmaceutical industries. Many studies have confirmed that EOs and plant extracts possess *in vitro* antibacterial, antifungal, and antiviral activities. However, it is necessary to consider issues of safety and toxicity when EOs and extracts are to be used for food preservation or medicinal purposes. The scientific studies on antimicrobial activity in plant products show a good correlation with the reported traditional medicinal uses. The medical applications of plant EOs further highlights the importance of preserving species of native flora for the maintenance of biodiversity; thus, there is a need to preserve native plant populations in their natural habitats [57].

References

[1] Dorrell S. Overcoming drug-resistant yeast infections. Drug Discov Today 2002;7(6):332–3.

[2] Harris R. Synergism in the essential oil world. Int J Aromatherapy 2002;12:179–86.

[3] Jin Y, Samaranayake LP, Samaranayake Y, Yip HK. Biofilm formation of Candida albicans is variably affected by saliva and dietary sugars. Arch Oral Biol 2004;49:789–98.

[4] Tangarife-Castaño V, Correa-Royero J, Zapata-Londoño B, Durán C, Stanshenko E, Mesa-Arango AC. Anti-Candida albicans activity, cytotoxicity and interaction with antifungal drugs of essential oils and extracts from aromatic and medicinal plants. Infectio 2011;15(3):160–7.

[5] Vataru-Nakamura C, Ishida K, Faccin LC, Dias Filho BP, Garcia Cortez DA, et al. *In vitro* activity of essential oil from Ocimum gratissimum L. against four Candida species. Res Microbiol 2004;155:579–86.

[6] Warnke PH, Becker ST, Podschun R, Sivananthan S, Springer IN, Russo PAJ, et al. The battle against multi-resistant strains: Renaissance of antimicrobial essential oils as a promising force to fight hospital-acquired infections. J Cranio-Maxill Surg 2009;37:392–7.

[7] Sullivan DJ, Moran GP, Coleman DC. Candida dubliniensis: Ten years on. FEMS Microbiol Lett 2005;253:9–17.

[8] Panda SK, Brahma S, Dutta SK. Selective antifungal action of crude extracts of Cassia fistula L.: A preliminary study on Candida and Aspergillus species. Malaysian J Microbiol 2010;6(1):62–8.

[9] Perlin DS. Antifungal drug resistance: Do molecular methods provide a way forward? Curr Opin Infec Dis 2009;22:568–73.

[10] Cardoso AMR, Cavalcanti YW, Dantas de Almeida LF, Alves de Lima Pérez AL, Padilha WWN. Antifungal activity of plant-based tinctures on Candida. RSBO 2012;9(1):25—30.

[11] Bona da Silva C, Guterres SS, Weisheimer V, Schapoval EES. Antifungal activity of the lemongrass oil and citral against Candida spp. The Brazil J Infect Dis 2008;12(1):63—6.

[12] Jacobsen ID, Wilson D, Wächtler B, Brunke S, Naglik JR, Hube B. Candida albicans dimorphism as a therapeutic target. Expert Rev Anti Infect Ther 2012;10(1):85—93.

[13] Chen LM, Xu YH, Zhou CL, Zhao J, Li CY, Wang R. Over expresion of CDR1 and CDR2 genes plays an important role in fluconazol resistance in Candida albicans with G487T and T916C mutations. J Int Med Res 2010;38:536—45.

[14] Chami N, Chami F, Bennis S, Trouillas J, Remmal A. Antifungal treatment with carvacrol and eugenol of oral candidiasis in immunosuppressed rats. The Brazil J Infect Dis 2004;8(3):217—26.

[15] Sullivan DJ, Moran GP, Pinjon E, Al-Mosaid A, Stokes C, Vaughan C, et al. Comparison of the epidemiology, drug resistance mechanisms, and virulence of Candida dubliniensis and Candida albicans. FEMS Yeast Res 2004;4:369—76.

[16] Jewtuchowicz VM, Mujica MT, Brusca MI, Sordelli N, Malzone MC, Pola SJ, et al. Phenotypic and genotypic identification of Candida dubliniensis from subgingival sites in immunocompetent subjects in Argentina. Oral Microbiol Immun 2008;23:505—9.

[17] Collin B, Clancy CJ, Hong Nguyen M. Antifungal resistance in non-albicans candida species. Drug Resist Updates 1999;2:9—14.

[18] Tscherner M, Schwarzmüller T, Kuchler K. Pathogenesis and antifungal drug resistance of the human fungal pathogen Candida glabrata. Pharmaceuticals 2011;4:169—86.

[19] Kumar J, Fish D, Burger H, Weiser B, Ross JS, Jones D, et al. Successful surgical intervention for the management of endocarditis due to multidrug resistant Candida parapsilosis: case report and literature review. Mycopathologia 2011;172:287—92.

[20] Heimark L, Shipkova P, Greene J, Munayyer H, Yarosh-Tomaine T, Di Domenico B, et al. Mechanism of azole antifungal activity as determined by liquid chromatographic/mass spectrometric monitoring of ergosterol biosynthesis. J Mass Spectrom 2002;37(3):265—9.

[21] Kanafani ZA, Perfect JR. Resistance to antifungal agents: mechanisms and clinical impact. Clinic Infec Dis 2008;46:120—8.

[22] Faria NCG, Kim JH, Goncalves LAP, Martins ML, Chan KL, Campbell BC. Enhanced activity of antifungal drugs using natural phenolics against yeast strains of Candida and Cryptococcus. Lett Appl Microbiol 2011;52:506—13.

[23] Pfaller MA. Antifungal drug resistance: mechanisms, epidemiology, and consequences for treatment. Am J Med 2012;125(1A):S1—13.

[24] Odds FC. Antifungal Agents: Resistance and Rational Use. In: Gould IM, van der Meer JW, editors. Antibiotic Policies: Theory and Practice. New York: Kluwer Academic/Plenum Publishers; 2005. p. 311—30.

[25] Perea S, Lopez-Ribot JL, Kirkpatrick WR, Mcatee RK, Santillan RA, Martinez M, et al. Prevalence of Molecular Mechanisms of Resistance to Azole Antifungal Agents in Candida albicans Strains Displaying High-Level Fluconazole Resistance Isolated from Human Immunodeficiency Virus-Infected Patients. Antimicrob. Agents Ch 2001;45(10):2676—84.

[26] Watamoto LP, Samaranayake H, Egusa H, Yatani, Seneviratne CJ. Transcriptional Regulation Of Drug-Resistance Genes In Candida albicans Biofilms In Response To Antifungals. J Med Microbiol 2011;60:1241—7.

[27] Sher A. Antimicrobial Activity of Natural Products From Medicinal Plants. Gomal J Med Sc 2009;7(1):72—8.

[28] Ratera EL, Ratera M. Plantas de la Flora Argentina Empleadas en Medicina Popular. 1st ed. Buenos Aires, Argentina: Editoral Hemisferio Sur Sociedad Anónima; 1980.

[29] Cowan M. Plant products as antimicrobial agents. Clin Microbiol Rev 1999;12(4):564—82.

[30] Hammer K, Carson C, Riley T. Antimicrobial activity of essential oils and other plant extracts. J Appl Microbiol 1999a;86:985—90.

[31] Cos P, Vlietinck AJ, Vanden Berghe D, Maes L. Anti-infective potential of natural products: How to develop a stronger *in vitro* 'proof-of-concept'. J Ethnopharmacol 2006;106:290—302.

[32] Texeira Duarte MC, Figueira GM, Sartoratto A, Garcia Rehder VL, Delarmelina C. Anti-Candida activity of Brazilian medicinal plants. J Ethnopharmacol 2005;97:305—11.

[33] Martinez Guerra MJ, Lopez Barreiro M, Morejón Rodriguez Z, Rubalcaba Y. Actividad antimicrobiana de un extracto fluido al 80% de Schinus terebinthifolius Raddi (Copal). Rev Cub Pl Med 2000;5(1):23—5.

[34] Quiroga EN, Sampietro AR, Vattuone MA. Screening antifungal activities of selected medicinal plants. J Ethnopharmacol 2001;74:84—96.

[35] Oskay M, Ûsame Tamer A, Ay G, Sari D, Aktas K. Antimicrobial activity of the leaves of Lippia triphylla (L'Her) O Kuntze (Verbenaceae) against on bacteria and yeasts. J Biol Sc 2005;5(5):620—2.

[36] Akroum S, Satta D, Lalaoui K. Antimicrobial, Anti-oxidant, Cytotoxic Activities and Korrichi Lalaoui Phytochemical Screening of Some Algerian Plants. Eur J Sc Res. 2009;31(2):289−95.

[37] Dorman HJD, Deans SG. Antimicrobial agents from plants: antibacterial activity of plant volatile oils. J Appl Microbiol 2000;88:308−16.

[38] Carson CF, Mee BJ, Riley TV. Mechanism of Action of Melaleuca alternifolia (Tea Tree) Oil on Staphylococcus aureus Determined by Time-Kill, Lysis, Leakage, and Salt Tolerance Assays and Electron Microscopy. Antimicrob Ag Ch 2002;46(6):1914−20.

[39] Hammer K, Carson C, Riley T. *In Vitro* Susceptibilities of Lactobacilli and Organisms Associated with Bacterial Vaginosis to Melaleuca alternifolia (Tea Tree) Oil. Antimicrob Ag Ch 1999b;43(1):196.

[40] Hammer K, Carson C, Riley T. Antifungal effects of Melaleuca alternifolia (tea tree) oil and its components on Candida albicans, Candida glabrata and Saccharomyces cerevisiae. J Antimicrob Chemoth 2004;53:1081−5.

[41] Cobos MI, Rodriguez JL, Oliva M, de las M, Demo M, Faillaci S, et al. Composition and antimicrobial activity of the essential oil of Bacharis notosergila. Planta Médica 2001;67:84−5.

[42] Vila R, Mundina M, Tomi F, Burlan R, Zacchino S, Casanova J, et al. Composition and antifungal activity of the essential oil of Solidago chilensis. Planta Médica 2002;68:164−7.

[43] Gonzalez S, Guerra PE, Bottaro H, Molares S, Demo MS, Oliva Ma, et al. Aromatic plants from Patagonia. Part I. Composition and antimicrobial activity of Schinus polygamus (Cav) Cabrera essential oil. Flavour Frag J 2004;19:36−9.

[44] Oliva Demo, M, Ma de las M, Lopez ML, Zunino MP, Zygadlo JA. Antimicrobial activity of essential oils obtained from aromatic plants of Argentina. Pharm Biol 2005;43:129−34.

[45] Sartoratto A, Machado ALM, Delarmelina C, Figueira GM, Duarte MCT, Rehder VLG. Composition and antimicrobial activity of essential oils from aromatic plants used in Brazil. Braz J Microbiol 2004;35:275−80.

[46] Oliva M, de las M, Beltramino E, Gallucci N, Casero C, Zygadlo J, et al. Antimicrobial activity of the essential oils of Aloysia triphylla (L'Her.) Britton from different regions of Argentina. BLACPMA 2010;9(1):29−37.

[47] Moura Mendes J, Sarmento Guerra FQ, De Oliveira Pereira F, Pereira De Sousa J, Nogueira Trajano. V, De Oliveira Lima E. Actividad antifúngica del aceite esencial de Eugenia caryophyllata sobre cepas de Candida tropicalis de aislados clínicos. BLACPMA 2012;11(3):208−17.

[48] Burt S. Essential oils: antibacterial properties and potential applications in food. A review. Int J Food Microbiol 2004;94:223−53.

[49] Carvalho CCR, Fonseca MMR. Carvone: Why and how should one bother to produce this terpene. Food Chem 2006;95(3):413−22.

[50] Zunino M, Novillo-Newton M, Maestri D, Zygadlo J. Composition of Essential Oil of Bacharys crispa Spreng and Baccharys salicifolia pers. Grown in Córdoba (Argentina). Flavour Frag J 1997;12:405−7.

[51] Demo M, Oliva M, Ramos B, Zygadlo J. Determinación de Actividad Antimicrobiana de Componentes Puros de Aceites Esenciales. Rev Higiene alimentar 2001;15(85):87−90.

[52] Skaltsaa HD, Demetzos C, Lazarib D, Sokovicc M. Essential oil analysis and antimicrobial activity of eight Stachys species from Greece. Phytochemistry 2003;64:743−52.

[53] Cox SD, Mann CM, Markham JL, Bell HC, Gustafson JE, Warmington JR, et al. The mode of antimicrobial action of the essential oil of Melaleuca alternifolia (tea tree oil). J Appl Microbiol 2000;88:170−5.

[54] Oliva MM, Zunino MP, López ML, Soria YA, Ybarra FN, Sabini L, et al. Variation in the essential oil composition and antimicrobial activity of Baccharis spartioides (H. et A.) J. Rimy from different region of Argentina. J Essent Oil Res 2007;19:509−13.

[55] Tyagi A, Malik A. Liquid and vapour-phase antifungal activities of selected essential oils against Candida albicans: microscopic observations and chemical characterization of Cymbopogon citrates. BMC Complementary and Alternative Medicine 2010;10:65.

[56] Oliva M, Carezzano E, Gallucci N, Casero C, Demo M. Antimycotic effect of the essential oil of Aloysia triphylla against Candida species obtained from human pathologies. Nat Prod Commun 2011;6(7):1039−43.

[57] Demo M, Oliva MM. Antimicrobial Activity of Medicinal Plants from South America. Pub. Botanical Medicine in Clinical Practice 2008. Ed. Prof. R. R. Watson and V. R. Preedy. UK.

5

Plants and Tuberculosis:
Phytochemicals Potentially Useful in the Treatment of Tuberculosis

Philip G. Kerr

School of Biomedical Sciences, Charles Sturt University,
WaggaWagga, New South Wales, Australia

TUBERCULOSIS: ONE OF THE FORGOTTEN DISEASES

Global tuberculosis control is facing major challenges today. In general, much effort is still required to make quality care accessible without barriers of gender, age, type of disease, social setting, and ability to pay. Coinfection with *Mycobacterium tuberculosis* and HIV (TB/HIV), especially in Africa, and multi-drug-resistant and extensively drug-resistant (MDR-TB and XDR-TB, respectively) tuberculosis in all regions, make control activities more complex and demanding. Several risk groups need special attention and these are discussed below. These challenges need to be addressed by national tuberculosis programs with focused attention and tailored approaches. *World Health Organization [1]*

There can be no doubt that tuberculosis, let alone MDR-TB and XDR-TB strains of the organism, poses severe health threats to much of the population of the developing world, and indeed to the world as a whole. The World Health Organization (WHO) has published an annual global tuberculosis report since 1997 [2].

Like its cousin leprosy, tuberculosis has been known, and feared since Biblical times (c. 2000 BC), and probably even earlier. These disease states are caused by the generally slow-growing organisms, *Mycobacterium leprae* and *M. tuberculosis*, respectively [3,4]. This feature of slow growth and the unusually high lipid content of the cell wall and envelope [5–7] have contributed to difficulties in diagnosing and treating mycobacterial infections [8].

It is a generally acknowledged estimate that around one-third of the world's population is infected with *M. tuberculosis* [9–11]. Although this is an alarming statistic, the vast majority will not develop tuberculosis. The WHO paints a relatively rosy picture in its web brochure "10 facts about tuberculosis" [12]. Despite this, the WHO currently publishes 38 books dedicated to tuberculosis, thus highlighting the significance of the disease [13].

Fighting Multidrug Resistance with Herbal Extracts,
Essential Oils and Their Components
http://dx.doi.org/10.1016/B978-0-12-398539-2.00005-7

The literature related to tuberculosis continues to grow, with reviews of medicinal plant research being published at regular intervals [14,15], often buried within reviews of plant use for other diseases, e.g., malaria [16]. However, there is a very uneven approach to the problem and much of the published work appears rather *ad hoc*. This I suspect is due to the vagaries of research funding for medicinal plant research.

Tuberculosis is sometimes referred to as the neglected disease, since a majority of sufferers live in the poorer socioeconomic areas of developed or prosperous countries such as South Africa, where the incidence rate is around 948 per 100,000 population [17]. In developing countries, the situation is also severe. Unfortunately, there is no incentive for *Big Pharma* to develop antitubercular drugs, since the affected population cannot pay for them. However, MDR-TB is not restricted to developing countries. In the 1990s, there were great concerns about New York drug users displaying a high incidence of MDR-TB [18]. There was a strong association with HIV coinfection and this brought home to roost the fact that tuberculosis is also alive and well in developed nations.

Systems of traditional medicine from around the world include a considerable number of plant species listed as being used for the treatment of tuberculosis and related respiratory diseases. Identifying these plants and cataloging their phytochemistry is an important start in the search for new antitubercular drugs. A second arm of attack is to study the offending organisms thoroughly. Knowledge of their structure and biochemistry may reveal significant pathways, e.g., mycolic acid synthesis [19,20], that will provide clues for the medicinal chemist to exploit for the effective treatment of tuberculosis. Unlike the majority of fatty acids, which tend to have only a single alkyl chain attached to a carboxyl moiety, mycolic acids possess a distinctive double alkyl strand structure with cyclopropyl groups attached to a β-hydroxy acid moiety. The cyclopropyl groups of the α-mycolic acids have a *cis* conformation, while the keto and methoxymycolic acids exist in both *cis* and *trans* forms (see Fig. 5-1).

Like most plant-based science, there are numerous journals in which the research findings are scattered. Not only are published reports dispersed widely among the dominant English language journals but they also often appear in regional languages e.g., Chinese [20,21], Turkish [21], and Russian [22], and are thus unnoticed by many researchers. These factors contribute to the rich but nonetheless sometimes infuriatingly complex task of keeping abreast of developments in this field, requiring the patient unraveling of numerous threads of research findings.

ORTHODOX PHARMACEUTICALS USED IN THE TREATMENT OF TUBERCULOSIS

The treatment of tuberculosis using Western orthodox therapies has remained relatively stagnant since most of the agents used currently were introduced over 50 years ago [8]. A major difficulty in the treatment of the disease is that of patient compliance. The treatments are unpleasant and daily dosing for an average 6−8 months is not exactly a delightful treat! Much of the drug resistance (up to 36%) appears to have arisen as a consequence of unfinished drug therapy regimens [8].

Among the relatively few compounds that make up the pharmaceutical armamentarium against mycobacterial infections, a range of chemical classes is represented. These include aminoglycosides, bis-amino alcohols (ethylenediamines), pyridines (and pyrazines), and rifamycins as first-line agents, and *p*-aminosalicylic acid, cyclic amino acids and peptides, and the synthetic 4-quinolones [23,24] as second-line agents. In the main, second-line agents exhibit greater toxicity and so their use

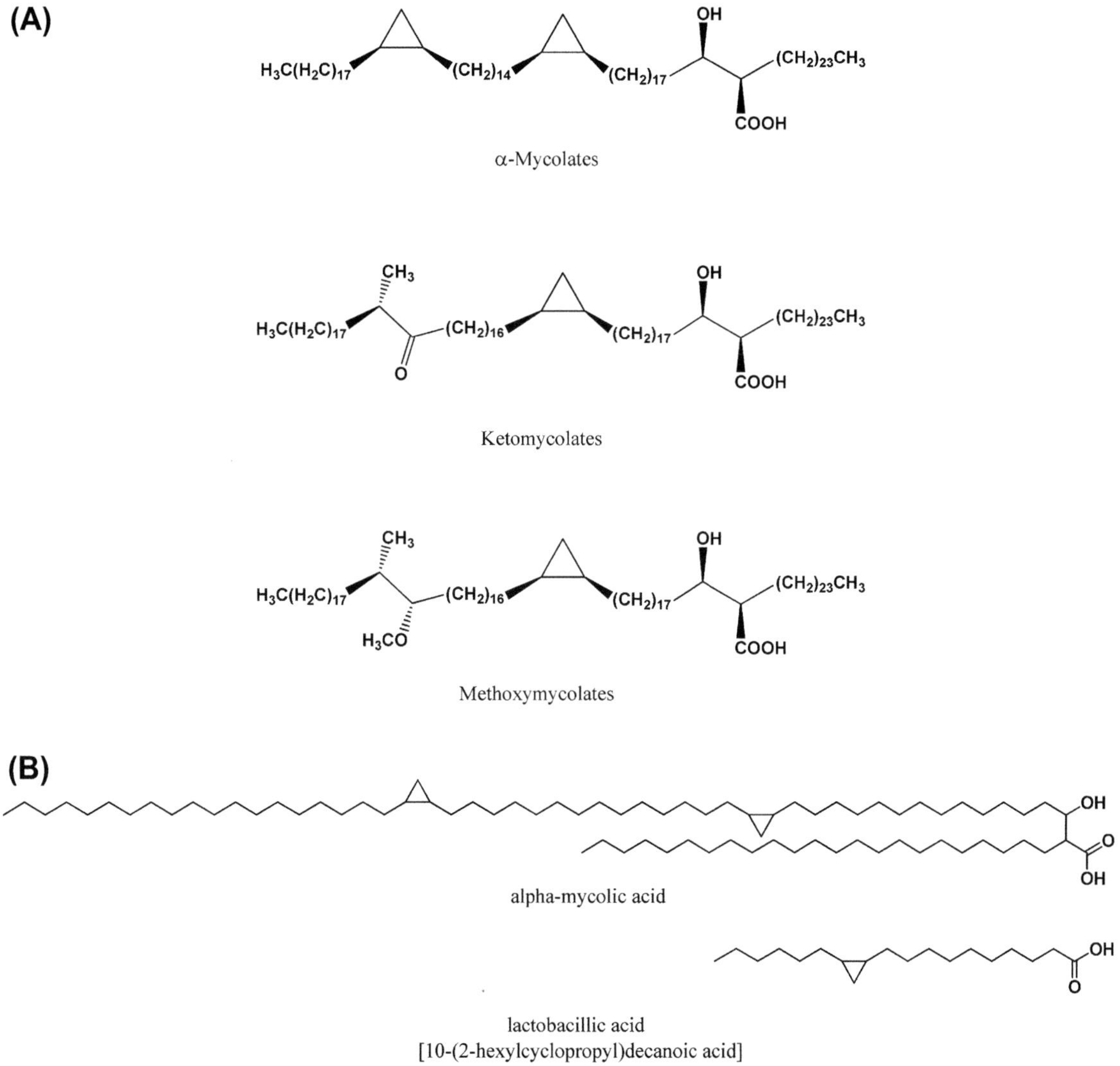

FIGURE 5-1 **Structures of mycobacterial mycolic acids.** (A) Representative mycolic acids. (B) Comparison of an alpha-mycolic acid with lactobacillic acid to highlight the differences between the long twin-chain mycolic acids and the more common shorter single-chain bacterial fatty acids. *Drawn by P. G. Kerr, after Lemke* [14].

is less palatable to the tuberculosis patient, who has to suffer the rigors of both the disease and treatment!

The therapeutic molecules range from very small (cycloserine; MW = 102 Da) to quite large (rifampin; MW approximately 823 Da) and beyond. Compounds with a variety of chemical functionalities are also well represented among antitubercular drugs. Structures of selected agents along with comments regarding their putative mechanisms of action are shown in Table 5-1.

TABLE 5-1   Current Orthodox Tuberculosis Treatments

Chemical class (example)	Structure of representative example	Mechanism of action (where known)
Pyridines (isoniazid)		Inhibition of mycolic acid synthesis
Pyrazines (pyrazinamide)		Prodrug of pyrazinoic acid, which acts by an unknown mechanism; may inhibit synthesis of fatty acids
Rifamycins (rifampin)		Binds to the β subunit of bacterial DNA-dependent RNA polymerase, thereby inhibiting RNA synthesis
Bis-amino alcohols (ethylenediamines and ethambutol)		Inhibits arabinosyl transferases involved in polymerization of arabinoglycan (cell wall)
Aminoglycosides (streptomycin)		Protein synthesis inhibition by binding to the 30S ribosomal subunit

TABLE 5-1   Current Orthodox Tuberculosis Treatments (*cont'd*)

Chemical class (example)	Structure of representative example	Mechanism of action (where known)
Phenolic acids (*p*-aminosalicylic acid)		Folate synthesis antagonist
Cyclic amino acids (cycloserine)		Cell wall synthesis inhibitor acting as a D-alanine mimic
Cyclic peptides (capreomycins)		Protein synthesis inhibition
4-quinolones (moxifloxacin)		Blocks DNA synthesis via topoisomerase II/IV inhibition

An examination of the structures of the pharmacophores and structure-activity correlations of synthetic agents is likely to be instructive for identifying natural leads that share structural similarities with the synthetic drugs in current use. Identifying new therapeutic targets based on the biochemistry of *M. tuberculosis* is a major priority in the search for promising effective agents.

Having said that, it is perhaps instructive to consider the following statement from 1963:

Nearly all bactericidal and fungicidal compounds which are in practical use today have been discovered by empirical methods. They have arisen either as a result of the 'screening' of large numbers of synthetic compounds, or they have been obtained as natural products of the fermentation of microorganisms (antibiotics). Bactericides or fungicides which have been synthesized by deliberate design based upon biochemical knowledge have rarely become of practical importance, though there have been results of great theoretical interest as a consequence of the hypothesis of metabolite antagonism. Nevertheless, the biochemical approach to new agents by a study of metabolism of the organism concerned is becoming increasingly popular, and will doubtless in due course yield practical results of the highest value. The object of such metabolic studies is to discover specific biochemical mechanisms whereby molecules having chemotherapeutic activity may be designed. *W.A. Sexton [25]*

I think this quotation is very instructive in that it illustrates the fact that medicinal chemists were certainly thinking along similar lines, perhaps not too optimistically, but at least suggesting future possibilities, in the early 1960s. It appears that not a great deal has changed since then!

Pyridine Derivatives

Currently, the most important first-line antitubercular agents are pyridine derivatives and isosteric pyrazines. Considering their structural features, it would appear that they act as false cofactors in the biosynthesis of mycolic acids by *Mycobacterium* spp. They possess a hydrazine side chain that appears to be oxidized to generate free radicals and/or reactive oxygen species. These, in turn, act as acylating agents to inactivate the NADH-dependent enzyme system responsible for the synthesis of long chain unsaturated fatty acids that are essential for the synthesis of mycolic acids, the major lipid constituents of the *Mycobacterium* cell wall, which contributes to the permeability barrier of the organism. Unlike those of many organisms, mycobacterial mycolic acids have uncommon structures (see Fig. 5-1). In contrast to the straight chain saturated or unsaturated fatty acids present in the phospholipid membrane, mycolic acids are *alpha branched*, possessing short (20–24 carbon) and long (50–60 carbon) arms [24]. This feature leads to

the formation of a highly impervious lipophilic barrier that prevents the entry of drugs to the organism.

The principal pyridine compounds used are isoniazid (an iso-nicotinic acid derivative) and pyrazinamide (an isostere of nicotinic acid). However, the mechanism of action of the latter compound has not been elucidated. A second-line pyridine derivative, ethionamide, which is not well tolerated by patients, is believed to act via a similar mechanism to that of isoniazid, with ethionamide sulfoxide being the putative active metabolite [23,24].

Rifamycins

Rifamycins, a subclass of ansamycins, are bacterial, macrocyclic products derived from the acylpolymalonate (acetate-malonate) pathway that act by inhibiting the bacterial DNA-dependent RNA polymerase (DDRP). The suggested mechanism for rifamycin binding to DDRP is thought to involve π-π stacking interactions between the naphthalene ring of the drug and an aromatic moiety in the DDRP protein. In addition, because DDRP is a metalloenzyme, it has been postulated that the phenolic -OH groups of the naphthalene ring function as bidentate ligands for zinc atoms in the enzyme [24].

Ethylenediamines

Ethambutol is the sole representative of this class currently in use as a therapeutic agent. The site of action of this drug is unknown but several studies, noted by Lemke [24], have reported inhibition of arabinosyltransferase. This would lead to reduced synthesis of the arabinofuranose-galactose peptidoglycan in the *Mycobacterium* cell outer envelope, which is the point of attachment of the mycolic acids, thus enabling greater penetration of the mycobacterial cell wall by coadministered rifamycins.

Aminoglycosides

Only streptomycin and, to a lesser extent, kanamycin and amikacin are used as injectable antitubercular treatments due to their poor oral bioavailability. The mechanism of action has not been fully elucidated but is believed to involve inhibition of protein synthesis by inducing mRNA misreading that prevents translational initiation [24].

p-Aminosalicylic Acid

Like the antibacterial sulfonamides, p-aminosalicylic acid provides a false substrate for folate synthesis, but bacterial resistance to this second-line antitubercular treatment has diminished its usefulness [24], as have the major unpleasant side effects experienced by patients [23]. It is believed to act via reducing isoniazid acetylation, thus increasing isoniazid levels in plasma. p-Aminosalicylic effectively acts as a suicide substrate for the mycobacterial acetylase that would otherwise deactivate the isoniazid.

Cycloserine

D-cycloserine is considered to be the active form of the drug, which inhibits two enzymes, D-alanine racemase and D-alanine ligase. The naturally occurring, related amino acid is L-alanine, but the *Mycobacterium* converts this to D-alanine for incorporation into its cell wall [24]. Cycloserine thus acts as a false substrate for the enzyme and a competitive inhibitor. Chemically this molecule is a 1,2-oxazolidine-3-one, and this structural feature is potentially useful in the search for antitubercular drugs [26].

Capreomycins

The capreomycin used for the treatment of tuberculosis is a mixture of four cyclopeptide compounds and the mechanism of action is unknown. Based on structure-activity considerations, the drug is likely to inhibit protein synthesis dependent on mRNA at the 70S ribosome, leading to cessation of chain elongation [24]. Side effects are unpleasant and include hepatotoxicity, nephrotoxicity, and ototoxicity [23]. Johansen et al. [27] have reported that isolates of *M. tuberculosis* exhibit resistance to these drugs.

Fluoroquinolones

These are synthetic drugs based on the structure of nalidixic acid (a naphthyridone rather than a quinolone). Their mechanism of action involves binding to DNA gyrase (topoisomerase), thereby inhibiting DNA replication [24]. There is chiral discrimination for levofloxacin (the L-isomer of ofloxacin, which is the racemate) [23]. The naphthyridone pharmacophore may provide structural clues for finding naturally occurring leads in this class of compound.

WORLD TRADITIONS AND ETHNOPHARMACOLOGY OF TUBERCULOSIS

Ancient indigenous traditions are steeped in their particular philosophical systems regarding the causes and treatment of illnesses and, accordingly, until quite recently there has been a lack of modern *scientific* validation of these treatments. This is changing, as illustrated by the rapidly increasing number of papers being published in the expanding number of journals devoted to ethnomedicine, ethnopharmacology, phytomedicine, and traditional and complementary medicines. The literature has become vast and is extensively spread across a range of publishing houses originating in Africa, China, India, and Latin America. It is exciting that many fledgling journals are providing an ever-increasing resource for the chemical and biologic study of natural products.

In my own country of Australia, there has been very limited research into tuberculosis and plant products [28]. It is important to note that before early white settlement (1788) tuberculosis was probably unknown in Australia and therefore indigenous phytomedicines specific for this disease were of no concern [29]. Earlier literature cataloging Australian indigenous medicines is not readily accessible and many of the medicines have been collected and published in secondary sources. These often depend on collections made and studies carried out in the nineteenth and early twentieth centuries. Particularly useful in this regard are the books by Lassak and McCarthy [30], Cribb and Cribb [31], and more recently the compendia of Isaacs [32], Low [33], Latz [34], and the Aboriginal communities of the Northern Territory of Australia [35,36]. All of these sources contain material collated from scattered Australian phytochemical and ethnobotanical studies and so form good starting points for the avid researcher. Unfortunately, the chemistry in these collections is scant, apart from the composition of the essential oils, which are tabulated in the Aboriginal pharmacopoeias under the direction and editorship of Barr [35,36]. Despite the lack of investigation of Australian plants for the treatment of tuberculosis, it is possible that they contain novel compounds active against this disease.

Within the scope of this chapter, a preliminary catalog of some of the plant species that were or are being used as treatments for tuberculosis is provided (see Table 5-2). Where possible, I have sought to organize these plant uses in such a way as to suggest potential new directions in the search for drugs to combat MDR- and XDR-TB. The table is by no means exhaustive, but I hope it is arranged in such a way that it gives a worldwide perspective ranging from anecdotal reports describing how plants are used for treating tuberculosis to the phytochemical and laboratory-based tests that have been performed. It is clear that certain plant families

contribute most to the botanical medicines used against tuberculosis.

Some of the literature surveyed can be viewed as region focused, i.e., the uses of plants from a specific geographical location are reported, including India [37,46] and the Madhupur forest region of Bangladesh, in which the plants used by the Garo tribe are noted [43]. In a survey of 65 plants, only *Piper longum* L. (Piperaceae) was listed as being used to treat tuberculosis. However, an abstract from a 2009 conference presentation by Rahman et al. [38] reports 23 species used against tuberculosis in the Bogra district of Bangladesh.

Journals are no longer being exclusively published by US, UK and European publishing houses and, increasingly, reports of medicinal plants from South American countries are appearing. Surveys have included plants from Colombia, where Lopez et al. [39] surprisingly reported anti-*Mycobacterium phlei* activity almost incidentally for 19 species, using a disk diffusion assay with 2 mg of extract per disk. The study by Bueno-Sanchez et al. [5] found antitubercular activity in the essential oils of eight species (two of which comprised five chemovars). Only two species (see Table 5-2) displayed antitubercular activity below concentrations of around 100 μg/mL. Although it reports modest activity, the study indicates the need for an analysis to identify correlations between activity and test material composition. A study of 37 Brazilian plants used by the people of the Cerrado region reported antitubercular activity from both nonpolar (chloroform) and polar (methanol) solvent extracts [41] at concentrations < 125 μg/mL. However, perhaps somewhat surprisingly, no details of the chemical composition of the extracts were furnished.

I have noted papers dealing, for example, with collections of North African [40,47,45] Mexican [48–54], Peruvian [55–60], Argentinian and Chilean [61,62], and Puerto Rican [42] plant species. The majority of these studies have

TABLE 5-2 Selected Traditional Plant-Based Tuberculosis Remedies from Around the World

Family	Genus	Species	Countries	Reference(s)
Acanthaceae	*Adhatoda*	*vasica*	India, Bangladesh	[37,38]
	Andrographis	*paniculata*	Bangladesh	[38]
Aizoaceae	*Carpobrotus*	spp.	South Africa	[11]
	Galenia	*africana*		
Alliaceae	*Tulbaghia*	*alliacea*		
		violacea		
Apiaceae	*Centella*	*asiatica*	Bangladesh	[38]
Apocynaceae	*Catharanthus*	*roseus*		
	Holarrhena	*antidysenterica*		
Araceae	*Colocasia*	*esculenta*		
	Pistia	*stratiotes*		
Asclepiadaceae	*Asclepias*	*fruticosa*	South Africa	[11]
Asparagaceae	*Protasparagus*	*africanus*		
Asphodelaceae	*Aloe*	*vera*	Bangladesh	[38]
Asteraceae	*Achyrocline*	*alata*	Colombia	[5]
	Calendula	*officinalis*	Bangladesh	[38]
	Conyza	*bonariensis*	Colombia	[39]
	Dittrichia	*graveolens*	South Africa	[11]
	Eupatorium	*glutinosum*	Colombia	[39]
	Helichrysum	*crispum*	South Africa	[11]
		melanacme		
		nudifolium		
		odoratissimum		
		pilosellum		
	Nidorella	*anomala*		
	Nidorella	*auriculata*		
	Senecio	*serratuloides*		
	Tagetes	*erecta*	Colombia	[39]
	Vernonia	*mespilifolia*	South Africa	[11]
Buddlejaceae	*Buddleja*	*saligna*	South Africa	[11]

(Continued)

TABLE 5-2 Selected Traditional Plant-Based Tuberculosis Remedies from Around the World (*cont'd*)

Family	Genus	Species	Countries	Reference(s)
Clusiaceae	*Garcinia*	*kola*	Nigeria	[40]
	Symphonia	*globulifera*	Colombia	[39]
	Vismia	*macrophylla*		
Costaceae	*Costus*	*afer*	Nigeria	[40]
Dipterocarpaceae	*Shorea*	*robusta*	Bangladesh	[38]
Euphorbiaceae	*Acalypha*	*indica*	India	[37]
	Ricinus	*communis*	Bangladesh	[38]
Fabaceae	*Senna*	*reticulata*	Colombia	[39]
Gentianaceae	*Swertia*	*chirata*	Bangladesh	[38]
Juglandaceae	*Juglans*	*neotropica*	Colombia	[39]
Lamiaceae	*Ocimum*	*sanctum*	Bangladesh	[38]
Lecythidaceae	*Eschweilera*	*rufifolia*	Colombia	[39]
Leguminosae	*Indigofera*	*suffruticosa*	Brazil [m]	[41]
Liliaceae	*Allium*	*sativum*	Bangladesh	[38]
Loganiaceae	*Strychnos*	*pseudoquina*	Brazil [c]	[41]
Malphighiaceae	*Byrsonima*	*verbascifolia*	Colombia	[39]
	Byrsonima	*basiloba*	Brazil [c]	[41]
	Byrsonima	*crassa*		
	Byrsonima	*fagifolia*		
Malvaceae	*Hibiscus*	*rosa-sinensis*	Bangladesh	[38]
Melastomataceae	*Miconia*	*cabuku*	Brazil [m]	[41]
	Miconia	*rubiginosa*		
Meliaceae	*Swietenia*	*mahagoni*	Bangladesh, Puerto Rico	[38,42]
Menispermaceae	*Iryanthera*	*megistophylla*	Colombia	[39]
	Iryanthera	*tricornis*	Colombia	[39]
	Tinospora	*cordifolia*	Bangladesh	[38]
Monimiaceae	*Siparuna*	*guianensis*	Colombia	[39]
Moraceae	*Ficus*	*citrifolia*	Puerto Rico	[42]
Myristicaceae	*Virola*	*multinervia*	Colombia	[39]
Myrtaceae	*Eucalyptus*	*globulus*	Bangladesh	[38]
	Myrteola	*nummularia*	Colombia	[39]

TABLE 5-2 Selected Traditional Plant-Based Tuberculosis Remedies from Around the World (*cont'd*)

Family	Genus	Species	Countries	Reference(s)
Nyctaginaceae	*Guapira*	*noxia*	Brazil [m]	[41]
	Neea	*theifera*	Brazil [c]	[41]
	Pisonia	*borinquena*	Puerto Rico	[42]
Piperaceae	*Piper*	*lanceaefolium*	Colombia	[39]
	Piper	*longum*	Bangladesh	[38,43]
Poaceae	*Cymbopogon*	*citratus*	Bangladesh	[38]
Polygonaceae	*Polygonum*	*punctatum*	Colombia	[39]
Pteridophyta	*Adiantum*	*latifolium*	Colombia	[39]
Rhamnaceae	*Ziziphus*	*mauritiana*	Bangladesh	[38]
Rubiaceae	*Duroia*	*hirsuta*	Colombia	[39]
	Morinda	*citrifolia*	Bangladesh	[38]
Rutaceae	*Swinglea*	*glutinosa*	Colombia	[5]
Sapotaceae	*Vitellaria*	*paradoxa*	Nigeria	[40]
Solanaceae	*Solanum*	spp.	Colombia	[39]
Sterculiaceae	*Cola*	*acuminata*	Nigeria	[40]
Verbenaceae	*Acampe*	*papillosa*	India	[44]
	Aerides	*odorata*		
	Bletilla	*striata*		
	Dendrobium	*nobile*		
	Lantana	*camara*	Uganda	[45]
	Lippia	*alba*	Puerto Rico	[42]
	Malaxis	*acuminata*	India	[44]
Vitaceae	*Cissus*	*suscicaulis*	Brazil [c]	[41]
	Vitis	*vinifera*	Bangladesh	[38]
Vochysiaceae	*Qualea*	*grandiflora*	Brazil [c]	[41]
	Qualea	*multiflora*		
Zingiberaceae	*Aframomum*	*melegueta*	Nigeria	[40]

[c] *chloroform extract.*
[m] *methanol extract.*
For Brazil, both nonpolar and polar fractions were tested.

focused on chemical classes, and some pertinent points are briefly discussed in the section devoted to the chemical classes as promising leads.

Of the papers focusing on plant species, Hossain [44] provides an extensive review of the therapeutic uses of orchids, including a large number of Orchidaceae from around the world. Unfortunately, the chemical information provided, although quite extensive, is limited to lists of compound names. This feature is not especially helpful for determining tentative directions for lead drug development. The use of structures for the presentation of chemical information is to be preferred.

When reporting on targeted Peruvian species containing anti-cancer activity Aponte et al. [55] mentioned that some also contained antitubercular activity, virtually as an aside. They isolated eight compounds from five species using bioassay-guided fractionation, evaluated their cytotoxic activities, and found antitubercular activity for the dimeric tertiary indole alkaloid, bisnordihydrotoxiferine. These investigators only considered compounds to be active when used at concentrations < 25 μg/mL. The use of a compound-focused approach was followed by Begum et al. [63], who tested and found a range of moderate activities in a series of β-carboline derivatives at 6.25 μg/mL (see Fig. 5-2). Similarly, flavonoids from *Lantana camara* L. were also investigated [64]. In their review of phthalides present in plants of the Apiaceae family, Beck and Chou [65] only briefly addressed the matter of tuberculosis. However, their extensive catalog of structures offers a tantalizing smorgasbord of compounds from which antitubercular testing and structural modification of leads could be performed. Unfortunately, there did not appear to be any specific references to antitubercular activity in their paper. However, it may be worth following up on their report on senkyunolides A, N, and J (see Fig. 5-2), which exhibit topoisomerase I and II activity, as this is also one of the mechanisms

of action for fluoroquinolones that display antitubercular activity.

Barnes et al. [66] have described the bioassay-guided fractionation of an antitubercular "resin glycoside" from *Ipomoea leptophylla*. This report is an example of the wide array of complex structures available to the natural product chemist and of the extraordinary intellectual effort expended to elucidate the structures of two closely related antitubercular glycolipid compounds.

Following these short descriptions of the various approaches used for identifying promising leads for new antitubercular drugs, it is now pertinent to summarize the major chemical classes that have been identified. As there is insufficient space in this chapter to provide more detailed information about alkaloids, the interested reader may wish to obtain important reviews addressing this large family of compounds [7,15,67]. Figure 5-2 illustrates exemplars of alkaloid skeleta that have been reported to possess antitubercular activity.

CHEMICAL CLASSES AS PROMISING LEADS

An extremely diverse chemistry is exhibited by antitubercular plant compounds. This is illustrated in a review of the potential of isatin (indoline-2,3-dione) derivatives, either natural or synthetic, as antitubercular drugs [9]. Another class of compounds studied is the phenylethanoids from *Buddleja cordata* subsp. *cordata* stem bark extracts [68]. These highly lipophilic simple esters of 2-(4'-hydroxyphenyl)-ethanol derivatives with long chain saturated fatty acids yield compounds that can potentially act as surfactants and lead to cell membrane disruption in *Mycobacterium* spp.

Among the selected reviews of natural products with antimycobacterial potential, those of Garcia et al. [67] and Okunade et al. [7] are especially outstanding. They describe

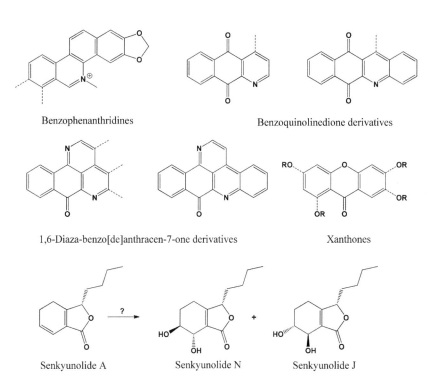

Carbazole and Carboline derivatives

Benzophenanthridines

Benzoquinolinedione derivatives

1,6-Diaza-benzo[de]anthracen-7-one derivatives

Xanthones

Senkyunolide A

Senkyunolide N

Senkyunolide J

FIGURE 5-2 **Selected classes of alkaloids exhibiting antitubercular activity and senkyunolides (phthalides) exhibiting topoisomerase I and II activity.**

a comprehensive range of structures that have been found to possess significant antimycobacterial activity. The latter paper investigated bioactivity that interferes with the glycolysis pathway [7]. Such a focused biochemical and structural chemical approach appears to be an extremely useful way of getting at the *big picture* when searching for new antitubercular drugs. Another, earlier, review by Newton et al. [15] is comprehensive and suggestive of future research directions but it lacks the

pictorial structural detail provided by Okunade et al. The comprehensive table provided in their review contains reports derived from the literature that are often stated only in relation to an extract (e.g., *Alnus rubra* Bong. (Betulaceae); methanol extract of bark; significant activity) and where activity is reported to be due to a specific compound, it is named without reference to structure [e.g., *Amyris elemifera* L. (Rutaceae); parts not specified; taxalin (oxazole) significantly active]. However, despite these limitations, the review [15] is a useful resource.

The review of antitubercular terpenoids by Cantrell et al. [26] is also very instructive. For example, a plot of log P values versus antitubercular activity is a simple way of identifying partition coefficients that are optimal for lead compounds. Here, I have highlighted only selected classes of compounds that appear to be promising sources of drug leads. The various structural categories are grouped according to common biosynthetic origins and some hypothetical explanations for bioactivity against *M. tuberculosis* have been added.

Lower Terpenoids

The mono- and sesquiterpenoids usually comprise the majority of constituents of the volatile oils of medicinal plants. In terms of their antitubercular potential, these molecules have not been well investigated. Cantrell, Franzblau, and Fischer's review of terpenoids reports only three linear monoterpenoid alcohols, viz, citronellol, geraniol, and nerol [26], as exhibiting activity. However, they note the structural similarities between these molecules and their sesquiterpenoid and diterpenoid homologs.

In a review of great mullein (*Verbascum thapsus*), a herb traditionally used in Ireland for chest and lung complaints, McCarthy and O'Mahony [69] suggest that the iridoid glycosides in mullein share aspects of their molecular architecture with streptomycin. This provides another incentive for examining structure-activity relations, and even the use of molecular docking studies, in the pursuit of new antitubercular drugs.

Sesquiterpenoids abound and exhibit varying degrees of antitubercular activity. As per monoterpenoids, the linear molecules nerolidol and farnesol are reported to have amongst the highest activities, along with sesquiterpene lactones of the eudesmanolide, germacranolide,

and guaianolide series [26]. See Figure 5-3 for illustrative structures.

Diterpenoids

Diterpenoids of the abietane and pimarane series, together with aromatized norabietane diterpenoids, have been identified as having antitubercular activity [70,71]. All compounds that were investigated comprise a fused, angular, three six-membered carbocyclic ring structure. A number of ring oxygenations (e.g., phenolic at C-12 or ketonic at C-11, C-12 and/or C-14; see Fig. 5-3) furnish sites of ready oxidation that can generate radicals, which might provide some explanation of their antitubercular behavior. Seco-abietanes also display significant antitubercular activity [72]; these compounds share a similar molecular architecture with their tricyclic cousins. In addition, labdanes were reported to have only weak antitubercular activity [71]. The polyacylated jatrophane diterpenoids from Thai medicinal plants [73] possess mild-to-moderate activity but their structures are very different, consisting of a bicyclic architecture (cyclopentane fused with a dodecadiene or modified moiety). At first glance, their architecture approximates that of the rifamycins.

Triterpenoids

The active pentacyclic triterpenoids isolated from Argentinian and Chilean plants [60] belong to the lupane/hopane or oleanane/ursane skeletal types (see Fig. 5-3). These skeletal types appear in many species used in traditional medicines and, although not mentioned specifically, the Ecuadorian species *Minquartia guianensis* [74] contains not only triterpenoids of this skeletal type but also xanthones, which have also been found in plants attributed with tuberculosis curative properties. Additional triterpenoids, as well as sesquiterpenoids and glycolipids isolated from four Mexican species

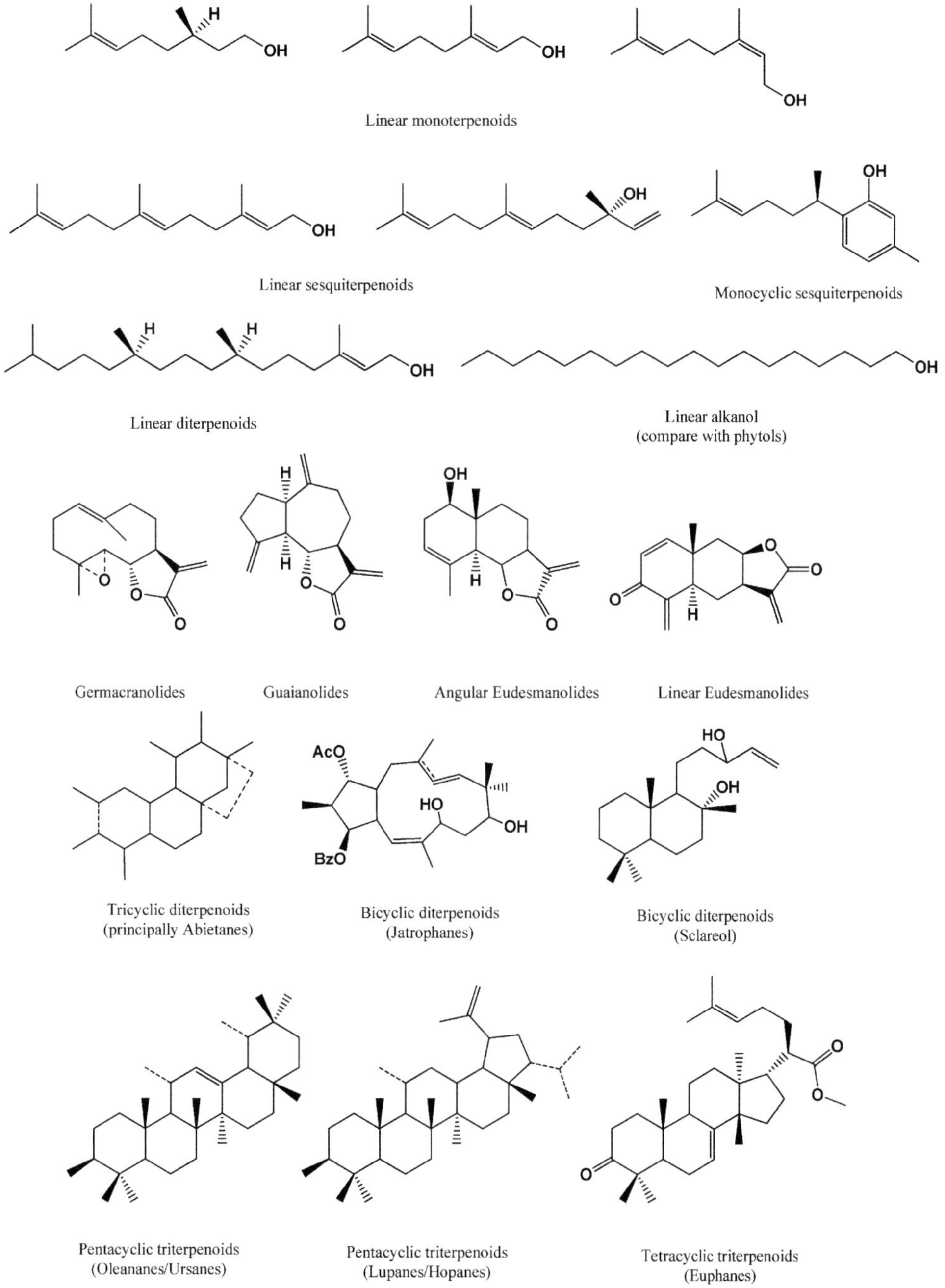

FIGURE 5-3 Some common skeletal features of selected antitubercular active terpenoids.

[52] and the tetracyclic euphanes (see Fig. 5-3) isolated from the East African species *Melia volkensii* [75], further indicate that triterpenoids are significant players in the search for new antitubercular drugs.

Phenolics: Coumarins and Lignans

The Zulu medicine derived from *Pelargonium sidoides* and used to treat *heavy coughs*, is reported to be rich in coumarins and phenolic acids (e.g., caffeic, chlorogenic, and *p*-coumaric acids), but only one coumarin and its 7-glucoside derivative are mentioned by name [76]. This is the 5,6,7-trioxygenated *simple* coumarin known as umckalin (7-hydroxy-5,6-dimethoxycoumarin). The more complete profile of coumarins identified earlier by Kayser and Kolodziej [77] were all highly oxygenated representatives, namely, 6,8-dihydroxy-7-methoxycoumarin, 6,7,8-trihydroxycoumarin, 6,8-dihydroxy-5,7-dimethoxycoumarin, 7-acetoxy-5,6-dimethoxycoumarin, scopoletin, 5,6,7-trimethoxycoumarin, and 5,6,7,8-tetramethoxycoumarin (artelin). It is believed that the oxygenation pattern contributes to the antibacterial activity of these compounds.

Shikimates: Stilbenoids, Xanthones, and Flavonoids

Flavonoids abound in nature and are reported frequently as plant active compounds. Synergism of chalcones with isoniazid has been reported [78] and this might be a useful adjunct to conventional antitubercular pharmacotherapy. Stilbenes have not had a high profile in antitubercular studies, although some have been reported, with the most well-known stilbene, resveratrol, making a brief appearance as a template for the isosteric synthetic azastilbenes [79].

Antitubercular xanthones are quite prominent, with structure-activity studies done in the mid-to-late 1970s reporting activities comparable with that of streptomycin [80,81]. More recent research might lead to a revival in the study of these molecules [6].

Polyketides: Quinones and Naphthoquinones

Various solvent extracts of the South African species *Euclea natalensis* were investigated for their antitubercular activity: the chloroform extract, along with at least three out of four named naphthoquinones, had the *best* activity [82]. Since the naphthoquinones share a structural moiety in common with rifampin, it is quite feasible that these molecules coordinate with the zinc atoms in the metalloenzyme DDRP. As these compounds possess both quinone and phenolic moieties, it is very likely that redox chemistry plays a significant role in their mechanism of action.

THE WAY FORWARD

The future is bright for medicinal and natural product chemists researching antitubercular compounds. A wealth of anecdotal reports is now available online and, with the increasing trend toward open access and electronic journals, virtually limitless information is becoming available to researchers.

Increasingly, researchers are targeting biochemical pathways unique to *M. tuberculosis*, seeking to identify sites of attack (e.g., by enzymes) [83] for natural products. In the main, the studies still involve plant extracts and/or pure compounds, using *in vitro* assays to determine activity.

In conclusion, I want to stress the following points:

(1) The need to look for targets in the organism or disease state under scrutiny by exploiting the unusual cell wall characteristics of

M. tuberculosis and investigating the biochemistry involved.

(2) Look for patterns! For instance, plant use can provide a key. So, ask the following questions

- Which plant families have the greater number of species (e.g., Asteraceae and Verbenaceae) with activity against a particular disease state and which have the greater activity?
- Is there something special about this family or this genus?

(3) Examine the molecular architecture of the active compounds (e.g., triterpenoids). Look for structural trends that correlate with activity.

(4) Do we stick with *in vitro* assays (so often preferred) or should we go *in vivo*?

A caveat for those of us using only *in vitro* studies is that it might prove fruitful to devise and use model systems that mimic *in vivo* metabolic processes. For example, consider the historical case of prontosil, an inactive dyestuff, that exhibits activity only after metabolic activation to yield sulfanilamide, an active antibiotic [84].There are bound to be compounds in *prodrug* forms and, unless metabolic transformation is considered, a potentially useful antitubercular drug might be missed.

References

[1] Anonymous. Address TB/HIV, MDR/XDR-TB and other Challenges. WHO, http://www.who.int/tb/challenges/en/; 2012 [accessed 15.06.12].

[2] Floyd K, Raviglione M, editors. Global Tuberculosis Control: WHO Report 2011. Geneva: Switzerland; 2011. WHO.

[3] Gibbons S. Phytochemicals for Bacterial Resistance—Strengths, Weaknesses and Opportunities. Planta Med 2008;74:594—602.

[4] Tripathi RP, Tewari N, Dwivedi N, Tiwari VK. Fighting Tuberculosis: an Old Disease with New Challenges. Med Res Rev 2005;25:93—131.

[5] Bueno-Sánchez JG, Martínez-Morales JR, Stashenko EE, Ribón W. Anti-tubercular Activity of Eleven Aromatic and Medicinal Plants Occurring in Colombia. Biomédica 2009;29:51—60.

[6] Gautam R, Saklani A, Jachak SM. Indian Medicinal Plants as a Source of Antimycobacterial Agents. J Ethnopharmacol 2007;110:200—34.

[7] Okunade AL, Elvin-Lewis MP, Lewis WH. Natural Antimycobacterial Metabolites: Current Status. Phytochemistry 2004;65:1017—32.

[8] Anonymous. Tuberculosis New Faces of an Old Disease. Médecins Sans Frontières, http://www.msfaccess.org; 2009 [accessed 23.05.12].

[9] Aboul-Fadl T, Bin-Jubair FAS. Anti-Tubercular Activity of Isatin Derivatives. Int J Res Pharm Sci 2010;1:113—26.

[10] Anthony KG, Strych U, Yeung KR, Shoen CS, Perez O, Krause KL, et al. New Classes of Alanine Racemase Inhibitors Identified by High-Throughput Screening Show Antimicrobial Activity against Mycobacterium tuberculosis. PLoS ONE 2011;6: e20374.

[11] McGaw LJ, Lall N, Meyer JJ, Eloff JN. The potential of South African Plants Against Mycobacterium infections. J Ethnopharmacol 2008;119:482—500.

[12] Anonymous. 10 Facts about Tuberculosis. WHO. Retrieved [June 15] from, http://www.who.int/features/factfiles/tuberculosis/en/; 2012.

[13] Anonymous. Who Publications: Tuberculosis. WHO. Retrieved [June 15] from, http://www.who.int/publications/en/; 2012.

[14] Arya V. A Review on Anti-Tubercular Plants. Int J Pharm Tech Res 2011;3:872—80.

[15] Newton SM, Lau C, Wright CW. A Review of Antimycobacterial Natural Products. Phytother Res 2000; 14:303—22.

[16] Cocquyt K, Cos P, Herdewijn P, Maes L, Van den Steen PE, Laekeman G. Ajuga Remota Benth.: from Ethnopharmacology to Phytomedical Perspective in the Treatment of Malaria. Phytomedicine 2011;18: 1229—37.

[17] Anonymous. South Africa and Tuberculosis. Center for Strategic & International Studies. Retrieved [May 16] from, http://www.smartglobalhealth.org/blog/entry/south-africa-and-tuberculosis/; 2012.

[18] Frieden T, Sterling T, Pablos-Mendez A, Kilburn JO, Cauthen GM, Dooley SW. The Emergence of Drug-Resistant Tuberculosis in New York City. N Engl J Med 1993;328:521—6.

[19] Garcia A, Bocanegra-Garcia V, Palma-Nicolas JP, Rivera G. Recent Advances in Antitubercular Natural Products. Eur J Med Chem 2012;49:1—23.

[20] Lakshmanan D, Werngren J, Jose L, Suja KP, Nair MS, Varma RL, et al. Ethyl p-methoxycinnamate Isolated from a Traditional Anti-Tuberculosis Medicinal Herb

Inhibits Drug Resistant Strains of Mycobacterium Tuberculosis *in vitro*. Fitoterapia 2011;82:757—61.

[21] Değer M. [The Ottoman Pamphlet on Rosmary "Risale-i Büberiye"]. Tıp tarihi araştırmaları = History of medicine studies 1993;5:152—70.

[22] Kurilenko NA. [Use of Aloe in Complex Treatment of Patients with Focal Pulmonary Tuberculosis]. Vrachebnoe delo August 1974:110—1.

[23] Chambers HF. Antimycobacterial Drugs. In: Katzung BG, editor. Basic and clinical pharmacology 9th edit. New York: McGraw-Hill; 2004.

[24] Lemke TL. Antimycobacterial Agents. In: Lemke TL, Williams DA, editors. Foye's principles of medicinal chemistry. 7th ed. Philadelphia: Lippincott, Williams & Wilkins; 2013.

[25] Sexton WA. Chemical Constitution and Biological Activity. 3rd ed. Princeton, New Jersey: D. Van Nostrand Company Inc; 1963.

[26] Cantrell CL, Franzblau SG, Fischer NH. Antimycobacterial Plant Terpenoids. Planta Med 2001;67: 685—94.

[27] Johansen SK, Maus CE, Plikaytis BB, Douthwaite S. Capreomycin Binds Across the Ribosomal Subunit Interface Using TlyA-Encoded 2'-O-methylations in 16S and 23S rRNAs. Mol Cell 2006;23:173—82.

[28] Meilak M, Palombo EA. Anti-Mycobacterial Activity of Extracts Derived from Australian Medicinal Plants. Res J Microbiol 2008;3:535—8.

[29] Australian Indigenous HealthInfoNet. Summary of tuberculosis among Indigenous peoples. Retrieved [September 15] from, http://www.healthinfonet.ecu.edu.au/infectious-conditions/tuberculosis/reviews/our-review; 2008.

[30] Lassak EV, McCarthy T. Australian Medicinal Plants. Sydney: Methuen; 1983.

[31] Cribb AB, Cribb JW. Wild Medicine in Australia. Sydney: Fontana/Collins; 1981.

[32] Isaacs J. Bush Food: Aboriginal Food and Herbal Medicine. Sydney: Lansdowne Publishing Pty Ltd; 1987.

[33] Low T. Bush Medicine: a Pharmacopoeia of Natural Remedies. North Ryde, NSW: Angus & Robertson; 1990.

[34] Latz PK. Bushfires and Bushtucker: Aboriginal Plant use in Central Australia. Alice Springs, NT: IAD Press; 1995.

[35] Aboriginal Communities of the Northern Territory. Traditional Aboriginal Medicines in the Northern Territory of Australia. In: Barr A, Chapman J, Smith N, Wightman G, Knight T, Mills L, Andrews M, Alexander V, editors. Conservation Commission of the Northern Territory of Australia; 1993. Darwin.

[36] Aboriginal Communities of the Northern Territory of Australia. Traditional Bush Medicines: an Aboriginal Pharmacopoeia. In: Barr A, Chapman J, Smith N, editors. Darwin: Greenhouse Publications Pty Ltd; 1988.

[37] Gupta R, Thakur B, Singh P, Singh HB, Sharma VD, Katoch VM, et al. Anti-tuberculosis Activity of Selected Medicinal Plants Against Multi-Drug Resistant Mycobacterium Tuberculosis Isolates. The Indian Journal of Medical Research 2010;131:809—13.

[38] Rahman F, Hossan S, Mollik A, Islam T, Jahan R, Taufiq-Ur-Rahman M, et al. Medicinal Plants Used Against Tuberculosis by Traditional Medicinal Practitioners of Bogra District, Bangladesh. In: Hostettmann K, editor. 57th International Congress and Annual Meeting of the Society for Medicinal Plant and Natural Product Research. Geneva, Switzerland: Georg Thieme Verlag; 2009.

[39] Lopez A, Hudson JB, Towers GH. Antiviral and Antimicrobial Activities of Colombian Medicinal Plants. J Ethnopharmacol 2001;77:189—96.

[40] Ogbole OO, Ajaiyeoba EO. Traditional Management of Tuberculosis in Ogun State of Nigeria: The Practice and Ethnobotanical Survey. Afr J Tradit, Complementary Altern Med 2010;7:79—84.

[41] Pavan FR, Sato DN, Higuchi CT, Santos ACB, Vilegas W, Leite CQF. *In vitro* Anti-Mycobacterium Tuberculosis Activity of Some Brazilian "Cerrado" Plants. Brazilian Journal of Pharmacognosy 2009;19: 204—6.

[42] Antoun MD, Ramos Z, Vazques J, Oquendo I, Proctor GR, Gerena L, et al. Evaluation of the Flora of Puerto Rico for in vitro Antiplasmodial and Antimycobacterial Activities. Phytother Res 2001;15: 638—42.

[43] Mia M-u-k, Kadir MF, Hossan S, Rahmatullah M. Medicinal Plants of the Garo Tribe Inhabiting the Madhupur Forest Region of Bangladesh. American Eurasian Journal of Sustainable Agriculture 2009;3: 165—71.

[44] Hossain MM. Therapeutic Orchids: Traditional Uses and Recent Advances—an Overview. Fitoterapia 2011;82:102—40.

[45] Kirimuhuzya C, Waako P, Joloba M, Odyek O. The Anti-mycobacterial Activity of Lantana Aamara a Plant Traditionally Used to Treat Symptoms of Tuberculosis in South-western Uganda. African Health Sciences 2009;9:40—5.

[46] Gupta VK, Shukla C, Bisht GR, Saikia D, Kumar S, Thakur RL. Detection of Anti-tuberculosis Activity in Some Folklore Plants by Radiometric BACTEC Assay. Lett Appl Microbiol 2011;52:33—40.

[47] Adeleye IA, Onubogu CC, Ayolabi CI, Isawumi AO, Nshiogu ME. Screening of Crude Extracts of Twelve Medicinal Plants and " Wonder-cure " Concoction Used in Nigeria Unorthodox Medicine for Activity Against Mycobacterium Tuberculosis. Afr J Biotechnol 2008;7:3182−7.

[48] Bocanegra-Garcia V, Camacho-Corona MdR, Ramirez-Cabrera M, Rivera G, Garza-Gonzalez E. The Bioactivity of Plant Extracts Against Representative Bacterial Pathogens of the Lower Respiratory Tract. BMC Res Notes 2009;2:95.

[49] Camacho-Corona M d R, Ramirez-Cabrera MA, Santiago OG, Garza-Gonzalez E, Palacios Ide P, Luna-Herrera J. Activity Against Drug Resistant-tuberculosis Strains of Plants Used in Mexican Traditional Medicine to Treat Tuberculosis and Other Respiratory Diseases. Phytother Res 2008;22:82−5.

[50] Cruz-Vega DE, Verde-Star MJ, Salinas-Gonzalez N, Rosales-Hernandez B, Estrada-Garcia I, Mendez-Aragon P, et al. Antimycobacterial Activity of Juglans Regia, Juglans Mollis, Carya Illinoensis and Bocconia Frutescens. Phytother Res 2008;22:557−9.

[51] Jimenez-Arellanes A, Meckes M, Ramirez R, Torres J, Luna-Herrera J. Activity Against Multidrug-resistant Mycobacterium Tuberculosis in Mexican Plants Used to Treat Respiratory Diseases. Phytother Res 2003;17: 903−8.

[52] León-Díaz R, Meckes M, Said-Fernández S, Molina-Salinas GM, Vargas-Villarreal J, Torres J, et al. Antimycobacterial Neolignans Isolated from Aristolochia Taliscana. Mem Inst Oswaldo Cruz 2010;105:45−51.

[53] Molina-Salinas GM, Perez-Lopez A, Becerril-Montes P, Salazar-Aranda R, Said-Fernandez S, de Torres NW. Evaluation of the Flora of Northern Mexico for in vitro Antimicrobial and Antituberculosis Activity. J Ethnopharmacol 2007;109:435−41.

[54] Rivero-Cruz I, Acevedo L, Guerrero JA, Martinez S, Bye R, Pereda-Miranda R, et al. Antimycobacterial Agents from Selected Mexican Medicinal Plants. J Pharm Pharmacol 2005;57:1117−26.

[55] Aponte JC, Vaisberg AJ, Rojas R, Caviedes L, Lewis WH, Lamas G, et al. Isolation of Cytotoxic Metabolites from Targeted Peruvian Amazonian Medicinal Plants. J Nat Prod 2008;71:102−5.

[56] Aponte JC, Yang H, Vaisberg AJ, Castillo D, Málaga E, Verástegui M, et al. Cytotoxic and Anti-infective Sesquiterpenes Present in Plagiochila Disticha (Plagiochilaceae) and Ambrosia Peruviana (Asteraceae). Planta Med 2010;76:705−7.

[57] Graham JG, Pendland SL, Prause JL, Danzinger LH, Schunke Vigo J, Cabieses F, et al. Antimycobacterial Evaluation of Peruvian Plants. Phytomedicine 2003;10:528−35.

[58] Lewis WH, Lamas G, Vaisberg A, Corley DG, Sarasara C. Peruvian Medicinal Plant Sources of New Pharmaceuticals (International Cooperative Biodiversity Group-Peru). Pharm Biol 1999;37:69−83.

[59] Rojas R, Bustamante B, Bauer J, Fernández I, Albán J, Lock O. Antimicrobial Activity of Selected Peruvian Medicinal Plants. J Ethnopharmacol 2003;88:199−204.

[60] Rojas R, Bustamante B, Ventosilla P, Fernádez I, Caviedes L, Gilman RH, et al. Larvicidal, Antimycobacterial and Antifungal Compounds from the Bark of the Peruvian Plant Swartzia Polyphylla DC. Chem Pharm Bull 2006;54:278−9.

[61] Wächter GA, Franzblau SG, Montenegro G, Suarez E, Fortunato RH, Saavedra E, et al. A New Antitubercular Mulinane Diterpenoid from Azorella Madreporica Clos. J Nat Prod 1998;61:965−8.

[62] Wächter GA, Valcic S, Flagg ML, Franzblau SG, Montenegro G, Suarez E, et al. Antitubercular Activity of Pentacyclic Triterpenoids from Plants of Argentina and Chile. Phytomedicine 1999;6:341−5.

[63] Begum S, Hassan SI, Siddiqui BS. Synthesis and Antimycobacterial Activity of Some β-carboline Alkaloids. Nat Prod Res 2004;18:341−7.

[64] Begum S, Wahab A, Siddiqui BS. Antimycobacterial Activity of Flavonoids from Lantana Camara Linn. Nat Prod Res 2008;22:467−70.

[65] Beck JJ, Chou S-C. The Structural Diversity of Phthalides from the Apiaceae. J Nat Prod 2007;70: 891−900.

[66] Barnes CC, Smalley MK, Manfredi KP, Kindscher K, Loring H, Sheeley DM. Characterization of an Antituberculosis Resin Glycoside from the Prairie Medicinal Plant Ipomoea Leptophylla. J Nat Prod 2003;66: 1457−62.

[67] García A, Bocanegra-García V, Palma-Nicolás JP, Rivera G. Recent Advances in Antitubercular Natural Products. Eur J Med Chem 2012;49:1−23.

[68] Acevedo L, Martínez E, Castaæeda P, Franzblau S, Timmermann BN, Linares E, et al. New Phenylethanoids from Buddleja Cordata Subsp. Cordata. Planta Med 2000;66:257−61.

[69] McCarthy E, O'Mahony JM. What's in a Name? Can Mullein Weed Beat TB Where Modern Drugs Are Failing? Evid. Based Complement. Alternat Med 2011;2011:239237.

[70] Ulubelen A, Topcu G, Johansson CB. Norditerpenoids and Diterpenoids from Salvia Multicaulis with Antituberculous Activity. J Nat Prod 1997;60:1275−80.

[71] Hussein AA, Meyer JJM, Jimeno ML, Rodríguez B. Bioactive Diterpenes from Orthosiphon Labiatus and Salvia Africana-lutea. J Nat Prod 2007;70:293−5.

[72] Chen J-J, Wu H-M, Peng C-F, Chen I-S, Chu S-D. Seco-Abietane Diterpenoids, a Phenylethanoid Derivative,

and Antitubercular Constituents from Callicarpa Pilosissima. J Nat Prod 2009;72:223—8.

[73] Mongkolvisut W, Sutthivaiyakit S. Antimalarial and Antituberculous Poly-O-acylated Jatrophane Diterpenoids from Pedilanthus Tithymaloides. J Nat Prod 2007;70:1434—8.

[74] El-Seedi HR, Hazell AC, Torssell KBG. Triterpenes, Lichexanthone and an Acetylenic Acid from Minquartia Guianensis. Phytochemistry 1994;35:1297—9.

[75] Cantrell CL, Rajab MS, Franzblau SG, Fischer NH. Antimycobacterial Triterpenes from Melia Volkensii. J Nat Prod 1999;62:546—8.

[76] Bladt S, Wagner H. From the Zulu Medicine to the European Phytomedicine Umckaloabo. Phytomedicine 2007;14(Suppl. 6):2—4.

[77] Kayser O, Kolodziej H. Highly Oxygenated Coumarins from Pelargonium Sidoides. Phytochemistry 1995;39:1181—5.

[78] Mativandlela SP, Meyer JJ, Hussein AA, Houghton PJ, Hamilton CJ, Lall N. Activity Against Mycobacterium Smegmatis and M. tuberculosis by Extract of South African Medicinal Plants. Phytother Res 2008;22: 841—5.

[79] Pavan FR, de Carvalho GS, da Silva AD, Leite CQ. Synthesis and Anti-Mycobacterium Tuberculosis Evaluation of Aza-Stilbene Derivatives. Scientific World Journal 2011;11:1113—9.

[80] Ghosal S, Biswas K, Chaudhuri RK. Chemical Constituents of Gentianaceae XXIV Anti-Mycobacterium Tuberculosis Activity of Naturally Occurring Xanthones and Synthetic Analogs. J Pharm Sci 1978;67:721—2.

[81] Ghosal S, Chaudhuri RK. Chemical Constituents of Gentianaceae XVI Antitubercular Activity of Xanthones of Canscora Decussata Schult. J Pharm Sci 1975;64:888—9.

[82] McGaw LJ, Lall N, Hlokwe TM, Michel AL, Meyer JJM, Eloff JN. Purified Compounds and Extracts from Euclea Species with Antimycobacterial Activity Against Mycobacterium Bovis and Fast-growing Mycobacteria. Biol Pharm Bull 2008;31: 1429—33.

[83] Guzman JD, Gupta A, Evangelopoulos D, Basavannacharya C, Pabon LC, Plazas EA, et al. Anti-tubercular Screening of Natural Products from Colombian Plants: 3-methoxynordomesticine, an Inhibitor of MurE Ligase of Mycobacterium Tuberculosis. J Antimicrob Chemother 2010; 65:2101—7.

[84] Zavod RM, Knittel JJ. Drug Design and Relationship of Functional Groups to Pharmacological Activity. In: Lemke TL, Williams DA, editors. Foye's principles of medicinal chemistry. 7th ed. Philadelphia: Lippincott, Williams & Wilkins; 2013. p. 29—60.

6

Use of Essential Oils and Their Components against Multidrug-Resistant Bacteria

M.L. Faleiro[1], M.G. Miguel[2,3]

[1]Centro de Biomedicina Molecular e Estrutural, Instituto de Biotecnologia e Bioengenharia, Faculdade de Ciências e Tecnologia, Universidade do Algarve, Campus de Gambelas 8005–139 Faro, Portugal, [2]Centro Biotecnologia Vegetal, Instituto de Biotecnologia e Bioengenharia, Faculdade Ciências Lisboa, Universidade de Lisboa, Portugal and [3]Departamento de Química e Farmácia, Faculdade de Ciências e Tecnologia, Universidade do Algarve, Portugal

INTRODUCTION

Human health has greatly benefit from the use of antimicrobial chemotherapy, in particular during the twentieth century; however, the successful results of saving millions of lives also led to the false idea that infectious diseases would be defeated within a short time. In spite of this, extremely antibiotic resistant bacteria have emerged and there is great concern over their effects on human and animal welfare. The environmental spread of antibiotic-resistant bacteria and the extent of antibiotic resistance to food microbiota are extremely worrying [1,2]. Besides becoming exceptionally resistant to antibiotics, the virulence potential of both Gram-negative (e.g., *Acinetobacter* spp., *Escherichia coli*, *Klebsiella* spp., and *Pseudomonas aeruginosa*) and Gram-positive (e.g., *Staphylococcus aureus*, *Streptococcus pneumoniae*, *Clostridium difficile*) bacteria can be enhanced, as seen in the dramatic *E. coli* 0104:H4 outbreak in Germany [3,4]. In addition to dramatic events of this type is the ongoing struggle to identify new and effective antibiotics, especially against Gram-negative bacteria; new classes of effective antibiotics have not yet emerged and the prospects of this happening in the near future are low [5,6].

Clinicians frequently reach a therapeutic limit when treating patients with serious bacterial infections and efforts are being made worldwide to find new treatments for resistant bacteria [7,8]. New approaches include educating the public, improving sanitation infrastructures, imposing strict

Fighting Multidrug Resistance with Herbal Extracts,
Essential Oils and Their Components
http://dx.doi.org/10.1016/B978-0-12-398539-2.00006-9

regulations on antibiotic prescription, developing new antibiotics, reevaluating old and rejected antibiotics, abolishing the nontherapeutic use of antibiotics (e.g., in agriculture and animal feed), and developing nonantibiotic approaches that can successfully prevent and protect against infectious diseases [7,8]. The last category of new approaches for the combat of antibacterial resistance include the development of antibacterial vaccines, bacteriophages, probiotics, and substances of plant origin such as essential oils (EOs) [7,8]. EOs and their components have shown excellent results against multidrug-resistant (MDR) bacteria such as methicillin-resistant *S. aureus* [9–11]. However, efforts should be made to improve the use of these promising substances in daily practice. This chapter summarizes recent advances in our knowledge of the activity of EOs and their components against MDR bacteria.

ESSENTIAL OIL COMPOSITION

EOs (or volatile oils) are complex mixtures of volatile constituents that are biosynthesized by aromatic plants. They can be synthesized in buds, flowers, leaves, stems, twigs, seeds, fruits, roots, wood, or bark, and stored in secretory cells, cavities, canals, epidermal cells, or glandular trichomes. Some of the volatile compounds may be toxic at high concentrations to the plant itself. One way to overcome this problem is to store such volatile compounds as inactive precursors, for instance as glycosides, or in extracellular compartments, as in the case of glandular trichomes [12,13].

Generally, EOs are liquid at room temperature but few are solid or resinous. The oils are lipid soluble and soluble in organic solvents and generally have a lower density than that of water. They can be of different colors, ranging from pale yellow to emerald green and from blue to dark brownish red [14,15].

EOs are obtained by hydrodistillation, steam distillation, or dry distillation of a plant or plant part, or by a suitable mechanical process without heating (e.g., for citrus fruits). Vacuum distillation; solvent extraction combined offline with distillation; simultaneous distillation-extraction; supercritical fluid extraction; microwave-assisted extraction and hydrodistillation; and static, dynamic, and high concentration capacity headspace sampling are other techniques used for extracting the volatile fraction from aromatic plants, although the products of these processes cannot be termed EOs [16]. The yield and chemical profile of the extraction product depends on the type of extraction, climate, soil composition, and the plant organ, age, and vegetative cycle stage [14,17,18].

The components of the EOs include regular terpenoids (mono-, sesqui-, and diterpenes), terpenoids with irregular carbon skeletons (homoterpenes and norisoprenoids), products of the lipoxygenase pathway (oxylipins), volatile fatty acid derivatives, indoles, and phenolics, including methyl salicylate and aromatic compounds [13,19–24].

Terpenoid Compounds of Volatile Oils

Monoterpenes (C_{10}) have two isoprene units ($2 \times C_5$), whereas sesquiterpenes (C_{15}) have three isoprene units ($3 \times C_5$). Although considerable amounts of monoterpenes and sesquiterpenes are produced in leaf glandular trichomes, some authors suggest that monoterpenes are typical leaf products, whereas sesquiterpenes are typical flower fragrances [21].

All terpenoids originate from the isopentenyl diphosphate (IPP) and its allylic isomer dimethylallyl diphosphate (DMAPP), which are derived from two alternative pathways: IPP is derived from three molecules of acetyl-CoA via the mevalonic acid pathway in the cytosol; and DMAPP is derived from pyruvate and glyceraldehyde3-phosphate via the methylerythritol phosphate pathway (also known as the deoxy-xylulose-5-phosphate pathway) in plastids [25,26]. The two pathways are not completely autonomous and there is a unidirectional cross talk between the two IPP biosynthetic pathways from the plastids to the cytoplasm mediated by specific metabolite transporters [27,28].

In plastids, the condensation of one molecule of IPP and one molecule of DMAPP is catalyzed by geranyl pyrophosphate synthase, giving rise to GPP (C_{10}), the universal precursor of all monoterpenes. In the cytosol, the condensation of two molecules of IPP and one molecule of DMAPP, catalyzed by the enzyme farnesyl pyrophosphate synthase (FPPS), forms farnesyl pyrophosphate (C_{15}), the natural precursor of sesquiterpenes [29]. The condensation of one molecule of DMAPP and three molecules of IPP is catalyzed by geranylgeranyl pyrophosphate synthase, yielding geranylgeranyl pyrophosphate, the C_{20} diphosphate precursor of diterpenes [30]. The conversion of the various prenyl pyrophosphates, DMAPP (C_5), GPP (C_{10}), FPP (C_{15}), and geranylgeranyl diphosphate (C_{20}), to hemiterpenes (isoprene and 2-methyl-3-buten-2-ol), monoterpenes, sesquiterpenes, and diterpenes, respectively, is carried out by a large family of enzymes known as terpene synthases/cyclases (TPSs) [31].

Many terpene volatiles are direct products of terpene synthases, but others are formed through transformation of the initial products by acylation, dehydrogenation, oxidation, and other reaction types [32]. Examples include the 3-hydroxylation of limonene by a P450 enzyme to yield the *trans*-isopiperitenol, a volatile compound in mint [33], or the 6-hydroxylation of limonene by another P450 enzyme, resulting in the formation of *trans*-carveol, which undergoes further oxidation by nonspecific dehydrogenase to form carvone, an aroma volatile of caraway fruits [34,35]. Geraniol may undergo oxidation to the corresponding aldehyde geranial by dehydrogenases in sweet basil [36] or reduction to (*S*)-citronellol [37].

Acetylation may also occur, catalyzed by acetyltransferase to produce geranyl acetate [38].

Irregular acyclic C_{16} (4,8,12-trimethyltrideca-1,3,7,11-tetraene) and C_{11} (4,8-dimethylnona-1,3,7-triene) carbons skeletons, known as homoterpenes, result from modification reactions of geranyl linalool (C_{20}) and (3*S*)-(*E*)-nerolidol (C_{15}), respectively, by oxidative degradation, possibly catalyzed by cytochrome P450 enzymes [21,39]. These homoterpenes are the most typical compounds relevant to herbivore feeding [13].

Plants also produce irregular volatiles with carbon skeletons ranging from C_8 to C_{18}, derived from carotenoids (C_{40}). Examples include β-ionone, β-damascenone and dihydroactinidiolide found in the scent of flowers of *Rosa hybrida*, *Osmanthus fragrans*, or *Freesia hybrida*, or in the aroma of fruits and vegetables of *Averrhoa carambola* (star fruit) or *Lycopersicon esculentum* (tomato) [21,40,41]. The biosynthesis of carotenoid-derived volatile compounds (norisoprenoids) occurs via three steps: an initial dioxygenase cleavage, yielding apocarotenoids; followed by enzymatic transformations leading to the formation of polar aroma precursors; and finally acid-catalyzed conversions of these precursors to volatile compounds [41]. In some cases, a volatile product is the result of the initial dioxygenase cleavage step, as reported for β-ionone in *Arabidopsis* [42,43].

Volatiles of the Lipoxygenase Pathway (Oxylipins) or Volatile Fatty Acid Derivatives

Oxylipins are derived from polyunsaturated fatty acids (linoleic and linolenic acids) released from chloroplast membranes by lipase activity. These fatty acids can undergo dioxygenation by lipoxygenases (LOX) that catalyze the oxygenation of polyenoic fatty acids at the C_9 and C_{13} positions (9-LOX and 13-LOX, respectively). Oxidation at these positions yields the

9-hydroperoxy and the 13-hydroperoxy derivatives of polyenoic fatty acids. These derivatives can be the substrates of allene oxide synthase (AOS) and hydroperoxide lyase (HPL) to form volatile compounds [21,44,45]. Octadecanoids and jasmonates and formed by 13-allene oxide synthase (13-AOS) activity, whereas short chain C_6- or C_9- volatile aldehydes (3-hexenal or 3,6-nonadienal) and the corresponding C_{12} or C_9 ω-oxo fatty acids (12-oxo-dodecenoic acid or 9-oxo-nonanoic acid) and alcohols are formed by the activity of hydroperoxy lyases (13-HPLs) [13,21].

In addition, green leaf volatiles are generally defined as saturated and unsaturated C_6 alcohols (3-hexenol), aldehydes (3-Z-hexenal and 2-hexenal), and esters (3-hexenyl acetate), as well as C_5 compounds (2-pentenyl acetate and 2-pentenol). These volatiles are also derived from the cleavage of C_{18} acids (linoleic and linolenic acid) by HPLs to form C_{12} and C_6 compounds [13].

Phenylpropanoids and Benzenoids

Phenylpropanoids are derived from L-phenylalanine after conversion to trans-cinnamic acid catalyzed by L-phenylalanine ammonia-lyase. Hydroxycinnamic acids, aldehydes, and alcohols are metabolites obtained after hydroxylation and methylation of hydroxycinnamic acid esters and their corresponding aldehydes and alcohols. Some of these intermediates can form volatile compounds such as eugenol and isoeugenol, formed from coniferyl acetate. Methylation of those compounds forms methyleugenol and isomethyleugenol, respectively. Estragole or chavicol is another aromatic compound derived from the phenylpropanoid pathway via p-coumaryl alcohol [46–48].

The C_6-C_1 benzenoids are derived from C_6-C_3 phenylpropanoids through the elimination of two carbons of the three-carbon chain attached to the phenyl ring of phenylpropanoids; therefore, these benzenoids are also originated from trans-cinnamic acid. Several routes have been proposed for shortening the trans-cinnamic side chain: via a CoA-dependent β-oxidative pathway (this β-oxidative pathway is analogous to that underlying the β-oxidation of fatty acids); a CoA-independent non-β-oxidative pathway (involving hydration of the free trans-cinnamic acid to 3-hydroxy-3-phenylpropionic acid, side chain degradation via a reverse aldol reaction with formation of benzaldehyde, and its oxidation to benzoic acid), or via a combination of these two mechanisms [21,49].

The biosynthesis of C_6-C_2 compounds does not occur via trans-cinnamic acid [21,49]. For example, in petunia, phenylacetaldehyde is formed directly from phenylalanine via an unusual combined decarboxylation-amine oxidation reaction catalyzed by phenylacetaldehyde synthase, whereas in tomato its biosynthesis occurs via two separate steps, starting with the decarboxylation of phenylalanine to phenylethylamine [21,50,51].

Other Volatile Aromatic Compounds

The volatile indole is made in maize by the cleavage of indole-3-glycerol phosphate, an intermediate in tryptophan biosynthesis [13]. This volatile can be found in the EO of jasmine but only when obtained by the enfleurage process [52]; methyl anthranilate has been detected in jasmine oil [52]. Indole may also derive from chorismate, via anthranilate, an intermediate of phenylalanine.

PERSPECTIVE ON THE USE OF ESSENTIAL OILS

Volatile compounds synthesized by plants, such as EOs, have several properties and are consequently exploited by humans for diverse purposes, e.g., for flavoring food and drink; as food preservatives; in perfumes and cosmetics; in massage oils mixed with vegetable oils;

added to in bathwater; in household cleaning products; in dentistry; in agriculture; and as natural remedies and aromatherapy [53]. In addition, EOs have also been used by different human populations for other purposes (summarized in Table 6-1). Therapeutic properties have also been attributed to the EOs, and they have been used as agents for preventing and/or treating cardiovascular diseases, analgesics, antibacterial, antifungal, and antiviral treatments, anticancer treatments, anti-inflammatories, antinociceptives, antioxidants, antiphlogistics, antipyretics, antispasmodics, cholinesterase inhibitors, insect repellents, penetration enhancers for improving transdermal drug delivery, among others [56–62].

TABLE 6-1   Some Examples of Essential Oils and Their Properties

Common name	Botanical name	Family	Main components	Applications
Aniseed oil	*Pimpinella anisum* L.	Apiaceae	*trans*-Anethole	Carminative, stimulant, expectorant, condiment, and flavoring agent
Cinnamon oil	*Cinnamomum zeylanicum* Nees, sin. *C. verum*, J. S. Presl	Lauraceae	*trans*-Cinnamaldehyde	Carminative, stomachic, astringent, stimulant, and antiseptic
Clove oil	*Syzygium aromaticum* (L.) Merill et L. M. Perry sin. *Eugenia caryophyllus* (C. Spreng) Bull et Harr.	Myrtaceae	Eugenol	Dental analgesic, carminative, stimulant, and antiseptic
Eucalyptus oil	*Eucalyptus globulus* Labillardière	Myrtaceae	1,8-Cineole	Counterirritant, antiseptic, expectorant, and cough reliever
Geranium oil	*Pelargonium graveolens* L.	Geraniaceae	Citronellol, geraniol	Flavoring agent and stimulant
Lavender oil	*Lavandula angustifolia* P. Miller F., sin. *L. officinalis* Chaix	Labiatae	Linalool, linalyl acetate	Stimulant and flavoring agent
Lemon oil	*Citrus limon* (L.) Burm.f.	Rutaceae	Limonene	Carminative, stimulant, perfuming, and flavoring agent
Lemongrass oil	*Cymbopogon citratus* D.C.	Poaceae	Neral, geranial, citronella	Flavoring agent, antiseptic, and deodorant
Orange oil	*Citrus sinensis* (L.) Osbeck (*Citrus aurantium* L. var. *dulcis* L.)	Rutaceae	Limonene	Carminative, stimulant, perfuming, and flavoring agent
Peppermint oil	*Mentha* × *piperita* L.	Labiatae	Menthol, menthone	Stomachic, stimulant, and tonic
Rosemary oil	*Rosmarinus officinalis* L.	Labiatae	1,8-Cineole, camphor, verbenone	Carminative, stimulant, and flavoring agent

Sources: references [54,55]

The antiseptic qualities of aromatic and medicinal plants and their EOs have been known since antiquity. The antimicrobial properties of EOs against a wide range of Gram-positive and Gram-negative bacteria are the reason why they were used for embalming in ancient Egypt [63]. Nevertheless, it was only in the late nineteenth century that the first efforts were made to characterize these properties in the laboratory [64,65]. During this period, Buchholtz studied the capacity of caraway oil, thyme oil, phenol, and thymol to inhibit bacterial growth and showed that thymol is a more effective antimicrobial than phenol. As consequence, thymol started to be used as surgical antiseptic at that time [65].

However, over the years it has become evident that evaluation of the antimicrobial activities of EOs is a very difficult task because they are complex mixtures that are poorly soluble in water and generally volatile [66].

Pauli and Schilcher [65] compiled the results of three different antimicrobial tests (agar diffusion test, dilution test, and vapor phase test) on 28 EOs (anise, bitter-fennel fruit, caraway, cassia, Ceylon cinnamon bark, Ceylon cinnamon leaf, citronella, clary sage, clove, coriander, dwarf pine, eucalyptus, juniper, lavender, lemon, mandarin, matricaria, mint, neroli, nutmeg, peppermint, *Pinus sylvestris*, turpentine, rosemary, star anise, sweet orange, tea tree, and thyme oils) listed in the *European Pharmacopoeia 6th edition*. This comprehensive study by the authors generated great variability in the results, which may be explained by the following factors: natural variability in the composition of EOs, natural variability in the susceptibility of microorganisms, different parameters of the microbiologic testing methods, the unknown history of the EOs tested (their production, storage conditions, and age), and insufficient knowledge about exact phytochemical composition [65].

Most EO research is directed toward analyzing *in vitro* antimicrobial activity and there is a marked absence of corresponding *in vivo* human studies, despite good results apparently being obtained by some therapists treating infections with topical and systematic applications of EOs [67]. Some examples of the *in vivo* antimicrobial activity of EOs follow. EOs of *Ocimum gratissimum* (thymol chemotype) and *Melaleuca alternifolia* can significantly reduce the severity of acne caused by *Propionibacterium acnes*. *M. alternifolia* oil is also reported to be effective as a topical decolonization agent for methicillin-resistant *S. aureus* (MRSA) carriers. In addition, bitter orange oil has been shown to be useful for treating patients with tinea corporis; *M. alternifolia* oil has shown efficacy in the treatment of toenail onychomycosis and intertriginous tinea pedis; *Eucalyptus pauciflora* oil has been reported to combat dermatophytosis; and *Zataria multiflora* oil has been reported to reduce vaginal burning, vaginal pruritus, vulvar pruritus, and urinary burning caused by *Candida albicans* [67].

BACTERIAL CELL TARGETS AND ANTIBIOTIC MODE OF ACTION

Currently, increasing knowledge about the mode of action of EOs against bacterial pathogens means that it is now possible to have a more complete picture of the bacterial cellular targets of EOs. The mode of action of EOs was recently reviewed [68,69] and is summarized in Fig. 6-1.

To date, studies have shown that EO bacterial cell targets include the cell wall and membrane, thereby disturbing ATP production and pH homeostasis. EOs can affect the cellular transcriptome, proteome, and the quorum-sensing system. The following sections summarize the EO target sites that have been identified.

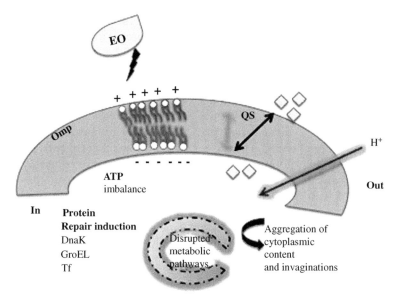

FIGURE 6-1 **Bacterial cell structures and cellular processes disrupted by the activity of essential oils or their components.** Essential oil-treated cells are more permeable to protons, experience an ATP imbalance, and induce the synthesis of chaperones. Metabolic pathways can be affected. EO, essential oil; Omp, outer membrane protein; QS, quorum sensing. *Reprinted from Faleiro [68] with permission from the Formatex Research Center.*

Effects on the Bacterial Cell Wall and Membrane Structure

Treatment with EOs leads to increased bacterial cell permeability, resulting in loss of cellular material [19,70–73], and it is important to note that injury to a cellular structure can result in many downstream effects on cellular functions [74]. Ultrastructural analysis by atomic force microscopy, scanning electron microscopy, and transmission electron microscopy (TEM) has shown severe-to-moderate changes in the surface of several bacterial pathogens, including *Aeromonas hydrophila* exposed to *Origanum vulgare* and *Rosmarinus officinalis* EOs [73] and both Gram-positive and Gram-negative foodborne pathogens treated with carvacrol (3.3 mM) [75]. Following treatment with the same EO (oregano and clove), Gram-positive and Gram-negative bacteria display different types of injuries on their cell surface [76,77]. *E. coli* cells exposed to oregano and clove EOs showed the formation of holes in the cell surface, in contrast to *Bacillus subtilis*, which merely exhibited cell surface malformation

[76]. Variation in the effects of EOs on different bacteria can be explained by differences in their cell wall structure. Other studies reported similar results, for example Hafedh et al. [78] showed that treatment with *Mentha longifolia* EO led to the formation of holes on surface of *E. coli* and *Salmonella enterica* serovar Typhimurium cells; in contrast, *Micrococcus luteus* and *S. aureus* only exhibited a slight disturbance of the cell surface under the same conditions.

The main constituents of *Cuminum cyminum* EO are cumin aldehyde (25%), γ-terpinene (19%), *p*-mentha-1,4-dien-7-al (16.6%), *p*-mentha-1,3-dien-7-al (13%), β-pinene (10.4%), and *p*-cymene (7.2%). Treatment with *C. cyminum* EO reduced capsule formation and induced the formation of filamentous *Klebsiella pneumoniae* cells [79].

Treatment of *P. aeruginosa* and *S. aureus* with *Origanum compactum* EO resulted in a more significant injury to *P. aeruginosa* cells than to *S. aureus*. At the minimum inhibitory concentration (MIC) and at 1.5 × MIC. *P. aeruginosa* cells were unable to grow, cell permeability was disturbed, and membrane potential was disrupted. In contrast, no such detrimental effects were observed in *S. aureus* cells; in particular at the MIC, the membrane potential and permeability were not significantly disturbed and only at 1.5 × MIC was there a significant reduction in viability and membrane potential [70]. When exposed to sublethal concentrations of the terpenes carvacrol and 1,8-cineole (2.5 μL/mL and 10 μL/mL, respectively) *P. fluorescens* experienced a series of morphologic alterations and increased membrane permeability that resulted on shrinkage of the protoplasm and leakage of cell contents [80]. In the study conducted by Bouhdid et al., the activity of *Cinnamomum verum* EO (*E*-cinnamaldehyde is the major component at 73.35%) against *P. aeruginosa* and *S. aureus* cells was monitored by plate counts, potassium outflow, flow cytometry and TEM [71]; there were marked differences in response between the two bacteria although they showed a similar MIC [0.125% (v/v)]: *P. aeruginosa* cell membrane permeability was more disturbed following treatment with the MIC of *C. verum* EO, but no significant effects on *S. aureus* membrane permeability were observed. There were also different effects on membrane potential between the two bacteria species: at the MIC and at 1.5 × MIC, a large percentage of *P. aeruginosa* cells showed altered membrane potential, whereas no effect on membrane potential was observed in *S. aureus* cells. TEM observations showed the greatest structural alterations to *P. aeruginosa* cells, manifested by coagulated material and projections from the cell wall and cell leakage; in contrast, *S. aureus* cells produced fibers around the cell surface. Furthermore, flow cytometry analysis indicated that in the presence of *C. verum* EO *S. aureus* enters in a viable but nonculturable state (VBNC) [71]. This physiologic state is the consequence of a set of metabolic alterations, including diminished nutrient transport, lower respiration rates, and slower biosynthesis, which contributes to their inability to grow on medium that normally supports *in vitro* growth; however, under these conditions the cells can remain viable for long periods of time [81]. When a pathogen is in the VBNC state, it is not expected to cause disease although its virulence potential is preserved and the cycle of infection restarts when the cells are resuscitated and recover their full metabolic capacity. Similar to many other stressor agents, EOs can induce the VBNC state, which poses a risk to both the food industry and health institutions. Therefore, it is important that this physiologic response is taken into account when EO activity is being evaluated.

Salmonella ser. Typhimurium exposed to eugenol (4-allyl-2-methoxyphenol) at 1% and 5% (v/v) developed increased membrane permeability followed by leakage of the cell contents [82]. In contrast, the activity of eugenol (5 mM) on *Listeria monocytogenes* cells results in inhibition of the uptake and utilization of glucose [83]. The production of invaginations in the cytoplasmic membrane, together with the development of a thicker cell wall and aggregation of the cytoplasmic content, was observed in *S. aureus* cells exposed to *Inula graveolens* EO [bornyl acetate (43.3%) and borneol (26.2%) are the main components] or *Santolina corsica* EO [myrcene (34.6%) and santolina triene (13.5%) are present in significant amounts] at the MICs (5 mg/mL for both) [84]. Geraniol, a component of lemongrass oil, also disrupts the *S. aureus* cell membrane, accompanied by cytoplasmic outflow [85]. The same bacterial target structures were identified when geraniol and citronellol were tested against *S. pneumoniae* [86]. In addition, the exposure of *P. acnes* cells to citronella oil (citronellal 34.40%, geraniol 23.40%, and citronellol 11.10%) caused cell elongation, rupture of the cell wall, and cytoplasmic membrane damage, with the consequent loss of the cellular constituents [87].

A series of vital cellular functions rely on an intact cell membrane, including the energy production, nutrient uptake and processing, synthesis of macromolecules, and secretion. Once the cell membrane is disturbed by EOs, or by any other agent, those processes will be compromised. In prokaryotes, the production of ATP takes place in the membrane and the cytosol by glycolysis. Therefore, one of the consequences of EO activity on the cell membrane is disturbance of ATP production, both intracellular and extracellular. Indeed, ATP losses caused by treatment with EOs have been reported [88–93].

Pathogens secrete their toxins through the cell membrane and EOs therefore indirectly affect toxin secretion by disturbing the cell membrane. For example, enterotoxin production of *S. aureus* was completely eliminated following treatment with oregano EO at 0.3–15 μL/mL [72], and study showed that treatment with an EO rich in α-linolenic acid (perilla oil) inhibited the production of α-toxin, enterotoxins A and B, and toxic shock syndrome toxin 1 by both methicillin-sensitive *S. aureus* (MSSA) and MRSA [94].

The indirect targeting action against MDR bacteria is extremely interesting, since the virulence potential of the pathogen can be diminished without killing the pathogen, thus reducing the pressure to develop resistance mechanisms [94].

pH Homeostasis

The exposure of bacterial cells to conditions that compromise their survival induce the cells to adopt a series of strategies to accomplish the most essential processes, in particular DNA transcription, protein synthesis, and enzymatic activity; maintaining intracellular pH (pH_{in}) is crucial to reach this goal [95]. A significant reduction of pH_{in} has been reported in bacterial cells exposed to EOs [88,96]. The action of EOs in disrupting pH homeostasis is comparable to the action of weak acids [88], since both EOs and weak acids are lipophilic and, similar to weak organic acids, EOs are highly lipid permeable and can cause proton influx [88,97–99]. If proton levels exceed the cytoplasmic buffering capacity or overwhelm proton efflux systems, pH_{in} decreases and

essential cellular functions are compromised [100–102]. Disruption of pH homeostasis by EOs is a consequence of their action on bacterial membranes, which become incapable of blocking proton entry [88,96,103]. The use of EOs from Spanish oregano at 0.025% (v/v), Chinese cinnamon at 0.025% (v/v), and savory at 0.05% against *E. coli* 0157:H7 caused a reduction in pH_{in}. However, this became more pronounced when Chinese cinnamon was used, i.e., the pH_{in} decreased from 7.25 ± 0.20 to $5.16\,0.05$, compared to a reduction from 7.25 ± 0.20 to 6.68 ± 0.37 when the same concentration of Spanish oregano EO was applied [75]. Carvacrol at 0.25–0.5 mM caused a decrease in the pH gradient of *B. cereus* cells, and an increase in the concentration to 1 mM caused complete disruption of the pH gradient [90].

Overexpression of Chaperone and Cell Surface Proteins

As mentioned above, several vital cellular processes can be compromised when bacterial cells face stress conditions; in particular, protein synthesis can be severely affected. Two studies reported the action of EO components on protein synthesis [104,105]. One group of researchers observed heat shock protein induction in *E. coli* 0157:H7 cells treated with carvacrol [104]; moreover, flagellin synthesis was inhibited, with the consequent production of nonmotile cells [104]. In the second study [105], *Salmonella enterica* serovar Thompson cells were treated with a sublethal concentration of thymol and their proteome was analyzed by two-dimensional electrophoresis. The proteomics data showed that exposure of *S. enterica* serovar Thompson to 0.01% thymol caused overexpression of a group of molecular chaperone proteins (DnaK, GroEL, HtpG, and the Trigger factor Tf) and outer membrane-associated proteins (OmpX and two OmpA proteins). Moreover, they also observed upregulation of proteins related to citrate metabolism and ATP synthesis.

The results of the proteomics experiment suggest that thymol has a broad range of activity, affecting different cellular processes [105].

The Quorum-Sensing Target

In Nature, bacteria live in communities and can modulate their population density by using small signaling molecules to assess the outside environment and their physiologic response. This capacity is termed *quorum sensing* (QS) and involves cell-cell communication. QS molecules are known as autoinducers, and different types can be found in Gram-negative and Gram-positive bacteria: autoinducers produced by Gram-negative bacteria are acyl homoserine lactones (HSLs) and those produced by Gram-positive bacteria are modified oligopeptides [106]. Important cellular processes such as biofilm production, motility, stress resistance, swarming, and virulence are regulated by QS [107,108]. Due to its importance to bacterial survival, QS has become an interesting target for applications in different areas, including medicine [109,110]. The potential use of plant extracts, including EOs, as anti-QS drug has been investigated [111–118]. Methods employed to evaluate inhibitory effects on QS include evaluating the bioluminescence of *Vibrio harveyi*, which is mediated by 3-hydroxy-C4-HSL and the autoinducer-2 (AI-2); measuring transcription levels of HSLs [113]; and monitoring the production of violacein, which is induced by N-hexanoyl homoserine lactone in the bacterium *Chromobacterium violaceum* CV026 [116,118]. Niu et al. [113] observed that cinnamaldehyde affects transcription of two HSLs, 3-oxo-C$_6$-HSL and 3-oxo-C$_{12}$-HSL, and the bioluminescence of *V. harveyi*, which is mediated by 3-hydroxy-C$_4$-HSL and the autoinducer-2 (AI-2). In the study by Szabó et al. [116], the sensor strain *C. violaceum* CV026 and the N-acyl homoserine lactone-producing *E. coli* strain ATCC 31298 were used to test the anti-QS activity of several EOs. The most active EOs against QS were found

to be clove, geranium, lavender, rose, and rosemary; in contrast, chamomile, juniper, and orange EOs showed no anti-QS activity. Olivero et al. [118] also used the sensor strain *C. violaceum* CV026 to evaluate the QS inhibitory activity of EOs of different *Piper* sp. The EO containing the best anti-QS activity was found to be that isolated from *P. bredemeyeri* [118].

Other Cellular Alterations

Diminished enzymatic activity and the appearance of coagulated material have been reported to result from EO activity [79,119]. Urease activity is important to the survival of pathogens such as *Helicobacter pylori* and *K. pneumoniae*, as these bacteria can hydrolyze urea to ammonia and carbon dioxide through their urease activity. Urease inhibitory activity was reported in the study of Derakhshan et al. [79]: the urease activity of *K. pneumoniae* was impaired by the action of *C. cyminum* EO at 4–32%. Considering the emergence of MDR *K. pneumoniae*, the results obtained by Derakhshan et al. [79] should stimulate the search for urease inhibitory activities in EOs and their components. Oregano and cinnamon EOs have been reported to cause intracytoplasmic changes to *E. coli* and *S. aureus* [79].

MECHANISMS OF ANTIBIOTIC RESISTANCE

The ability of bacteria to develop antibacterial resistance has largely exceeded our expectations. It started to be noticed soon after the clinical application of penicillin in the emerging penicillinase-producing *S. aureus* and evolved with the development of other antimicrobial agents and their abuse [120,121]. The various mechanisms of antibiotic resistance have been the subject of recent reviews [122–124] and we present an overview of EO applications against MDR bacteria in this section. It is important to distinguish between the different mechanisms through which bacteria show resistance: intrinsic resistance and acquired resistance.

Intrinsic Resistance

Some bacterial species are naturally resistant to antibiotics owing to the absence of antibiotic targets or the presence of natural barriers to antibiotic activity. The most common example of intrinsic resistance is that of Gram-negative bacteria to penicillin or vancomycin, which is a consequence of the antibiotic being unable to penetrate the outer bacterial membrane, and the resistance of enterococci to cephalosporins, which is a result of the expression of low affinity penicillin-binding proteins (PBPs) by these bacteria [125–126]. On the other hand, acquired resistance results from a modification to the bacterial genome that prevents the function of a previously effective antibiotic. In this case, antibiotic activity may still be detected, although at a diminished level that is no longer useful for combating the infectious disease [126]. The exceptional resistance phenotype represents another type of resistance mechanism exhibited by some bacterial species that are resistant to specific antimicrobial agents that are very rare or not yet described [126]. The resistance of members of the Enterobacteriaceae family to carbapenems (which, although rare, is increasing) and of *S. aureus* to vancomycin are examples of such exceptional phenotypes [126]. The main mechanisms of acquired resistance in bacteria will be examined in the following section and are summarized in Table 6-2.

Acquired Resistance

The cellular elements involved in the developed resistance include:

(1) Enzymes such as β-lactamases that inactivate antibiotics by disabling the mechanism of action;

(2) Efflux pumps that force out the antibiotic from the cell, thus preventing its activity;

(3) Modifications to bacterial targets that prevent the antibiotic function;

(4) Decreased membrane permeability and transport mechanisms that prevent the antibiotic reaching the cytoplasm; and

(5) Overproduction of the target [123,124,127] (see Table 6-2).

Genes associated with antibiotic resistance or that affect the expression levels of bacterial cellular components are mainly acquired by horizontal gene transfer. The molecular elements that carry these genes can be the bacterial chromosome itself, as well as bacteriophages, integrons, plasmids, and transposons, (Table 6-3) [133]. Exchange of genetic material between bacteria can be performed by different mechanisms, including conjugation (by means of plasmids and transposons), transformation (by means of chromosomal DNA integration into the plasmids or chromosomal DNA of lysing bacterial cells), and transduction (by means of bacteriophages) [133].

In addition to these mechanisms, other cellular responses have been associated with the antibiotic resistance such as the SOS response and the resistance present in biofilm cells [142,143]. The SOS response is a conserved regulatory mechanism that is induced when bacterial cells are exposed to stress conditions that cause DNA damage; it is necessary for repairing the DNA damage and restoring DNA replication [144], thus allowing bacterial cells to survive different environmental stresses. This system contains two main molecular elements: the repressor LexA and the activator RecA. During normal environmental conditions, LexA is attached to SOS boxes through the promoter region of the SOS regulon, thus preventing transcription of the SOS regulon genes; however, as soon as bacterial cells experience DNA damage or DNA replication is interrupted, RecA binds to single-stranded DNA

TABLE 6-2 Recognized Mechanisms of Acquired Bacterial Resistance to Antibiotics

Resistance mechanism	Example	Reference(s)
Enzymes that destroy or inactivate the antibiotic	Acetyltransferase, adenyltransferase, β-lactamases (β-lactams), phosphoryltransferases (aminoglycosides, chloramphenicol), thioltransferase (fosfomycin)	[122,124,127–129]
Modified target	Altered peptidoglycan cross-link (glycopeptides), altered penicillin-binding proteins (β-lactams), methylation of adenine residues prevents the antibiotic binding to the 50S ribosomal subunit (macrolides-lincosamides-streptogramins B), mutation generating a reduction on binding to active site(s) (oxazolidinones and quinolones), production of proteins that attach to the ribosome and modify the conformation of the active site (tetracyclines)	[124,129–132]
Efflux	Efflux pumps (aminoglycosides, chloramphenicol, fluoroquinolones, β-lactams, macrolides, and tetracyclines), novel membrane transporters (chloramphenicol, quinolones, and tetracyclines)	[130,132]
Decreased membrane permeability	Altered membrane permeability due to porin loss or other barriers (aminoglycosides, fusidic acid, quinolones)	[123,124]
Overproduction of the target	Excess peptidoglycan (glycopeptides), overproduction of dihydrofolate reductase (trimethoprim)	[124,129]

(ssDNA) that results from the injury. RecA binding to ssDNA leads to the formation of a nucleoprotein filament (RecA-ssDNA) and RecA activation, which causes LexA removal from the SOS boxes by autoproteolysis and thus derepression (induction) of the SOS regulon ([142] and references therein). The SOS response is linked to the development of antibiotic resistance, as the increased number of mutations following activation of the SOS system ultimately results in the appearance of the resistance mechanisms described above ([142] and references therein).

The acquisition of antibiotic resistance by biofilm cells (sessile cells or attached cells) is associated with several mechanisms. For example, as sessile cells are embedded in a polymer matrix, antibiotic diffusion will be diminished within the bacterial biofilm population. Moreover, a subpopulation of cells known as persisters (i.e., phenotypic variants of the normal bacterial population that appear after exposure to antibiotics and display a noninherited bacterial resistance [145]) can appear, bacterial physiology can be modified toward the development of resistance by the presence of stress conditions within the biofilm (e.g., oxidative stress, nutrient starvation, and pH alteration), and the mutation frequency may increase ([123,143] and references therein).

CONTROLLING MULTIDRUG-RESISTANT BACTERIA

As described in the section "*Bacterial Cell Targets and Antibiotic Mode of Action*" and illustrated in Figure 6-1, EOs can compromise bacterial cell integrity due to their action on one or more cell components, leading to growth inhibition or, at appropriate concentrations, cell death. Thus, the use of EOs to combat or control the

TABLE 6-3 Genetic Elements Involved in Bacterial Resistance Dissemination

Type	Description	Example	Reference(s)
Chromosomal	Bacterial chromosomal DNA is subject to mutations within drug targets	Quinolones—mutation in *gyrA* in enteric Gram-negative bacteria and *Staphylococcus aureus*; rifampin—mutation in *rpoB* of *Mycobacterium tuberculosis*	[134,135]
Plasmids	Circular DNA capable of self-replication that permits rapid synthesis of a large number of gene copies, including those of antibiotic resistance (1–100 Kb)	β-Lactams—*mecA* in *S. aureus*, *ampC* in *Escherichia coli*, *Klebsiella* spp., and *Salmonella* spp.	[136,137]
Transposons	DNA cassette that can move from one location to another in both chromosomes and plasmid. The antibiotic resistance sequence is flanked by insertion sequences (< 2.5Kb, containing terminal inverted repeats and encoding a transposase)	Tn5—kanamycin, streptomycin; Tn7—spectinomycin, streptomycin, trimethoprim; Tn1546—glycopeptides	[138,139]
Integrons	Integrons comprise a system that can capture unrelated genes and assemble them into arrays using site-specific recombination, which guarantees the correct expression of functional genes. An integrase, a primary recombination site, and a promoter that controls expression constitute the main parts of an integron	Four classes are recognized: Class 1—multiple single determinants and efflux pump of multidrug resistance; Class 2—spectinomycin, streptomycin, streptothricin, and trimethoprim; Class3—carbapenems; Class 4—*Vibrio* spp. superintegron	[133,140]
Bacteriophages	Bacterial virus that inserts antibiotic resistance genes into the bacterial chromosome	Mu and environmental phage DNA	[141]

development of MDR bacteria has been investigated and results show that they have an excellent potential to accomplish this purpose. In this section, relevant examples of the use of EOs against MDR bacteria will be described. A general overview is provided in Tables 6-4 and 6-5.

The majority of reports on the activity of EOs against MDR bacteria relate to *in vitro* studies. Although they constitute a strong body of evidence, the number of clinical reports is low. However, three clinical reports on the use of EOs as antibacterial agents are of considerable note [167–169]. One of the clinical cases relates to the application of a formulation containing

a mixture of EOs (clove, eucalyptus, lemongrass, tea tree, and thyme), named *polytoxinol*, in a MRSA infection of the lower tibia [167]. The second report [168] describes the successful use of tea tree oil (TTO) by inhalation to combat a *Mycobacterium tuberculosis* infection prior to the administration of conventional antitubercular medication (ethambutol, isoniazid, pyrazinamide, and rifampin). The use of TTO was reported by Edmondson et al. [169], who applied it to eliminate MRSA in wounds. Although the authors reported that 8 out of 11 patients had smaller wounds at the conclusion of the study, the application of TTO at 3.3%

TABLE 6-4 Essential Oils Active Against Multidrug-Resistant Bacteria, Identified Bacterial Cell Targets, and Mode of Action

Plant origin of essential oil (Common name)	Main components of essential oil (% v/v)	Target multidrug-resistant bacteria	MIC or inhibition zone	Bacterial cell target and/or mode of action[a]	Reference(s)
Allium sativum (Garlic)	Diallyl disulfide 38.9 ± 1.2, diallyl trisulfide 36.4 ± 2.2, 1-allyl-3-methyl trisulfide 18.7 ± 0.04	MRSA	8.3–530 μL/L[b]	Disorder of the cellular membrane, quorum sensing	[77,111,146]
Armoracia rusticana (Horseradish)	Allyl isothiocyanate 76.7 ± 1.9, phenylethyl isothiocyanate 23.3 ± 1.9	MRSA	8.3–17 μL/L[b]	N/A	[146]
Citrus sinensis (Orange)	Linalool 20.2, decanal 18.0, geranial 9.1, α-terpineol 5.8, valencene 5.2, neral 5, dodecanal 4.1, citronellal 3.9	MRSA; vancomycin-intermediate-resistant *Staphylococcus aureus*	31.50 ± 3.02–76.67 ± 2.74 mm; 78.8 ± 1.8 mm	Cell wall, malformed septum and cell lysis	[147–149]
Eucalyptus globulus (Tasmanian bluegum)	Fruit oil: aromadendrene 31.17, 1,8-cineole 14.55, globulol 10.69	MRSA; VRE; *Escherichia coli*; *Pseudomonas aeruginosa*; *Klebsiella pneumoniae*; *Acinetobacter baumannii*	0.25–1 mg/mL; 0.25–1 mg/mL; 8 mg/mL; > 8 mg/mL; > 8 mg/mL; 1 mg/mL	N/A	[10,150]
	Leaf oil: 1,8-cineole 86.51, α-pinene 4.74, γ-terpinene 2.57	MRSA; VRE; *E. coli*; *P. aeruginosa*; *K. pneumoniae*; *A. baumannii*	2–> 4 mg/mL; > 4 mg/mL; > 4 mg/mL; > 4 mg/mL; > 4 mg/mL; 2 mg/mL	N/A	[10,150]
Eucalyptus radiata (Narrow-leaf peppermint gum)	1,8-cineole 82.66, α-pinene 3.68, α-terpineol 7.03	MRSA; VRE; *E. coli*; *P. aeruginosa*; *K. pneumoniae*; *A. baumannii*	4–> 4 mg/mL; > 4 mg/mL; > 4 mg/mL; > 4 mg/mL; 1 mg/mL	N/A	[150]
Eucalyptus citriodora (Lemon-scented gum)	Citronellal 90.07, citronellol 4.32, β-caryophyllene 1.46	MRSA; VRE; *E. coli*; *P. aeruginosa*; *K. pneumoniae*; *A. baumannii*	2–> 4 mg/mL; > 4 mg/mL; > 4 mg/mL; > 4 mg/mL; 2 mg/mL	N/A	[150]

Plant (common name)	Major components	Bacteria	Concentration	Mechanism	Ref
Foeniculum vulgare (Sweet fennel)	trans-Anethole 80.73, estragole 5.18, limonene 8.03, fenchone 1.96	ESBL *K. pneumoniae*	0.008–0.064 mg/mL	N/A	[151]
Lantana achyranthiofolia (Brushland shrub verbena)	Carvacrol 30.64, 1,8-cineole 5.03, isocaryophyllene 10.73, β-bisabolol 11.23, β-bisabolene 5.68	Multiresistant *Staphylococcus epidermidis*; multiresistant *Bacillus subtilis*	0.50 mg/mL; 0.50 mg/mL	N/A	[152]
Lantana camara (Lantana)	Bicyclogermacrene 19.42, isocaryophyllene 16.70 valencene 12.94, germacrene D 12.34	Multiresistant *E. coli*; multiresistant *S. aureus*	0.51 mg/mL; 0.25 mg/mL	N/A	[153]
Melaleuca alternifolia (Tea tree)	Terpinen-4-ol 40.1, γ-terpinene 23.0, α-terpinene 10.4, 1,8-cineol 5.1	MRSA	0.25–2 % (v/v)	Coagulation and leakage of cellular contents, production of extracellular vesicles, inhibition of respiration, leakage of K^+	[154–160]
Mentha piperita (Peppermint)	Isomenthone 50.08, Menthol 21.77, p-menthone 4.19, 1,8-cineol 3.83	ESBL *K. pneumoniae*	0.008–0.064 mg/mL	N/A	[151]
Mentha spicata (Spearmint)	Carvone 75.07, limonene 7.84, cis-dihydrocarvone 4.08, 1,8-cineol 2.07	ESBL *K. pneumoniae*	0.008–0.064 mg/mL	NA	[151]
Ocimum basilicum (Basil)	Linalool 54.94, methyl chavicol 11.97, methylcinnamat 7.24	MRSA; Multiresistant *S. epidermidis*; multiresistant *Enterococcus faecalis*; multiresistant *P. aeruginosa*	0.0015–0.0030% v/v; 0.0015–0.0030% v/v; 0.0015–0.0030% v/v; 0.0030% v/v	Membrane permeabilization	[146,161,175]
	Linalool 75.94, 1,8-cineol 7.73, geraniol 2.40	ESBL *K. pneumoniae*	0.008–0.064 mg/mL	N/A	[151]

(Continued)

TABLE 6-4 Essential Oils Active Against Multidrug-Resistant Bacteria, Identified Bacterial Cell Targets, and Mode of Action (cont'd)

Plant origin of essential oil (Common name)	Main components of essential oil (% v/v)	Target multidrug-resistant bacteria	MIC or inhibition zone	Bacterial cell target and/or mode of action[a]	Reference(s)
Origanum vulgare (Oregano)	Thymol 24.7, p-cymene 14.6, carvacrol 14.0, γ-terpinene 11.7	MRSA; methicillin-resistant S. epidermidis	0.06–0.125% v/v; 0.06–0.125% v/v	Repression of enterotoxin production, increase in membrane permeabilization, leakage of PO_4^{3-} and K^+, and dissipation of pH gradient	[72,103,162]
	Carvacrol 81.25, γ-terpinene 6.03, p-cymene 2.81, caryophyllene 2.20	ESBL K. pneumoniae	0.008–0.064 mg/mL		[151]
Satureja cuneifolia (Apulian savory)	Thymol 41.66, p-cymene 22.27, linalool 3.03	ESBL K. pneumoniae	0.008–0.064 mg/mL	NA	[151]
Satureja hortensis (Summer savory)	Carvacrol 44.8 ± 2.7, γ-terpinene 36.2 ± 1.3, p-cymene 15.2 ± 1.1	MRSA	33–130 μL/L[b]	NA	[146]
Satureja montana (Winter savory)	Carvacrol 66.4 ± 1.8, p-cymene 22.3 ± 1.7, β-caryophyllene 5.0 ± 0.1, γ-terpinene 4.1 ± 0.2, linalool 2.3 ± 0.1	MRSA	33–130 μL/L[b]	Damaged cell wall, agglomeration of cellular content, cell lysis	[146,163]
Thymus vulgaris (Thyme)	Thymol 54.6, p-cymene 16.4 ± 1.2, γ-terpinene 11.2 ± 0.6	MRSA	66–260 μL/L[b]	Membrane permeabilization	[146,161]

[a] Identified bacterial cell target and/or mode of action in different bacteria but not determined in multidrug bacteria.

[b] Vapor phase.

ESBL, extended-spectrum β-lactamase; MIC, minimal inhibitory concentration; MRSA, methicillin-resistant Staphylococcus aureus; N/A, information not available; VRE, vancomycin-resistant enterococci.

v/v failed to completely eradicate MRSA [169]. However, it is possible that increasing the concentration of TTO to 5–10% v/v and allowing the product to remain on the wound instead of rinsing off would result in MRSA eradication [169,170]. The *in vitro* application of two blended EOs (*Motivation* and *Longevity*) led to a reasonable level of MRSA inhibition [170]. The addition of 33 μL of *Motivation* (the main components of which are isobutyl and isoamyl methacrylate, bornyl acetate, isoamyl angelate, camphene, and α-pinene) produced an inhibition zone of 55 mm and *Longevity* (a mixture of p-cymene, eugenol, limonene, α-pinene, and thymol) produced an inhibition zone of 46 mm. Both inhibition zones were larger than those obtained using the individual components, which indicates possible synergism between the EO components [170].

Carvacrol and thymol are the best-known EO components of *Lippia*, *Origanum*, *Satureja*, and *Thymus* spp. [172,173] and their activities against MRSA have been described [162]. All 25 MRSA and MSSA strains and all 22 methicillin-sensitive *Staphylococcus epidermidis* strains tested in the study by Nostro et al. [162] were equally susceptible to carvacrol and thymol. However, carvacrol showed a lower MIC value (0.015–0.03% v/v) than that of thymol (0.03–0.06% v/v). The *in vitro* activity of terpeneless cold-pressed Valencia orange oil (CPV) against MRSA strains resulted in large inhibition zones (from 31.50 ± 3.02 to 76.67 ± 2.74 mm), as determined by disk diffusion assays [147]. The major components of CPV oil are linalool (20.2%), decanal (18.8%), geranial (9.1%), α-terpineol (5.8%), valencene (5.2%), neral (5%), dodecanal (4.1%), citronellal (3.9%), and limonene (0.3%) [148]. Exposure of an MRSA strain to 0.1% CPV resulted in rapid cell lysis. Investigation of the mode of action of CPV using a transcriptomic approach and TEM revealed the cell wall to be a target of CPV. In particular, there was a 24-fold increase in expression of the *cwrA* gene (involved in the response to cell wall injury [174]) and, in addition to lysis, treated MSRA cells showed aggregation of cytoplasmic components, a collapsed septum, and loss of cellular material [147]. Moreover, when tested in an *in vitro* dressing model, CPV oil showed potential as a topical agent for combating MRSA infections [149].

Compared to other EOs (*Allium sativum*, *Origanum syriacum*, *Satureja hortensis*, *Satureja montana*, *Thymus serpyllum*, and *Thymus vulgaris*), the vapor phase application of EO isolated from the roots of horseradish (*Armoracia rusticana*) showed the lowest MIC for a single MRSA strain and six *S. aureus* clinical isolates, with values ranging from 8.3 to 17 μL/L [146]. The major constituents of *Armoracia rusticana* are allyl isothiocyanate and phenylethyl isothiocyanate, whereas the other EOs are rich in carvacrol and thymol [146]. In addition, linalool-rich EO from *Ocimum basilicum* (approximately 55%) is effective against several MDR strains of *Enterococcus faecalis*, *P. aeruginosa*, and *S. aureus* [175].

EO obtained from the leaves of *Lantana camara*, of which the main components are bicyclogermacrene (19.42%), isocaryophyllene (16.70%), valencene (12.94%), and germacrene D (12.34 %), was effective against MDR *E. coli* (MIC of 512 μg/mL) and *S. aureus* (MIC of 256 μg/mL) [153]. The potential use of *Lantana* EOs to combat MDR bacteria was indicated by the successful results obtained with EOs of other *Lantana* spp., namely *L. achyranthiofolia*, which was tested against MDR *S. epidermidis* and *B. subtilis* [152], and *L. montevidensis*, which was tested against MDR *S. aureus* and *E. coli* [176]. However, the EOs were found to differ in their composition: the main components of *L. achyranthiofolia* EO are carvacrol (30.64%), 1,8-cineole (5.03%), and the sesquiterpenes isocaryophyllene (10.73%), β-bisabolol (11.23%), and β-bisabolene (5.68%) [138]; and the main constituents of the EO of *L. montevidensis* are β-caryophyllene (31.50%), germacrene D (27.50%), and bicyclogermacrene (13.93%) [176].

TABLE 6-5 Essential Oil Components Active Against Multidrug-Resistant Bacteria, Identified Bacterial Cell Targets, and Mode of Action

Component (Origin)	Chemical group	Chemical structure	Target multidrug bacteria (MIC)	Bacterial cell target and/or mode of action[a]	Reference(s)
Aromadendrene (Eucalyptus)	Sesquiterpene hydrocarbon		MRSA (0.5–1 mg/mL); VRE (1 mg/mL); Escherichia coli, Pseudomonas aeruginosa, and Klebsiella pneumoniae (>8 mg/mL); Acinetobacter baumannii (2 mg/mL)	Disturbance of cellular membranes; incorrect protein conformation	[10,150,164,165]
Carvacrol (Oregano, savory, thyme)	Monoterpenoid phenol		MRSA(0.05–0.03% v/v); MRSA ATCC 25923 (15.25 mg/mL); methicillin-resistant S. epidermidis (0.03% v/v); K. pneumoniae (0.008–0.064 mg/mL)	Membrane damage, pH homeostasis disturbance, induction of heat shock proteins, and inhibition of flagellin synthesis	[80,74,75,162,151,166]
1,8-cineol (Basil, camphor tree coriander, eucalyptus, sage, rosemary)	Monoterpenoid cyclic ether		MRSA, VRE, E. coli; Pseudomonas aeruginosa; K. pneumoniae (> 8 mg/ mL), Acinetobacter baumannii (8 mg/mL)	Increase in permeability; contraction of protoplasm, and loss of cytoplasmic material	[10,80,150]
Citronellal (Eucalyptus, lemon, lemongrass)	Aliphatic monoterpenoid	(+) (−)	MRSA (0.5–8 mg/mL), VRE (8–> 8 mg/mL), E. coli, P. aeruginosa, K. pneumoniae (0.008–> 8 mg/mL), Acinetobacter baumannii (2–4 mg/mL)	Tumescence of cell wall, damage of cellular membrane, and leakage of cellular constituents	[87,150,151]

Compound (source)	Class	Activity/MIC	Mode of action	References
(+)-Citronellol, (-)-citronellol (Lemongrass, geranium)	Aliphatic monoterpenoid	MRSA (0.125–8 mg/mL); MRSA ATCC 25923 (not inhibited); VRE (2–8 mg/mL); E. coli (4 mg/mL); P. aeruginosa and K. pneumoniae (0.008-> 8 mg/mL); Acinetobacter baumannii (0.125–0.25 mg/mL)	Tumescence of cell wall, damage of cellular membrane, and leakage of cellular constituents	[86,87,150,151,166]
Eugenol (Clove)	Phenylpropanoid	MRSA ATCC 25923 (133.75 mg/mL); K. pneumoniae (0.008–0.064 mg/mL)	Alterations to membrane permeability and inhibition of uptake and utilization of glucose	[82,83,151,166]
Geraniol (Lemongrass)	Monoterpenoid alcohol	Methicillin-resistant S. aureus ATCC 25923 (55 mg/mL)	Disruption of the cellular membrane and leakage of cell constituents	[85,166]
Thymol (Thyme)	Monoterpenoid phenol	Methicillin-resistant S. aureus (0.06% v/v); MRSA ATCC 25923 (30.15 mg/mL); methicillin-resistant S. epidermidis (0.06% v/v); K. pneumoniae (0.008–0.064 mg/mL)	Increase in permeability and leakage of cell constituents	[19,103,162,151,166]

[a] The bacterial cell targets and mode of action were determined in different bacteria, not necessarily in multidrug-resistant bacteria. MIC, minimal inhibitory concentration; MRSA, methicillin-resistant Staphylococcus aureus; VRE, vancomycin-resistant enterococci

The control of antibiotic-resistant foodborne pathogens, such as *Campylobacter* spp. is of major importance, since the emergency of multi-drug resistance among thermophilic *Campylobacters* has been observed worldwide [177]. Aslim and Yucef [177] demonstrated that the EO of *Origanum minutiflorum* is effective against ciprofloxacin-resistant *Campylobacter* spp. (*C. coli*, *C. jejuni*, and *C. lari*) in *in vitro* assays. The main constituents of this EO are carvacrol (73.90%) and *p*-cymene (7.20%). It is therefore obvious that its activity against ciprofloxacin-resistant *Campylobacter* spp. is due to the main component, carvacrol. However, the contribution of minor constituents cannot be disregarded.

The activity of EOs of different *Eucalyptus* spp. (the fruit oil of *E. globulus* and leaf oils of *E. citriodora*, *E. globulus*, and *E. radiata*) and their main components (aromadendrene, 1,8-cineole, citronellal, and citronellol) were tested against various MDR bacteria [MRSA, vancomycin-resistant enterococci (VRE), *A. baumannii*, *E. coli*. *K. pneumoniae*, and *P. aeruginosa*] [150]. MRSA and VRE strains were inhibited by *E. globulus* EO (an aromadendrene-rich oil), the MRSA MIC values ranged from 250 to 1000 μg/mL, and the VRE MIC values ranged from 500 to 1000 μg/mL. EOs from *E. globulus* and *E. radiata* (rich in 1,8-cineole) showed higher MIC values for MRSA: those of *E. globulus* ranged between 2000 and >4000 μg/mL and those of *E. radiata* were >4000 μg/mL. The MIC values of the EO of *E. citriodora* (rich in citronellal and citronellol) against MRSA ranged from 1000 to >4000 μg/mL. The Gram-negative resistant bacteria (*E. coli*, *K. pneumoniae*, and *P. aeruginosa*), apart from *A. baumannii*, were less susceptible to the *Eucalyptus* oils. The diminished activity in these species may be simply explained by the presence of an outer membrane in Gram-negative bacteria that constitutes an obstacle to the passage of EO, although other resistance mechanisms (see the section "Bacterial Cell Targets and Antibacterial Mode of Action") may also make a contribution [150]. The component with the highest activity was aromadendrene and the lowest activity was displayed by 1,8-cineole. The efficacy of aromadendrene is likely to be linked to its structure, which contains an exocyclic methylene group and a cyclopropane ring that can cause alkylation of proteins, resulting in altered conformation [151,164,165]. In another study, Mulyaningsih et al. [10] found synergism to occur between aromadendrene and 1,8-cineole against MRSA, when aromadendrene was used at 0.1 mg/mL and 1,8-cineole at 16 mg/mL.

The inhibitory activity of several EOs has been evaluated against extended-spectrum β-lactamase (ESBL)-producing bacteria such as *E. coli* and *K. pneumoniae* [178,151]. The growth of ESBL-producing *K. pneumoniae* was inhibited by the EOs of *Foeniculum vulgare*, *Mentha piperita*, *Mentha spicata*, *Ocimum basilicum*, *Origanum majorana*, *O. onites*, *O. vulgaris*, and *Satureja cuneifolia*, which displayed MIC values in the range of 32−64 μg/mL [151]. The susceptibility of ESBL-producing *K. pneumoniae* strains to such diverse EOs shows the great potential of these products for combating the spread of bacteria exhibiting this mechanism of resistance.

SYNERGISTIC ACTIVITY BETWEEN ESSENTIAL OILS AND ANTIBIOTICS

A considerable number of studies have reported interactions between EOs and antibiotics [11,139,178−185]. Combinations of antimicrobial agents can have different type of effects, including indifferent, additive, synergistic, and antagonistic effects [186−188]. The indifferent effect is observed when a blend of antimicrobial agents or a combination of antibacterial agent and inactive substance has an identical effect to that of the most active constituent. Additive effects occur when a mixture of antibacterial agents has an effect equal to the sum of the effects of each component. The synergistic effect is observed when a combination of antibacterial agents has a greater effect than the added effects of each constituent. Antagonistic effects of a treatment combination arise when a reduced activity is observed relative to the effect of the most efficient individual constituent [186,187]. These effects can be quantified by the application of mathematical expressions: the fractional inhibitory concentration (FIC) and the fractional bactericidal concentration (FBC) [186].

For two antibacterial agents, A and B, acting individually or in combination:

$$FIC_A = \frac{MIC_{(A \text{ in the presence of } B)}}{MIC_{(A \text{ alone})}}$$

$$FIC_B = \frac{MIC_{(B \text{ in the presence of } A)}}{MIC_{(B \text{ alone})}}$$

The FIC index is the sum of FIC_A and FIC_B. The FBC index can be similarly determined by simply substituting the MIC values with the minimum bactericidal concentration values. A FIC index of < 0.5 indicates synergism, $> 0.5-1$ indicates additive effects, > 1 to < 2 indifference, and ≥ 2 is considered to be antagonism [186].

The effect of several terpenes (carvacrol, citronellol, eugenol, geraniol, menthol, menthone, myrcene, and thymol) in combination with penicillin against MRSA ATCC 25923 and an *E. coli* strain was evaluated in a study of Gallucci et al. [166]. The MICs of carvacrol, thymol, and eugenol for the MRSA strain were 15.25 mg/mL, 30.15 mg/mL, and 133.75 mg/mL, and the MICs of for carvacrol, eugenol, geraniol, and thymol for the *E. coli* strain were 7.62 mg/mL, 66.82 mg/mL, 222.25 mg/mL and 15.07 mg/mL, respectively; the other terpenes exhibited no inhibitory activities. The FIC for the terpene-penicillin combination was 0.0078 (i.e., synergism) for carvone and 3.98 (i.e., antagonism) for thymol; remainder had indifferent effects [166]. Veras et al. [185] observed an

increase in gentamycin activity against *S. aureus* when combined with the EO of *Lippia sidoides* (rich in thymol; 84.90%) of approximately 21.53% when the EO was added at a concentration of 6%; the antibiotic activity increased to 256.82 with an EO concentration of 12%. The combination of the antibiotic with thymol resulted in a lower increase in antibiotic activity.

The observation that many MDR bacteria are widely dispersed in the environment as a consequence of the intensive use of antibiotics in animal husbandry led Zhang et al. [183] to evaluate the inhibitory effect of carvacrol, cinnamaldehyde, eugenol, thymol, both individually and in combination with antibiotics, against two antibiotic-resistant bacteria isolated from animal feeds: *Klebsiella oxytoca* and *Sphingomonas paucimobilis*. The FIC of all EO components tested in combination with the antibiotic nitrofurantoin against *S. paucimobilis* was ≤ 0.5 (indicative of synergism) and, combined with the antibiotic ampicillin, only carvacrol and thymol exhibited a synergistic effect. Against *K. oxytoca*, synergism with both ampicillin and nitrofurantoin was only observed for carvacrol. Taken together, these studies show that the combined use of EOs with antibiotics can be highly effective in controlling the development of MDR bacteria.

DEVELOPMENT OF RESISTANCE UNDER CONTINUOUS EXPOSURE TO ESSENTIAL OILS?

The exposure of *E. coli* and *Salmonella* spp., MRSA, and methicillin-sensitive *S. aureus* (MSSA) to sublethal concentrations in a habituation mode (3 days in culture medium supplemented with TTO) is reported to result in the development of resistance to a group of antibiotics used in therapy [154,189]. However, in a recent study, Hammer et al. [155] determined single-step antibiotic-resistant mutation frequencies in *E. coli* and *S. aureus* exposed to $2-8 \times$ MIC of selected antibiotics: the presence or absence of TTO led to negligible differences (< 1 log) in mutation frequency for most antibiotics, and the combination of kanamycin and TTO against *E. coli* resulted in approximately 1 log fewer kanamycin-resistant mutants when TTO oil was present than when it was absent. The multistep antibiotic approach, comprising a daily treatment of *S. aureus* and *E. coli* with TTO or its main component terpinen-4-ol in combination with antibiotic, or antibiotic alone, for 6 days showed a fourfold increase in median MIC values for *S. aureus* treated with ciprofloxacin and vancomycin alone and an eightfold increase in the presence of mupirocin. However, the addition of TTO or terpinen-4-ol to the antibiotic treatments did not cause any significant change in MIC values [155]. When TTO or terpinen-4-ol was used alone for 6 days, there was a fourfold increase in *S. aureus* MIC values and a twofold increase for *E. coli* MIC compared to baseline values [155]. Hammer et al. [155] concluded that neither TTO nor terpinen-4-ol contributes to the development of multistep antibiotic resistance. The different results obtained in the three studies mentioned [154,155,189] can be explained by differences in their methodology. It will be interesting to explore how bacterial cells respond to continued exposure to other EOs and their components, since there is expected to be an increased use of EOs and/or their components against MDR bacteria as a consequence of the large body of evidence now available from published studies.

CONCLUSIONS AND FUTURE PERSPECTIVES

The analysis of studies on the use of EOs and their components against MDR bacteria support the conclusion that a wide range of EOs and/or their components constitute an excellent tool for combating the development of bacterial resistance worldwide. Since bacteria use different strategies to overcome the activity of different antimicrobial agents, it is possible that continued exposition to EOs may induce physiologic responses that may, at least, induce an adaptation (tolerance) response or, at most, the acquisition of a resistance phenotype. This type of responses must be investigated to ensure the safe use of these natural products.

References

[1] Popowska M, Rzeczycka M, Miernik A, Krawczyk-Balska A, Walsh F, Duffy B. Influence of soil use on prevalence of tetracycline, streptomycin, and erythromycin resistance and associated resistance genes. Antimicrob Agents Chemother 2012;56:1434—44.

[2] Devirgillis G, Barile S, Perozzi G. Antibiotic resistance determinants in the interplay between food and gut microbiota. Genes Nutr 2011;6:275—84.

[3] Buchholz U, Bernard H, Werber D, Böhmer MM, Remschmidt C, Wilking H, et al. German outbreak of Escherichia coli 0104:H4 associated with sprouts. New Engl J Med 2011;365:1763—70.

[4] Rubino S, Cappuccinelli P, Kevin DJ. Escherichia coli (STEC) serotype 0104 outbreak causing haemolytic syndrome (HUS) in Germany and France. J Infect Dev Cties 2011;5:437—40.

[5] Moellering RC. Discovering new antimicrobial agents. Int J Antimicrob Ag 2011;37:2—9.

[6] Piddock LJ. The crisis of no new antibiotics-what is the way forward? Lancet Infect. Dis 2011;12:249—53.

[7] Bush K, Courvalin P, Dantas G, Davies J, Eisenstein B, Huovinen P, et al. Tackling antibiotic resistance. Nat Rev Microbiol 2011;9:894—8.

[8] Carlet J, Jarlier V, Harbath S, Voss A, Goossens H, Pittet D. Ready for a world without antibiotics? The Pensières antibiotic resistance call to action. Antimicrob Res Inf Cont 2012;1:11.

[9] Warnke PH, Becker ST, Podschun R, Sivananthan S, Springer N, Russo PAJ, et al. The battle against multi-resistant strains: Renaissance of antimicrobial essential oils as a promising force to fight hospital-acquired infections. J Cranio Maxill Surg 2009;37: 392—7.

[10] Mulyaningsih S, Sporer F, Zimmermann S, Reichling J, Wink M. Synergistic properties of the terpenoids aromadendrene and 1,8-cineole from the essential oil of Eucalyptus globulus against antibiotic-susceptible and antibiotic-resistant pathogens. Phytomedicine 2010;17:1061—6.

[11] Fadli M, Chevalier J, Saad A, Mezriou N-E. Essential oils from Moroccan plants as potential chemosensitisers restoring antibiotic activity in resistant Gram-negative bacteria. Int J Antimicrob Ag 2011;38:325—30.

[12] Jerković I, Mastelić J. Composition of free and glycosidically bound volatiles of Mentha aquatic L. Croat Chem Acta 2001;74:431—9.

[13] Maffei ME. Sites of synthesis, biochemistry and functional role of plant volatiles. S Afr J Bot 2010;76: 612—31.

[14] Bakkali F, Averbeck S, Averbeck D, Idaomar M. Biological effects of essential oils — a review. Food Chem Toxicol 2008;46:446—75.

[15] Bassolé IHN, Juliani HR. Essential oils in combination and their antimicrobial properties. Molecules 2012;17:3989—4006.

[16] Rubiolo P, Sgorbini B, Liberto E, Cordero C, Bicchi C. Essential oils and volatiles: sample preparation and analysis. Flavour Frag J 2010;25:282—90.

[17] Masotti V, Juteau F, Bessière JM, Viano J. Seasonal and phenological variations of the essential oil from the narrow endemic species Artemisia molinieri and its biological activities. J Agric Food Chem 2003;51: 7115—21.

[18] Angioni A, Barra A, Coroneo V, Dessi S, Cabras P. Chemical composition, seasonal variability, and antifungal activity of Lavandula stoechas L. ssp. stoechas essential oils from stem/leaves and flowers. J Agric Food Chem 2006;54:4364—70.

[19] Dorman HJD, Deans SG. Antimicrobial agents for plants: antibacterial activity of plant volatile oils. J Appl Microbiol 2000;83:308—16.

[20] Arze JBL, Collin G, Garneau F-X, Jean F-I, Gagnon H. Essential oils from Bolivia. III. Asteraceae: Artemisia copa Philippi. J Essent Oil Res 2004;16:554—7.

[21] Dudareva N, Negre F, Nagegowda DA, Orlova I. Plant volatiles: recent advances and future perspectives. Crit Rev Plant Sci 2006;25:417—40.

[22] Miguel MG. Antioxidant and anti-inflammatory activities of essential oils: a short review. Molecules 2010;15:9252—87.

[23] Stefanello MEA, Pascoal ACRF, Salvador MJ. Essential oils from neotropical Myrtaceae: chemical diversity and biological properties. Chem Biodivers 2011;8:73—94.

[24] Rather MA, Hassan T. Analysis of the diterpene rich essential oil of Nepeta clarkei Hooke from Kashmir Himalayas by capillary GC-MS. Int J ChemTech Res 2011;3:959—62.

[25] McCaskill D, Croteau R. Monoterpene and sesquiterpene biosynthesis in glandular trichomes of peppermint (Mentha × piperita) rely exclusively on plastid-derived isopentenyl diphosphate. Planta 1995;197:49—56.

[26] Rodríguez-Concépcion M, Boronat A. Elucidation of the methylerythritol phosphate pathway for isoprenoid biosynthesis in bacteria and plastids. A metabolic milestone achieved through genomics. Plant Physiol 2002;130:1079—89.

[27] Bick JA, Lange BM. Metabolic crosstalk between cytosolic and plastidial pathways of isoprenoid biosynthesis: unidirectional transport of intermediates across the chloroplast envelope membrane. Arch Biochem Biophys 2003;415:146—54.

[28] Dudareva N, Andersson S, Orlova I, Gatto N, Reichelt M, Rhodes D, et al. The nonmevalonate pathway supports both monoterpene and sesquiterpene formation in snapdragon flowers. Proc Natl Acad Sci USA 2005;102:933—8.

[29] McGarvey DJ, Croteau R. Terpenoid metabolism. Plant Cell 1995;7:115—1026.

[30] Ogura K, Koyama T. Enzymatic aspects of isoprenoid chain elongation. Chem Rev 1998;98: 1263—76.

[31] Davis EM, Croteau R. Cyclization enzymes in the biosynthesis of monoterpenes, sesquiterpenes, and diterpenes. In: Topics in Current Chemistry, vol. 209. Berlin, Heidelberg, Germany: Springer-Verlag; 2000. p. 53—94.

[32] Dudareva N, Pichersky E, Gershenzon J. Biochemistry of plant volatiles. Plant Physiol 2004;135: 1893—902.

[33] Lupien S, Karp F, Wildung M, Croteau R. Regiospecific cytochrome P450 limonene hydroxylases from mint (Mentha) species: cDNA isolation, characterization, and functional expression of (-)-4S-limonene-3-hydroxylase and (-)-4S-limonene-6-hydroxylase. Arch Biochem Biophys 1999;368:181—92.

[34] Bouwmeester HJ, Gershenzon J, Konings MCJM, Croteau R. Biosynthesis of the monoterpenes limonene and carvone in the fruit of caraway-I. Demonstration of enzyme activities and their changes with development. Plant Physiol 1998;117:901—12.

[35] Bouwmeester HJ, Konings MCJM, Gershenzon J, Karp F, Croteau R. Cytochrome P-450 dependent (+)-limonene-6-hydroxylation in fruits of caraway (Carum carvi). Phytochemistry 1999;50:243—8.

[36] Iijima Y, Wang G, Fridman E, Pichersky E. Analysis of the enzymatic formation of citral in the glands of sweet basil. Arch Biochem Biophys 2006;448:141—9.

[37] Luan F, Mosandl A, Munch A, Wust M. Metabolism of geraniol in grape berry mesocarp of Vitis vinifera L. cv. Scheurebe: Demonstration of stereoselective reduction, E/Z-isomerization, oxidation and glycosylation. Phytochemistry 2005;66:295—303.

[38] Shalit M, Guterman I, Volpin H, Bar E, Tamari T, Menda N, et al. Volatile ester formation in roses: Identification of an acetyl-CoA:geraniol acetyltransferase in developing rose petals. Plant Physiol 2003;131:1868–76.

[39] Degenhardt J, Gershenzon J. Demonstration and characterization of (E)-nerolidol synthase from maize: A herbivore-inducible terpene synthase participating in (3E)-4,8-dimethyl-1,3,7-nonatriene biosynthesis. Planta 2000;210:815–22.

[40] Kaiser R. Carotenoid-derived aroma compounds in flower scents. In: P. Winterhalter P, Rouseff L, editors. Carotenoid-derived aroma compounds. Washington DC, USA: American Chemical Society; 2002. p. 160–82.

[41] Winterhalter P, Rouseff R. Carotenoid-derived aroma compounds: an introduction. In: P. Winterhalter P, Rouseff L, editors. Carotenoid-derived aroma compounds. Washington DC, USA: American Chemical Society; 2002. p. 1–17.

[42] Schwartz SH, Qin X, Zeevaart JAD. Characterization of a novel carotenoid cleavage dioxygenase from plants. J Biol Chem 2001;276:25208–11.

[43] Schwartz SH, Qin XQ, Loewen MC. The biochemical characterization of two carotenoid cleavage enzymes from Arabidopsis indicates that a carotenoid-derived compound inhibits lateral branching. J Biol Chem 2004;279:46940–5.

[44] Feussner J, Wasternack C. The lipoxygenase pathway. Ann Rev Plant Biol 2002;53:275–97.

[45] Wasternack C. Jasmonates: An Update on Biosynthesis, Signal Transduction and Action in Plant Stress Response. Growth and Development Ann Bot 2007;100:681–97.

[46] Gang DR, Wang JH, Dudareva N, Nam KH, Simon JE, Lewinsohn E, et al. An investigation of the storage and biosynthesis of phenylpropenes in sweet basil. Plant Physiol 2001;125:539–55.

[47] Koeduka T, Fridman E, Gang DR, Vassão DG, Jackson BL, Kish CM, et al. Eugenol and isoeugenol characteristic aromatic constituents of spices, are biosynthesized via reduction of a coniferyl alcohol ester. Proc Natl Acad Sci USA 2006;103:10128–33.

[48] Vassão DG, Gang DR, Koeduka T, Jackson B, Pichersky E, Davin LB, et al. Chavicol formation in sweet basil (Ocimum basilicum): cleavage of an esterified C9 hydroxyl group with NAD(P) H-dependent reduction. Org Biomol Chem 2006;4: 2733–44.

[49] Boatright J, Negre F, Chen XL, Kish CM, Wood B, Peel G, et al. Understanding *in vivo* benzenoid metabolism in petunia petal tissue. Plant Physiol 2004;135:1993–2011.

[50] Kaminaga Y, Schnepp J, Peel G, Kish CM, Ben-Nissan G, Weiss D, et al. Plant phenylacetaldehyde synthase is a bifunctional homotetrameric enzyme that catalyzes phenylalanine decarboxylation and oxidation. J Biol Chem 2006;281:23357–66.

[51] Tieman D, Taylor M, Schauer N, Fernie AR, Hanson AD, Klee HJ. Tomato aromatic aminoacid descarboxylases participate in synthesis of the flavor volatiles 2-phenylethanol and 2-phenylacetaldehyde. Proc Natl Acad Sci USA 2006;106:8287–92.

[52] Parry EJ. The essential oil in the plant. In: Chemistry of Essential Oils and Artificial Perfumes, vol. II. New York. D.: Scott, Greenwood and Son; 1922. p. 1–24. Van Nostrand Company, Eight Warren Street, USA.

[53] Bakkali F, Averbeck S, Averbeck D, Idaomar M. Biological effects of essential oils – a review. Food Chem Toxicol 2008;46:446–75.

[54] Cunha AP, Ribeiro JA, Roque OR. Plantas aromáticas em Portugal. Caracterização e utilizações. Lisboa, Portugal: Fundação Calouste Gulbenkian; 2007.

[55] Cunha AP, Roque OR. Plantas Medicinais da Farmacopeia Portuguesa. Constituintes, controlo, farmacologia e utilização. Lisboa, Portugal: Fundação Calouste Gulbenkian; 2008.

[56] Adorjan B, Buchbauer G. Biological properties of essential oils: an updated review. Flavour Frag J 2010;25:407–26.

[57] Karkabounas S, Kostoula OK, Daskalou T, Veltsistas P, Karamouzis M, Zelovitis I, et al. Anticarcinogenic and antiplatelet effects of carvacrol. Exp Oncol 2006;28:121–5.

[58] Magalhães PJC, Criddle DN, Tavares RA, Melo EM, Mota TL, Leal-Cardoso JH. Intestinal myorelaxant and antispasmodic effects of the essential oil of Croton nepetaefolius and its constituents cineole, methyleugenol and terpineol. Phytother Res 1998;12: 172–7.

[59] Perry NSL, Bollen C, Perry EK, Ballard C. Salvia for dementia therapy: review for pharmacological activity and pilot tolerability clinical trial. Pharmacol Biochem Behav 2003;75:651–9.

[60] Santos MRV, Moreira FV, Fraga BP, de Sousa DP, Bonjardim LR, Quintans-Junior LJ. Cardiovascular effects of monoterpenes: a review. Rev Brasil Farmacogn 2011;21:764–71.

[61] Silva J, Abebe W, Sousa SM, Duarte VG, Machado MIL, Matos FJA. Analgesic and antiinflammatory effects of essential oils of Eucalyptus. J Ethnopharmacol 2003;89:277–83.

[62] Stefanello MEA, Pascoal ACRF, Salvador MJ. Essential oils from neotropical Myrtaceae: chemical diversity and biological properties. Chem Biodivers 2011;8:73–94.

[63] Edris AE. Pharmaceutical and therapeutic potentials of essential oils and their individual volatile constituents: a review. Phytother Res 2007;21: 308−23.

[64] Dorman HJD, Deans SG. Antimicrobial agents from plants: antibacterial activity of plant volatile oils. J Appl Microbiol 2000;88:308−16.

[65] Pauli A, Schilcher H. In vitro antimicrobial activities of essential oils monographed in the European pharmacopoeia 6th Edition. In: Başer K, Buchbauer G, editors. Handbook of essential oils. Science, Technology and Application. Boca Raton, London, NY: CRC Press, Taylor & Francis Group; 2010. p. 353−548.

[66] Lahlou M. Methods to study the phytochemistry and bioactivity of essential oils. Phytother Res 2004;18:435−48.

[67] Harris B. Phytotherapeutic uses of essential oils. In: Başer K, Buchbauer G, editors. Handbook of essential oils. Science, Technology and Application. Boca Raton, London, NY: CRC Press, Taylor & Francis Group; 2010. p. 315−52.

[68] Faleiro ML. The mode of antibacterial action of essential oils. In: Méndez-Vilas A, editor. Science against microbial pathogens: communicating current research and technological advances, vol. 2. Badajoz, Spain: Edition Microbiology book series-2011, Formatex Research Center; 2011. p. 1143−56.

[69] Hyldgaard M, Mygind T, Meyer RL. Essential oils in food preservation: mode of action, and interactions with food matrix components. Front Microbiol 2012;3:12.

[70] Bouhdid S, Abrinin J, Zhiri A, Espuny MJ, Munresa A. Investigation of functional and morphological changes in Pseudomonas aeruginosa and Staphylococcus aureus cells induced by Origanum compactum essential oil. J Appl Microbiol 2009;106:1558−68.

[71] Bouhdid S, Abrinin J, Amensour M, Zhri A, Espuny MJ, Manresa A. Functional and ultrastructural changes in Pseudomonas aeruginosa and Staphylococcus aureus cells induced by Cinnamomum verum essential oil. J Appl Microbiol 2010;109: 1139−49.

[72] De Souza EL, Barros JC, Oliveira CEV, Conceição ML. Influence of Origanum vulgare L. essential oil on enterotoxin production, membrane permeability and surface characteristics of Staphylococcus aureus. Int. J. Food Microbiol 2010;137: 308−11.

[73] Azeredo GA, Stamford TLM, de Figueiredo RCBQ, de Souza EL. The cytotoxic effect of essential oils from Origanum vulgare L. and /or Rosmarinus officinalis L. on Aeromonas hydrophila. Foodborne Pathog Dis 2012;9:298−304.

[74] Carson CF, Mee BJ, Riley TV. Mechanism of action of Melaleuca alternifolia (tea tree) on Staphylococcus aureus determined by time-kill, leakage and salt tolerance assays and electron microscopy Antimicrob. Agents Chemother 2002;46:1914−20.

[75] La Storia A, Ercolini D, Marinello F, Di Pasqua R, Villani F, Mauriello G. Atomic force microscopy analysis shows surface structure changes in carvacrol-treated bacterial cells. Res Microbiol 2011;162: 164−72.

[76] Rhayour K, Bouchikhi T, Tantaoui-Elaraki A, Sendice K, Remmal A. The mechanism of bactericidal action of oregano and clove essential oils and of their phenolic major components on Escherichia coli and Bacillus subtilis. J Essent Oil Res 2003;15:356−62.

[77] Perry CC, Weatherly M, Beale T, Randriamahefa A. Atomic force microscopy study of the antimicrobial activity of aqueous garlic versus ampicillin against Escherichia coli and Staphylococcus aureus. J Sci Food Agric 2009;89:958−64.

[78] Hafedh H, Fethi BA, Mejdi S, Emira N, Amina B. Effect of Mentha longifolia L. ssp. longifolia essential oil on the morphology of four pathogenic bacteria visualized by atomic force microscopy. Afr J Microbiol 2010;4:1122−7.

[79] Derakhshan S, Sattari M, Bigdeli M. Effect of subinhibitory concentrations of cumin (Cuminum cyminum L.) seed essential oil and alcoholic extract on the morphology, capsule expression and urease activity of Klebsiella pneumoniae. Int J Antimicrob Ag 2008;32:432−6.

[80] De Sousa JP, Torres RA, Azerêdo GA, Figueiredo RCBQ, Vasconcelos MAS, de Souza EL. Carvacrol and 1,8-cineole alone or in combination at sublethal concentrations induce changes in the cell morphology and membrane permeability of Pseudomonas fluorescens in a vegetable-based broth. Int J Food Microbiol 2012;158:9−13.

[81] Oliver JD. Recent findings on the viable but nonculturable state in pathogenic bacteria. FEMS Microbiol Rev 2010;34:415−25.

[82] Devi KP, Nisha SA, Sakthivel R, Pandian SK. Eugenol (an essential oil of clove) acts as an antibacterial agent against Salmonella typhi by disrupting the cellular membrane. J Ethnopharmacol 2010;130:107−15.

[83] Gill AO, Holley RA. Mechanisms of bactericidal action of cinnamaldehyde against Listeria monocytogenes and eugenol against L. monocytogenes and Lactobacillus sakei. Appl Environ Microbiol 2004;70:5750−5.

[84] Guinoiseau E, Luciani A, Rossi PG, Quilichini Y, Ternengo S, Bradesi P, et al. Cellular effects induced by Inula graveolens and Santolina corsica essential oils on Staphylococcus aureus. Eur J Clin Microbiol Infec Dis 2010;29:873−9.

[85] Aiemsaard J, Aiumlamai S, Aromdee C, Taweechaisupapong S, Khunkitti W. The effect of lemongrass oil and its major components on clinical isolate mastitis pathogens and their mechanisms of action on Staphylococcus aureus DMST 4745. Res Vet Sci 2011;91:e31−7.

[86] Horne DS, Holm M, Oberg C, Chao S, Young DG. Antimicrobial effects of essential oils on Streptococcus pneumoniae. J Essent Oil Res 2001;13: 387−92.

[87] Lertsatitthanakorn P, Taweechaisupapong S, Arunyanart C, Aromdee C, Khunkitti W. Effect of citronella oil on time kill profile, leakage and morphological changes of Propionibacterium acnes. J Essent Oil Res 2010;22:270−4.

[88] Turgis M, Han J, Caillet S, Lacroix M. Antimicrobial activity of mustard essential oil against Escherichia coli 0157:H7 and Salmonella typhi. Food Control 2009;20:1073−9.

[89] Helander IM, Alakomi H-L, Latva-Kala K, Mattila-Sandholm T, Pol I, Smid EJ, et al. Characterization of the action of selected essential oil components on Gram-negative bacteria. J Agric Food Chem 1998;46:3590−5.

[90] Ultee A, Kets EPW, Smid EJ. Mechanisms of action of carvacrol on the food-borne pathogen Bacillus cereus. Appl Environ Microbiol 1999;65:4606−10.

[91] Ultee E, Smid J. Influence of carvacrol on growth and toxin production by Bacillus cereus. Int J Food Microbiol 2001:64,373−64,378.

[92] Caillet S, Lacroix M. Effect of gamma radiation and oregano Essential Oil on murein and ATP concentration of Listeria monocytogenes. J Food Protect 2006;69:2961−9.

[93] Caillet S, Ursachi L, Shareck F, Lacroix M. Effect of gamma radiation and oregano essential oil on murein and ATP concentration of Staphylococcus aureus. J Food Sci 2009;74:M499−508.

[94] Qiu J, Zhang X, Luo M, Li H, Dong J, Wang J, et al. Subinhibitory concentrations of Perilla oil affect the expression of secreted virulence factor genes in Staphylococcus aureus. PLoS ONE 2011;6:e16160.

[95] Iwami Y, Kawarada K, Kojima I, Miyasawa H, Kakuta H, Mayanagi H, et al. Intracellular and extracellular pHs of Streptococcus mutans after addition of acids: loading and efflux of a fluorescent pH indicator in streptococcal cells. Oral Microbiol Immun 2002;17:239−44.

[96] Oussalah M, Caillet S, Saucier L, Lacroix M. Antimicrobial effects of selected plant essential oils on the growth of a Pseudomonas putida strain isolated from meat. Meat Sci 2006;73:236−44.

[97] Ricke SC. Perspectives on the use of organic acids and short chain fatty acids as antimicrobials. Poultry Sci 2003;82:632−9.

[98] Carpenter CE, Broadbent JR. External concentration of organic acid anions and pH, key independent variables for studying how organic acids inhibit growth of bacteria in mildly acidic foods. J Food Sci 2009;74:12−5.

[99] Faleiro ML. Response of foodborne bacteria to acid shock. In: Wong H-C, editor. Stress Response of Foodborne Pathogens. New York: Nova Press; 2012. p. 35−70.

[100] Booth IR. Regulation of cytoplasmic pH in bacteria. Microbiol Rev 1985;49:359−78.

[101] Axe DD, Bailey JE. Transport of lactate and acetate through the energized cytoplasmic membrane of Escherichia coli. Biotechnol Bioeng 1995;47:8−19.

[102] Foster JW. Escherichia coli acid resistance: tales of an amateur acidophile. Nature Microbiol Rev 2004;2: 898−907.

[103] Lambert RJW, Skandamis PN, Coote PJ, Nychas GJE. A study of the minimum inhibitory concentration and mode of action of oregano essential oil, thymol and carvacrol. J Appl Microbiol 2001;91:453−62.

[104] Burt SA, van der Zee R, Koets AP, De Graaff AM, van Knapen F, Gaastra W, et al. Carvacrol induces heat shock protein 60 and inhibit synthesis of flagellin in Escherichia coli 0157:H7. Appl Environ Microbiol 2007;73:4484−90.

[105] Di Pasqua R, Mamone G, Ferranti P, Ercolini D, Mauriello G. Changes in the proteome of Salmonella enterica serovar Thompson as stress adaptation to sublethal concentrations of thymol. Proteomics 2010;10:1040−9.

[106] Camilli A, Bassler BL. Bacterial small-molecule signaling pathways. Science 2006;311:1113−6.

[107] Kjelleberg S, Molin S. Is there a role for quorum sensing signals in bacterial biofilms? Curr Opin Microbiol 2002;5:254−8.

[108] Gobbetti M, de Angelis M, di Cagno R, Minervini F, Limitone A. Cell-cell communication in food related bacteria. Int J Food Microbiol 2007;120:34−45.

[109] Hentzer M, Givskov M. Pharmacological inhibition of quorum sensing for the treatment of chronic bacterial infections. J Clin Invest 2003;112:1300−7.

[110] Rasmussen TB, Givskov M. Quorum-sensing inhibitors as anti-pathogenic drugs. Int J Med Microbiol 2007;296:149−61.

[111] Bjarnsholt T, Jensen PO, Rasmussen TB, Christophersen L, Calum H, Hentzer M, et al. Garlic blocks quorum sensing and promotes rapid clearing of pulmunary Pseudomonas aeruginosa infections. Microbiology 2005;151(Pt12):3873–80.

[112] Adonizio AL, Downum K, Bennet BC, Mathee K. Anti-quorum sensing activity of medicinal plants in southern Florida. J Ethnopharmacol 2006;105: 427–35.

[113] Niu S, Afre S, Gilbert ES. Subinhibitory concentrations of cinnamaldehyde interfere with quorum sensing. Lett Appl Microbiol 2006;43:489–94.

[114] Brackman G, Defoirdt T, Miyamoto C, Bossier P, Calenbergh SV, Nelis H, et al. Cinnamaldehyde and cinnamaldehyde derivatives reduce virulence in Vibrio spp. by decreasing the DNA-binding activity of the quorum sensing response regulator LuxR. BMC Microbiology 2008;8:149.

[115] Khan MSA, Zahin M, Hasan S, Husain FM, Ahmad I. Inhibition of quorum sensing regulated bacterial functions by plant essential oils with special reference to clove oil. Lett Appl Microbiol 2009;49: 354–60.

[116] Szabó IA, Varga GZ, Hohmann J, Schelz Z, Szegedi E, Amaral L, et al. Inhibition of Quorum-sensing Signals by Essential Oils. Phytother Res 2010;24:782–6.

[117] Brackman G, Celen S, Hillaert U, Calenbergh SV, Cos P, Maes L, et al. Structure-activity relationship of cinnamaldehyde analogs as inhibitors of AI-2 based quorum sensing and their effect on virulence of Vibrio spp. PLoS ONE 2011;6:e16084.

[118] Olivero JTV, Pájaro JTC, Stashenko E. Antiquorum sensing activity of essential oils isolated from different species of the genus Piper. Vitae Columbia 2011;18:77–82.

[119] Becerril R, Gómez-Lus R, Goñi P, López P. Combination of analytical and microbiological techniques to study the antimicrobial activity of a new active food packaging containing cinnamon or oregano against E. coli and S. aureus. Anal Bioanal Chem 2007;388:1003–11.

[120] French GL. The continuing crisis in antibiotic resistance. Int J Antimicrob Ag 2010;36S3:S3–7.

[121] Saga T, Yamaguchi K. History of antimicrobial agents and resistant bacteria. JMAJ 2009;52:103–8.

[122] Chen LF, Chopra T, Kaye KS. Pathogens resistant to antibacterial agents. Infect Dis Clin North Am 2009;23:817–45.

[123] Kyd JM, McGrath J, Krishnamurthy A. Mechanisms of bacterial resistance to antibiotics in infections of COPD patients. Curr Drug Targets 2011;12:521–30.

[124] Van Hoek AHAM, Mevius D, Guerra B, Mullany P, Roberts AP, Aarts HJM. Acquired antibiotic resistance genes: an overview. Front Microbiol 2011;2:203.

[125] Bush LM, Calmon J, Cherney CL, Wendeler M, Pitsakis P, Poupard J, et al. High-level penicillin resistance among isolates of enterococci: implications for treatment of enterococcal infections. Ann Intern Med 1989;110:515–20.

[126] Leclercq R, Canton R, Brown DF, Giske CG, Heisig P, MacGowan AP, et al. EUCAST expert rules in antimicrobial susceptibility testing. Clin Microbiol Infec, Oct 21 2011;. http://dx.doi.org/10.1111/j.1469-0691. 2011.03703. x. (Epub ahed of print).

[127] Wright GD. Bacterial resistance to antibiotics: enzymatic degradation and modification. Adv Drug Deliver Rev 2005;57:1451–70.

[128] Strateva T, Yordanov D. Pseudomonas aeruginosa—a phenomenon of bacterial resistance. J Med Microbiol 2009;58:1133–48.

[129] Chen CM, Huang M, Chen HF, Ke SC, Li CR, Wang JH, et al. Fusidic acid resistance among clinical isolates of methicillin-resistant Staphylococcus aureus in a Taiwanese hospital. BMC Microbiol 2011;11:98.

[130] Rice LB. Mechanisms of resistance and clinical relevance of resistance to β-lactams, glycopeptides, and fluoroquinolones. Mayo Clin Proc 2012;87: 198–208.

[131] Cox G, Gary S. Thompson GS, Jenkins HT, Peske F, Savelsbergh A, Marina V, et al. Ribosome clearance by FusB-type proteins mediates resistance to the antibiotic fusidic acid. Proc. Natl. Acad. Sci. U.S.A. 2012;109:2102–7.

[132] Bhardwaj AK, Mohanty P. Bacterial efflux pumps involved in multidrug resistance and their inhibitors: rejuvenating the antimicrobial chemotherapy. Recent Pat AntiInfec Drug Discov 2012;7:73–89.

[133] Alekshun MN, Levy SB. Molecular mechanisms of antibacterial multidrug resistance. Cell 2007;128: 1037–50.

[134] Hopkins KL, Davies RH, Threlfall EJ. Mechanisms of quinolone resistance in Escherichia coli and Salmonella: recent developments. Int J Antimicrob Ag 2005;25:358–73.

[135] Morris S, Bai GH, Suffys P, Gomez-Portillo L, Fairchok M, Rouse D. Molecular mechanisms of multiple drug resistance in clinical isolates of Mycobacterium tuberculosis. J Infect Dis 1995;71: 954–60.

[136] Fuda CC, Fisher JF, Mobashery S. Beta-lactam resistance in Staphylococcus aureus: the adaptive resistance of a plastic genome. Cell Mol Life Sci 2005;62:2617–33.

[137] Jacoby GA, Munoz-Price LS. The new beta-lacta-mases. New Engl J Med 2005;352:380—91.

[138] Weigel LM, Clewell DB, Gill SR, Clark NC, McDougal LK, Flannagan SE, et al. Genetic analysis of a high-level vancomycin-resistant isolate of Staphylococcus aureus. Science 2003;302:1569—71.

[139] Gueguen E, Rousseau P, Duval-Valentin G, Chandler M. The transpososome: control of trans-position at the level of catalysis. Trends Microbiol 2005;13:543—9.

[140] Boucher Y, Labbate M, Koenig JE, Stokes HW. Inte-grons: mobilizable platforms that promote genetic diversity in bacteria. Trends Microbiol 2007;15: 301—9.

[141] Colomer-Lluch M, Jofre J, Muniesa M. Antibiotic resistance genes in the bacteriophage DNA fraction of environmental samples. PLoS ONE 2011;6:e17549.

[142] Da Rea S, Ploy M-C. Antibiotiques et réponse SOS bactérienne. Une voie efficace d'acquisition des resistances aux antibiotiques. M/S 2012;28:179—84.

[143] Paraje MG. Antimicrobial resistance in biofilms. In: Méndez-Vilas A, editor. Science against microbial pathogens: communicating current research and technological advances, Microbiology book series-2011, vol. 2. Badajoz, Spain: Edition Microbiology book series-2011, Formatex Research Center; 2011. p. 736—44.

[144] Erill I, Campoy S, Barbe J. Aeons of distress: an evolutionary perspective on the bacterial SOS response. FEMS Microbiol Rev 2007;31:637—56.

[145] Lewis K. Multidrug tolerance of biofilms and persister cells. Curr Top Microbiol 2008;322:107—31.

[146] Nedorostova L, Kloucek P, Urbanova K, Kokoska L, Smid J, Urban J, et al. Antibacterial effect of essential oil vapours against different strains of Staphylo-coccus aureus, including MRSA. Flavour Frag J 2010;26:403—7.

[147] Muthaiyan A, Martin EM, Natesan S, Crandall PG, Wilkinson BJ, Ricke SC. Antimicrobial effect and mode of action of terpeneless cold-pressed Valencia orange essential oil on methicillin-resistant Staphy-lococcus aureus. J Appl Microbiol 2012;102:1020—33.

[148] Nannapaneni R, Chalova VI, Crandall PG, Ricke SC, Johnson MG, O'Bryan CA. Campylobacter and Arcobacter species sensivity to commercial orange oil fractions. Int J Food Microbiol 2009;129:43—9.

[149] Muthaiyan A, Biswas D, Grandall PG, Wilkinson BJ, Ricke SC. Application of orange essential oil as an antistaphylococcal agent in a dressing model. BMC Complem. Altern M 2012;12:125.

[150] Mulyaningsih S, Sporer F, Reichling J, Wink M. Antibacterial activity of essential oils from Eucalyptus and of selected components against multi-drug resistant bacterial pathogens. Pharm Biol 2011;49: 893—9.

[151] Orhan IE, Ozcelik B, Kan Y, Kartal M. Inhibitory effects of various essential oils and individual components against extended-spectrum beta-lacta-mase produced by Klebsiella pneumoniae and their chemical compositions. J Food Sci 2011;76:538—46.

[152] Hernández T, Canales M, Avila JG, García AM, Martínez A, Caballero J, et al. Composition and antibacterial activity of essential oil of Lantana achyranthifolia Desf. (Verbenaceae). J Ethno-pharmacol 2005;96:551—4.

[153] Sousa EO, Silva NF, Rodrigues FFG, Campos AR, Lima SG, da Costa JGM. Chemical composition and resistance-modifying effect of the essential oil of Lantana camara Linn. Pharmacogn Mag 2010;6: 79—82.

[154] McMahon M, Tunney MM, Moore JE, Blair IS, Gilpin DF, McDowell D. Changes in antibiotic susceptibility in staphylococci habituated to sub-lethal concentrations of tea tree oil (Melaleuca alternifolia). Lett Appl Microbiol 2008;47:263—8.

[155] Hammer KA, Carson CF, Riley TV. Effects of Mela-leuca alternifolia (Tea Tree) essential oil and the major monoterpene component terpinen-4-ol on the development of single- and multistep antibiotic resistance and antimicrobial susceptibility. Anti-microb Agents Chemother 2012;56:909—15.

[156] Brophy JJ, Davies NW, Southwell IA, Stiff A, Williams LR. Gas chromatographic quality control for oil of Melaleuca terpinen-4-ol type (Australian tea tree). J Agric Food Chem 1989;37:1330—5.

[157] Gustafson JE, Liew YC, Chew S, Markham J, Bell HC, Wyllie SG, et al. Effects of tea tree oil on Escherichia coli. Lett Appl Microbiol 1998;26:194—8.

[158] Cox SD, Gustafson JE, Mann CM, Markham JL, Liew YC, Hartland RP, et al. Tea tree oil causes K+ leakage and inhibits respiration in Escherichia coli. Lett Appl Microbiol 1998;26:355—8.

[159] Cox SD, Mann CM, Markham JL, Bell HC, Gustafson JE, Warmington JR, et al. The mode of antimicrobial action of the essential oil of Melaleuca alternifolia (tea tree oil). J Appl Microbiol 2000;80: 170—5.

[160] Brady A, Loughlin R, Gilpin D, Kearney P, Tunney M. In vitro activity of tea-tree oil against clinical skin isolates of methicillin-resistant and sensitive Staphylococcus aureus and coagulase-negative staphylococci growing planktonically and as biofilms. J Med Microbiol 2006;55:1375—80.

[161] Nguefack J, Budde BB, Jakobsen M. Five essential oils from aromatic plants of Cameroon: their anti-bacterial activity and ability to permeabilize the

cytoplasmic membrane of Listeria innocua examined by flow cytometry. Lett Appl Microbiol 2004;39: 395–400.

[162] Nostro A, Blanco AR, Cannatelli MA, Enea V, Flamini G, Morelli I, et al. Susceptibility of methicillin-resistant staphylococci to oregano essential oil, carvacrol and thymol. FEMS Microbiol Lett 2004;230: 191–5.

[163] De Oliveira TLC, Soares RA, Ramos EM, Cardoso MG, Alves E, Piccoli RH. Antimicrobial activity of Satureja Montana L. essential oil against Clostridium perfringens type A inoculated in mortadella-type sausages formulated with different levels of sodium nitrite. Int J Food Microbiol 2011; 144:546–55.

[164] Sikkema J, de Bont JA, Poolman B. Mechanisms of membrane toxicity of hydrocarbons. Microbiol Rev 1995;59:201–22.

[165] Wink M. Evolutionary advantage and molecular modes of action of multi-component mixtures used in phytomedicine. Curr Drug Metab 2008;9: 996–1009.

[166] Gallucci N, Casero C, Oliva M, Zygadlo J, Demo M. Interaction between terpenes and penicillin on bacterial strains resistant to beta-lactam antibiotics. Mol Med Chem 2006;10:30–2.

[167] Sherry E, Boeck H, Warnke PH. Percutaneous treatment of chronic MRSA osteomyelitis with a novel plant-derived antiseptic. BMC Surg 2001;1:1.

[168] Sherry E, Reynolds M, Sivananthan S, Mainawalala S, Warnke PH. Inhalational phytochemicals as possible treatment for pulmonary tuberculosis: two case reports. Am J Infect Control 2004;32:369–70.

[169] Edmondson M, Newall N, Carville K, Smith J, Riley TV, Carson CF. Uncontrolled, open-label, pilot study of tea tree (Melaleuca alternifolia) oil solution in the decolonization of methicillin-resistant Staphylococcus aureus positive wounds and its influence on wound healing. Int Wound J 2011;8:375–84.

[170] Bowler WA, Bresnahan J, Bradfish A, Fernandez C. An integrated approach to methicillin-resistant Staphylococcus aureus control in a rural, regional-referral healthcare setting. Infection Control and Hospital Epidemiology 2010;31:269–75.

[171] Chao S, Young G, Oberg C, Nakaoka K. Inhibition of methicillin-resistant Staphylococcus aureus (MRSA) by essential oils. Flavour Frag J 2008;23:444–9.

[172] Figueiredo AC, Barroso JG, Pedro LG, Salgueiro L, Miguel MG, Faleiro ML. Portuguese Thymbra and Thymus species volatiles: chemical composition and biological activities. Curr Pharm Des 2008;14: 3120–40.

[173] Baser KHC. Biological and pharmacological activities of carvacrol and carvacrol bearing essential oils. Curr Pharm Des 2008;14:3106–20.

[174] Balibar CJ, Shen X, McGuire D, Yu D, McKenney D, Tao J. cwrA, a gene that specifically responds to cell wall damage in Staphylococcus aureus. Microbiology 2010;156:1372–83.

[175] Opalchenova B, Obreshkova D. Comparative studies on the activity of basil—an essential oil from Ocimum basilicum L.—against multidrug resistant clinical isolates of the genera Staphylococcus, Enterococcus and Pseudomonas by using different tests methods. J Microbiol Methods 2003;54:105–10.

[176] de Sousa EO, Rodrigues FFG, Campos AR, Lima SG, da Costa JGM. Chemical composition and synergistic interaction between amino glycosides antibiotics and essential oils of Lantana montevidensis Briq. Nat Prod Res 2012. 2012 Apr 5. [Epub ahead of print].

[177] Aslim B, Yucel N. In vitro antimicrobial activity of essential oil from endemic Origanum minutiflorum on ciprofloxacin-resistant Campylobacter spp. Food Chem 2008;107:602–6.

[178] Si H, Hu J, Liu Z, Zeng Z-L. Antibacterial effect of oregano essential oil alone and in combination with antibiotics against extended-spectrum β-lactamase-producing Escherichia coli. FEMS Immunol Med Mic 2008;53:190–4.

[179] Nascimento GGF, Locatelli J, Freitas PC, Silva GL. Antibacterial activity of plant extracts and phytochemicals on antibiotic-resistant bacteria. Braz J Microbiol 2000;31:247–56.

[180] Sato Y, Shibata H, Arai T, Yamamoto A, Okimura Y, Arakaki N, et al. variation in synergistic activity by flavone and its related compounds on the increased susceptibility of various strains of methicillin-resistant Staphylococcus aureus to β-lactams antibiotics. Int J Antimicrob Ag 2004;24:28–35.

[181] Hemaiswarya S, Doble M. Synergistic interaction of eugenol with antibiotics against Gram negative bacteria. Phytomedicine 2009;16:997–1005.

[182] Guerra FQS, Mendes JM, de Sousa JP, Morais-Braga MFB, Santos BHC, Coutinho HDM, et al. Increasing antibiotic activity against multi-drug Acinetobacter spp. by essential oils of Citrus limon and Cinnamomum zeylanicum. Nat Prod Res 2011;26(23):2235–8.

[183] Zhang D, Hu H, Rao Q, Zhao Z. Synergistic effects and physiological responses of selected bacterial isolates from animal feed to four natural antimicrobials and two antibiotics. Foodborne Pathog Dis 2011;8:1055–62.

[184] Fadli M, Saad A, Sayadi S, Chevalier J, Mezrioui N-E, Pagès J-M, et al. Antibacterial activity of Thymus

maroccanus and Thymus broussonetii essential oils against nosocomial infection-bacteria and their synergistic potential with antibiotics. Phytomedicine 2012;19:464—71.

[185]   Veras HNH, Rodrigues FFG, Colares AV, Menezes IRA, Coutinho HDM, Botelho MA, et al. Synergistic antibiotic activity of volatile compounds from the essential oil of Lippia sidoides and thymol. Fitoterapia 2012;83:508—12.

[186]   European Committee for Antimicrobial Susceptibility Testing (EUCAST) of the European Society of Clinical Microbiology and Infectious Diseases (ESC-MID). Terminology relating to methods for the determination of susceptibility of bacteria to antimicrobial agents. Clin Microbiol Infec 2000;6:503—8.

[187]   Lis-Balchin M, Deans SG, Eaglesham E. Relationship between bioactivity and chemical composition of commercial essential oils. Flavour Frag J 1998; 13:98—104.

[188]   Hodges NA, Hanlon GW. Detection and Measurement of combined biocide action. In: Denyer SP, Hugo WB, editors. Mechanisms of action of chemical biocides. Their study and exploitation. Oxford, UK: Technical series of the Society for Applied Bacteriology. Blackwell Scientific Publications; 1991. p. 297—310.

[189]   McMahon M, Blair I, Moore J, McDowell D. Habituation to sublethal concentrations of tea tree oil (Melaleuca alternifolia) is associated with reduced susceptibility to antibiotics in human pathogens. J Antimicrob Chemother 2007;59:125—7.

In Vivo Antileishmanial Activity of Plant-Based Secondary Metabolites

Luiz Felipe Domingues Passero[1], Márcia D. Laurenti[1], Gabriela Santos-Gomes[2], Bruno Luiz Soares Campos[1], Patrícia Sartorelli[3], João Henrique G. Lago[3]

[1]Laboratório de Patologia de Moléstias Infecciosas, Departamento de Patologia, Faculdade de Medicina da Universidade de São Paulo, Brazil, [2]Unidade de Ensino e Investigação de Parasitologia Médica, Centro de Malária e outras Doenças Tropicais, Instituto de Higiene e Medicina Tropical, Universidade Nova de Lisboa, Lisboa, Portugal and [3]Laboratório de Química Bioorgânica Prof. Otto R. Gottlieb. Instituto de Ciências Ambientais, Químicas e Farmacêuticas, Universidade Federal de São Paulo, Brazil

INTRODUCTION

Leishmaniasis is an infectious disease caused by protozoa belonging to the genus *Leishmania*. According to World Health Organization, the global prevalence of leishmaniasis exceeds 12 million cases, with 350 million individuals currently residing in areas that are considered to confer a risk of contracting the infection [1]. This illness is endemic in 88 countries, mainly located in tropical and subtropical regions, where socioeconomic conditions promote the spread of the disease and thus increase the number of victims.

In the case of leishmaniasis, the *Leishmania* parasite is transmitted to the vertebrate host when an infected female sand fly (the *Lutzomyia* genus in the New World and the *Phlebotomus* genus in the Old World) bites the host, feeds on its blood, and then regurgitates metacyclic promastigote into the dermis together with saliva. The saliva of the vector attracts neutrophils and macrophages to the site of feeding. However, the saliva also inhibits macrophage activity, prevents antigen presentation to

T lymphocytes, and creates an environment that is advantageous to the survival of *Leishmania*. This allows the parasite to infect the host and thrive [2,3]. After the infection takes hold, the cutaneous or visceral form of the disease manifests itself.

The best prophylactic measure against human leishmaniasis would be vaccination; however, vaccines still are under development. For this reason, the control of illness relies on chemotherapy. The most commonly used drugs to treat leishmaniasis are amphotericin B, miltefosine pentavalent, and antimonials. However, in Eastern countries, a number of parasites have been found to be resistant to antimonials [4], while in the West, there have been only few reports of parasite factors being resistant to antimonials, as in the case of *L. (V.) panamensis* parasites [5]. Therefore, new antileishmanial drugs need to be developed.

Plants have a broad range of molecules whose therapeutic activities are being investigated and identified. In regions of broad biodiversity, it is possible to find a large variety of phytochemical, many of which may have antileishmanial properties and should be explored at the molecular level [6]. In spite of the great importance of plants to drug development, less than 10% of 250,000 most common plant species identified worldwide have actually been investigated for their potential pharmacologic and therapeutic properties. While this number indicates that many more plant species than are currently used may contain therapeutic activities, it also highlights the fact that we have been neglecting 90% of plant species, which has affected drug development.

In this chapter, we critically review the most relevant findings from the past 10 years (2002 to 2012) concerning natural compounds derived from plants with demonstrated leishmanicidal activity *in vivo*. The findings presented herein may offer interesting alternatives to replace the current toxic treatments for leishmaniasis,

which may lead to a revolution in therapeutic protocols and management of this disease.

TRADITIONAL ANTILEISHMANIAL DRUGS

The pentavalent antimonials, meglumine antimoniate (Glucantime) and sodium stibogluconate (Pentostam), have been used since the 1930s and are considered the front-line drugs of choice for all the clinical forms of leishmaniasis, although their mechanisms of action have remained essentially unknown. Evidence suggests that pentavalent antimony is a prodrug, which is converted to trivalent antimony within the organism, and subsequently becomes biologically active [7–9]. However, the exact site where reduction occurs and the underlying mechanisms of reduction remain unknown.

The trivalent antimony compounds seem to interact with thiol metabolism [10,11], thereby inducing DNA fragmentation [12–14]. Other reports have suggested that pentavalent antimonials specifically inhibit *Leishmania* type I DNA topoisomerase, which acts on the metabolism of purines [15,16] to promote ATP and GTP depletion, subsequently reducing or depleting the parasite's energy [17,18].

Antimonials also seem to promote parasite killing by inducing the production of proinflammatory cytokines, which can enhance the phagocytic capacity of monocytes and neutrophils, thus increasing the release of reactive oxygen species (ROS) by phagocytes [19].

Since 1945, pentavalent antimonials have been used in the treatment of cutaneous leishmaniasis. N-methyl-glucamine antimoniate is the most commonly used drug to treat human and canine visceral leishmaniasis. This drug requires hospitalization, daily parenteral administration for at least three consecutive weeks, and constant monitoring during treatment. Intramuscular administration of the drug is often accompanied by local pain and systemic side effects, which include nausea, vomiting, weakness, myalgia, abdominal colic, diarrhea, skin rashes, hepatotoxicity and even cardiotoxicity, and therefore this form of treatment requires intensive medical supervision. However, antimonials are sometimes not overly effective in treating leishmaniasis, which is a major concern as this can lead to the development of treatment-resistant *Leishmania* strains.

RECENT ANTILEISHMANIAL DRUGS

Amphotericin B is a polyene antimycotic, which was first isolated in 1955 from *Streptomyces nodosus* [20]. The drug (Fungizone) possesses a wide spectrum of activity and is used to target systemic infections that are caused by a large number of fungal species and some protozoan parasites [21]. It has a very limited solubility profile, being almost completely insoluble in water [22], leading to low bioavailability via the oral route. Its use is therefore restricted to intravenous infusion and local application. Both the therapeutic activity and toxic effects of amphotericin B are derived from its interaction with lipids of parasites and host cells, particularly with membrane sterols. The antibiotic can form self-aggregating complexes with ergosterol, the principal sterol of *Leishmania* cell membranes, and cholesterol in mammalian cell membranes, resulting in extensive cell damage [23]. Furthermore, the affinity of amphotericin B for lipids indicates that it is readily incorporated into plasma lipoproteins, particularly low-density lipoproteins [24]. The uptake of low-density lipoprotein-carrying amphotericin B by renal epithelial cells is one mechanism of toxicity. Amphotericin B was found to be an effective treatment for visceral leishmaniasis in the 1990s (approximately 100% cure rate) [25,26] and has been administered in cases where antimonials failed.

Given its affinity for biologic membranes, incorporation of amphotericin B into lipid-based nanosystems improved its therapeutic index and reduced toxicity. The liposomal formulation of amphotericin B (AmBisome), approved by the US Food and Drug Administration in 1997 [27], accumulates in macrophages, and it is therefore becoming an attractive alternative for treating leishmaniasis. However, the high cost of this treatment prevents its widespread use in many of leishmaniasis endemic regions, where even short courses of liposomal formulations are unaffordable [28].

Amphotericin B also affects the immune system and can modulate macrophage activity by inducing the production of proinflammatory cytokines [29] and ROS [30,31], and promoting chemotaxis and phagocytosis.

Miltefosine is an alkyl phospholipid compound that was originally developed for the topical treatment of cutaneous metastasis in mammary carcinomas, and is the most recent antileishmanial drug. The fact that miltefosine is administered orally is a particularly attractive quality, as it is an alternative means through which to treat leishmaniasis on an outpatient basis and is therefore suitable for use at primary healthcare centers. However, adverse effects include gastrointestinal symptoms, diarrhea, vomiting, moderate elevation of hepatic enzymes, and nephrotoxicity.

The drug possesses a direct toxic effect on the promastigote form of protozoa. Mitochondrial cytochrome *c* oxidase has been suggested as a miltefosine target [32]. Miltefosine causes perturbation in ether-lipid metabolism, glycosylphosphatidylinositol anchor biosynthesis, and signal transduction in *Leishmania* [33,34]. Inhibition of the glycosomal alkyl-specific acyl-CoA acyltransferase, an enzyme involved in lipid remodeling, was also reported to be a direct effect of this compound [35].

PARASITE RESISTANCE

In some geographical areas, the spread of resistance to classical leishmaniasis chemotherapy has become a clinical threat, as in the case of sodium antimony gluconate. The first instance of leishmaniasis drug resistance arose in Bihar, India, a hyperendemic area of anthroponotic visceral leishmaniasis, where around 30% of patients did not respond to antimony [36].

The rise of antimony resistance in this particular region can be attributed to the freely available antimony, the extensive use of which can lead to subtherapeutic dosages, promoting the development of resistant parasites [37]. In addition, the risk of transmission of resistant *Leishmania* is elevated, since the parasites are directly transmitted between humans via sand fly bites, without the need for animal reservoirs.

In vitro studies have been carried out using promastigote and amastigote macrophage assays, which have enabled the establishment of correlations between drug pressure and the selection of treatment-resistant parasites. These experimental studies have greatly improved our understanding of resistance and susceptibility at the molecular level; however, other intrinsic and extrinsic factors that influence drug efficacy must also be considered.

The first *in vitro* studies performed with *L. (L.) donovani* isolates from responder and nonresponder patients indicated a correlation between *in vitro* sensitivity to antimony and the clinical response [38]. Other studies have shown a direct correlation between the sensitivity of *Leishmania* isolates from cutaneous and mucocutaneous cases and clinical outcomes in these patients [39]. Further research demonstrated that the resistance to antimony is stage specific: intramacrophage amastigotes have been found to be more susceptible than promastigotes. These differences in the susceptibility of parasites based on their morphologic forms

relate to the mechanisms of drug uptake and bioreduction of antimony to its trivalent form [40].

In *Leishmania* parasites, the predominant mechanism being responsible for resistance to antimonials is gene amplification resulting in homologous recombination. It is thought that gene amplification leads to the overexpression of proteins [41,42]; for instance, gene amplification in *L. (L.) donovani* results in multiple overexpressed genes. Multidrug resistance protein A (*MRPA*, encoding a member of the multidrug-resistance protein family, a large family of ABC transporters), gamma-glutamylcysteine synthetase (γGCS), heat shock protein 83(*HSP83*), mitogen-activated protein kinase (*MAPK1*), and histones (*H1, H2A,* and *H4*) genes [43] are indicated as good biomarkers for antimony resistance. In fact, the three genes encoding the ABC transporter have been found to be consistently overexpressed, and their transfection into wild intracellular parasites has coincided with sodium stibogluconate resistance [44]. However, the levels at which these genes must be overexpressed to induce a resistant parasitic phenotype have yet to be determined.

Variations in the sensitivity of visceral *L. (L.) donovani,* cutaneous *L. (L) major, L. (L) tropica, L. (L) aethiopica, L. (L) mexicana,* and *L. (V) panamensis* to miltefosine have been observed [45]. However, a recent genome sequencing study of *L. major* indicated for the first time that an association among multiple genetic mutations is evident in this particular instance of miltefosine resistance [46].

NATURAL PRODUCTS AS SOURCES OF NEW ANTILEISHMANIAL COMPOUNDS

Several secondary metabolites of diverse molecular structures with potential antileishmanial activity have been isolated from plants. These represent interesting lead structures for the development of new prototype antileishmanial therapies [47,48].

The available literature includes wide-ranging studies involving bioguided fractionation procedures of active extracts that frequently lead to purification of different structural classes of compounds. Among the secondary metabolites that have shown significant leishmanicidal activity, the most notable are alkaloids, especially indoles [49]; naphtylisoquinolines [50]; bisbenzylisoquinolines [51] and benzoquinolizidines [52]; and terpenoids, including triterpenes [53], steroids [54], saponins [55–57], sesquiterpenes [58–61], and diterpenes [62,63]. In addition, flavonoids, especially isoflavones [64] and chalcones [65], have been found to show significant leishmanicidal activity.

Helietta apiculata

The *Helietta* genus belongs to the Rutaceae family and is primarily found in Mexico and nearby areas, including the northern Argentina, Brazil, Colombia, Cuba, Paraguay, Peru, and the United States (Texas), and Venezuela, [66]. In folk medicine, *H. apiculata* has been used to cure skin lesions and inflammatory diseases (reviewed in ref. 67).

Studies examining the therapeutic properties of the *H. apiculata* plant have demonstrated that furoquinoline alkaloids and coumarins are moderately active against the promastigote forms of *L. (L.) amazonensis*, *L. (L.) infantum*, and *L. (V.) braziliensis*. However, *in vivo* studies demonstrated that the alkaloid γ-fagarine (compound 1, Fig. 7-1), coumarins 3-(1'-dimethylallyl)-decursinol (compound 2, Fig. 7-1), and (-)-heliettin (compound 3, Fig. 7-1) are highly active, with effects comparable to those of the standard drug *N*-methylglucamine antimoniate. In this case, oral administration of 10 mg/kg of γ-fagarine plus polysorbate for 15 days to *L. (L.) amazonensis*-infected mice led to the suppression of 90.5% of the lesion weight,

accompanied by a 97.4% reduction in skin parasitism. Similar results were found after the subcutaneous administration of 10 mg/kg of 3-(1'-dimethylallyl)-decursinol and (-)-heliettin, which suggest that different coumarins with antileishmanial activity are present in this plant [67].

Kalanchoe pinnata

The *Kalanchoe* genus belongs to the Crassulaceae family and comprises about 125 species, including *K. pinnata* and *K. braziliensis*. These plants are employed in traditional medicine to treat gastric ulcers, respiratory infections, boils, wounds, and rheumatoid arthritis [68].

Thirty-day oral administration of the flavonoids quercetin 3-*O*-α-L-arabinopyranosyl (1→2)-α-L-rhamnopyranoside (compound 4, Fig. 7-1), quercetin (compound 5, Fig. 7-1), and kaempferol 3-*O*-α-L-arabinopyranosyl (1→2)-α-L-rhamnopyranoside (compound 6, Fig. 7-1) purified from *K. pinnata* was able to control lesion size in *L. (L.) amazonensis*-infected mice [69]. Similar effects were found after administration of *K. pinnata* aqueous extract and glucantime. In addition, although the parasite load seemed to be equal in mice treated with flavonoids, aqueous extract, and the standard drug, the parasitism was higher in untreated mice, suggesting that molecular exploration of the flavonoids and other components in the extract aid development of a new leishmanicidal molecule for humans. Indeed, [70] showed that oral administration of 21 mg/kg lyophilized aqueous extract of *K. pinnata* twice a day for 14 days could control the progression of human cutaneous leishmaniasis. However, it is important to note that the lesion progressed once this treatment was withdrawn, and administration of the standard drug was required to heal the lesion. Interestingly, treatment of human patients with this extract did not alter serum levels of alanine aminotransferase, aspartate aminotransferase, urea, creatinine, or alkaline

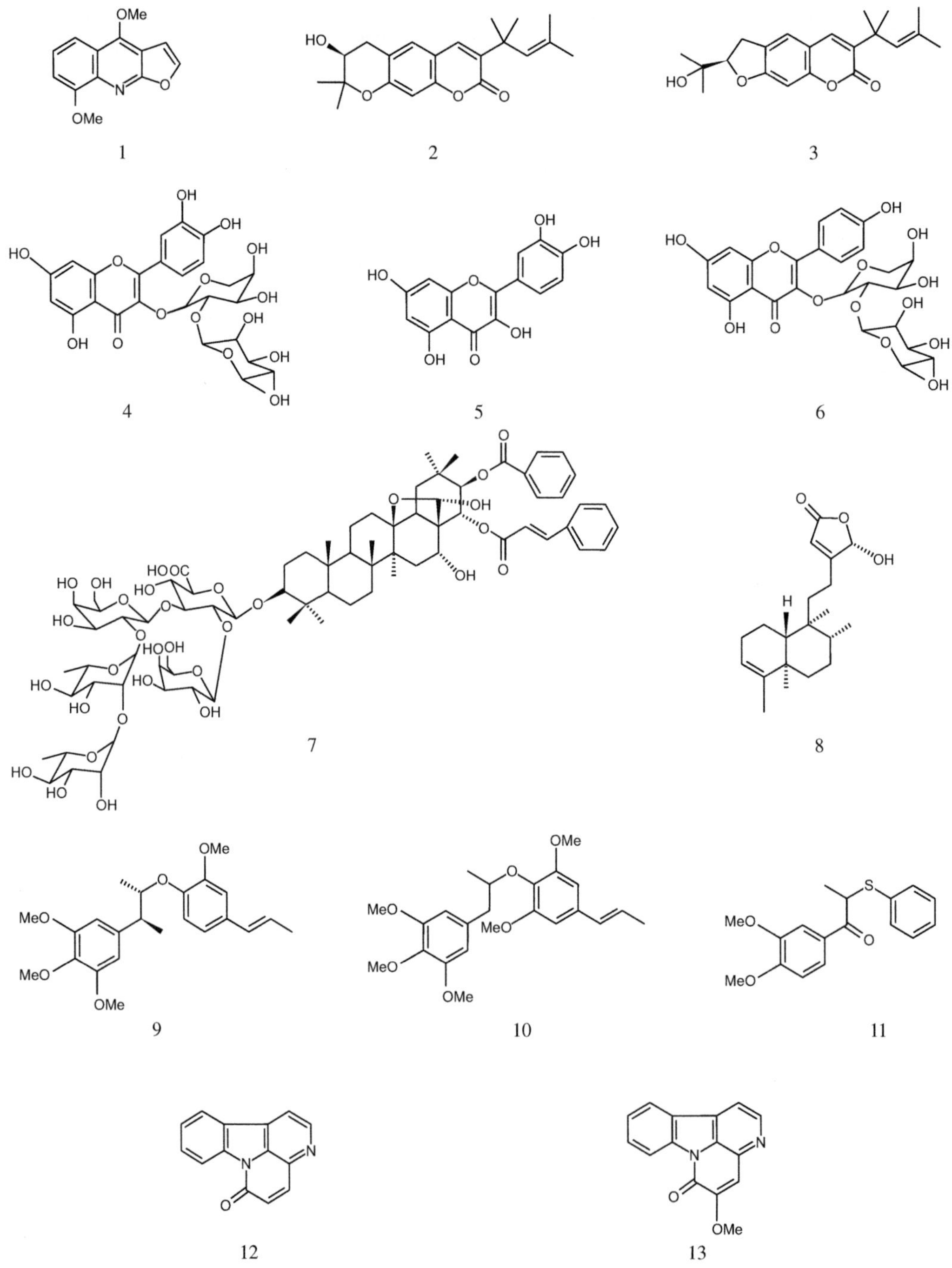

FIGURE 7-1 Structures of compounds 1–13 that exhibit leishmanicidal effects.

phosphatase. It is possible that the leishmanicidal effect of *K. pinnata* could be related to an immunomodulatory effect of the aqueous extract and that the compounds in this extract may act on innate immunity [71]. If this were the case, it could further encourage researchers to consider this plant as the source of a potential antileishmanial drug.

Maesa balansae

The genus *Maesa* (Myrsinaceae) comprises around 100 species from the Old World, and is commonly found in the paleotropics, i.e., Malaysia, New Guinea, and western Asia [72]. Ethnopharmacologically, this genus has been used against bacterial infections and for worm expulsion [73].

In this regard, Maes et al. [74] studied the effects of triterpene saponin maesabalide III (compound 7, Fig. 7-1) on the progression of experimental visceral leishmaniasis. Intraperitoneal administration of triterpene saponin (0.2—0.8 mg/kg) for five consecutive days abrogated the classical signs of kala-azar in hamsters, without causing any toxic effects. However, in spite of these results, the spleens of some experimental animals showed persistent parasites

Polyalthia longifolia

The *Polyalthia* genus belongs to the Annonaceae family and contains approximately 25 identified species. This genus is predominantly found in Asia, extending from southern India and Sri Lanka and spanning continental Asia to northern Australia and Melanesia. This plant is also commonly found in the lowland humid regions of East Africa and Madagascar [75]. In folk medicine, the bark of *P. longifolia* has long been used to treat skin diseases, fevers, diabetes, hypertension and worms [76]. In this species, a major constituent of the methanol extract is clerodane diterpene. Phytochemically, this species is composed mainly of diterpenes containing a clerodane skeleton [77].

Dose-dependent amelioration was confirmed when *L. (L.) donovani*-infected hamsters were treated with orally administered 25—250 mg/kg of the clerodane diterpene 16α-hydroxycleroda-3,13Z-dien-15,16-olide (compound 8, Fig. 7-1) for five consecutive days. In addition, treatment with 100 and 250 mg/kg of this compound led to diminished parasitism in the bone marrow, liver, and spleen of treated hamsters compared to those treated orally with 40 mg/kg of miltefosine for 5 days [78]. *In vitro*, this compound was found to inhibit *L. (L.) donovani* recombinant DNA topoisomerase I, which induced apoptosis in the parasites, detected by exposure of phosphatidylserine on the outer membrane of *L. (L.) donovani* promastigotes. Additionally, it is important to note that 16α-hydroxycleroda-3,13(14)Z-dien-15,16-olide displayed no cytotoxic effects on J774.1 macrophages [78]. This study suggests a potential interaction between the 16α-hydroxycleroda-3,13(14)Z-dien-15,16-olide diterpene and the DNA replication/repair machinery of promastigotes; however, the mechanism of action *in vivo* may be quite different to that proposed previously, since diverse cell subsets can be found *in vivo*. Moreover, in addition to amastigotes, the molecule may be highly selective for other cell populations, such as antigen-presenting cells. However, this detailed, extensive study showed both the mode of action of clerodane diterpenes in promastigote forms and their effects on chronically infected hamsters.

Virola Species

The *Virola* genus belongs to Myristicaceae, and has approximately 60 described species endemic to the lowland and cloud rainforests of Central and South America. *Virola* spp. are of considerable ethnobotanical importance to the traditional societies of Central and South America. Perhaps one of the most widespread applications of *Virola* is as hallucinogenic snuff, as it contains an abundance of tryptamine alkaloids [79]. In addition, sap obtained from *Virola* has been used to treat inflammation, skin mycosis, and malaria [80,81].

Furthermore, other compounds has also been shown to be biologically active, such as the neolignan surinamensin (3,4,5-trimethoxy-8-(2',6'-dimethoxy-4'E-propenylphenoxy)-phenylpropane; compound 9, Fig. 7-1) purified from *V. surinamensis*. Although synthetic analogs, the molecules 3,4,5-trimethoxy-8-[2',6'-dimethoxy-4'E-propenylphenoxy]-phenylpropane (compound 10, Fig. 7-1) of *V. pavonis* demonstrated comparable efficacy in eliminating *L. (L.) donovani* promastigotes when compared to stibogluconate. Moreover, it did so without cytotoxic effects on macrophages. However, an analogous form of 3,4-dimethoxy-8-(4'-methylthiophenoxy)-propiophenone (compound 11, Fig. 7-1) exhibited limited effects *in vivo*: subcutaneous administration of 100 mg/kg of this molecule for five consecutive days reduced parasitism in the liver by 42%, while the administration of 15 mg/kg of the standard drug for 5 days reduced the effects of this parasite on the liver by 74% [82]. The most important finding in this study was that it was possible to extract new analogs with leishmanicidal activity from this plant. In addition, it is possible that improvements to the structure of this molecule could increase its antileishmanial potential.

Zanthoxylum chiloperone

The *Zanthoxylum* genus belongs to the Rutaceae family and comprises about 549 species distributed worldwide, although it is mainly found in tropical and temperate regions [83,84]. In addition, this genus has been used as food and wood oils [83,85]. Ethnopharmacologically, *Zanthoxylum* spp. have been used worldwide to treat different conditions such as snakebites, stomach problems, skin lesions, inflammation, and parasitic diseases [86–89].

From a phytochemical perspective, the alkaloids are the most important class of secondary metabolite found in this genus, and these account for its diverse pharmacologic activities [90].

In fact, *in vitro* studies have demonstrated that the alkaloids canthin-6-one (compound 12, Fig. 7-1) and 5-methoxycanthin-6-one (compound 13, Fig. 7-1) are active against *L. (L.) amazonensis*, *L. (V.) braziliensis*, and *L. (L.) donovani*. In addition, four intralesional injections of 10 mg/kg canthin-6-one alkaloid in *L. (L.) amazonensis*-infected BALB/c mice inhibited parasite replication, although only meglumine antimoniate treatment (28 mg/kg) was found to significantly decrease parasite load, compared with untreated mice [91]. In spite of this finding, only 40 mg/kg of canthin-6-one was necessary to reduce skin parasitism by 77%, while 280 mg/kg of meglumine antimoniate was required to reduce tissue parasitism by approximately 91%, suggesting that some adjustments are necessary to improve the efficacy of this alkaloid. In another study using *Trypanosoma cruzi* (a parasite that also belongs to the Kinetoplastida family), canthin-6-one displayed effects comparable to benzonidazole (standard drug), thereby eliminating trypomastigote forms *in vivo* and improving the overall rate of mice survival to 80—100% [92].

The data summarized in this chapter are presented in Table 7-1.

CONCLUSIONS AND FUTURE PERSPECTIVES

Phytochemical studies of medicinal plants have resulted in the identification of different classes of secondary metabolites with interesting leishmanicidal activity *in vitro* [93–96]. These studies frequently use cellular models with relatively low complexity, and generally include initial trials with promastigotes,

TABLE 7-1 Class of Plant Compounds with Leishmanicidal Effects in Murine Experimental Leishmaniasis

Species (family)	*Leishmania* species	Experimental host	Class of compound	Dose	Parasite inhibition	Reference(s)
Helietta apiculata (Rutaceae)	*L. (L.) amazonensis*	BALB/c mouse	Alkaloids, coumarins	20 mg/kg	97,4–98,6%	[67]
Kalanchoe pinnata (Crassulaceae)	*L. (L.) amazonensis*	BALB/c mouse	Flavonoid	16 mg/kg	57–75%	[69]
Maesa balansae (Myrsinaceae)	*L. (L.) donovani; L. (L.) infantum*	Hamster; BALB/c mouse	Triterpene saponins	2.5–40 mg/kg; 0.2–0.4 mg/kg	>90%; 83–99.8%	[57,74]
Polyalthia longifolia (Annonaceae)	*L. (L.) donovani*	Hamster	Diterpene	25–250 mg/kg	74,9–93%	[78]
Virola pavonis (Myristicaceae)	*L. (L.) donovani*	BALB/c mouse	Lignans	100 mg/kg	42%	[82]
Zanthoxylum chiloperone (Rutaceae)	*L. (L.) amazonensis*	BALB/c mouse	Alkaloids	10 mg/kg	77.6%	[91]

followed by toxicity studies using macrophages or similar cell lineages. Finally the potency of compounds is assayed by the interaction of amastigotes with macrophages, which mimics, in part, an infection. Although this is an important experimental model used to guide a rational purification strategy, it cannot reflect the *in vivo* effects of compounds.

The few advances produced in the last 10 years using animal models to assay the efficacy of purified secondary metabolites have shown us that some of these plant extracts exhibit great potential for eliminating parasites without any detectable side effects. This is completely different to the results of chemotherapies used to treat human leishmaniasis, which are often associated with many side effects. Although this chapter has described the main advances in the search for new *in vivo* leishmanicidal drugs, it also highlighted the need for more interactions between phytochemical researchers and immunoparasitologists.

Although a number of studies investigating lead molecules using experimental models of cutaneous and visceral leishmaniasis have been discussed in this chapter, very few therapeutic developments have taken place in the last 10 years. By engaging in more research initiatives to explore human physiology by means other than simply assessing the interaction between parasites and macrophages *in vitro*, researchers might be able to find plant-based antileishmanial treatment options that reduce side effects and enhance the overall quality of life of affected individuals.

References

[1] WHO technical report series; no. 949 Control of the leishmaniasis: report of a meeting of the WHO Expert Committee on the Control of Leishmaniases. Geneva, 22−26; March 2010.

[2] Laurenti MD, Silveira VM, Secundino NF, Corbett CE, Pimenta PP. Saliva of laboratory-reared Lutzomyia longipalpis exacerbates Leishmania (Leishmania) amazonensis infection more potently than saliva of wild-caught Lutzomyia longipalpis. Parasitol Int 2009;58:220−6.

[3] Araújo-Santos T, Prates DB, Andrade BB, Nascimento DO, Clarêncio J, Entringer PF, et al.

Lutzomyia longipalpis saliva triggers lipid body formation and prostaglandin E2 production in murine macrophages. Plos Negl Trop Dis 2010;4:e873.

[4] Croft SL, Sundar S, Fairlamb AH. Drug resistance in leishmaniasis. Clin Microbiol Rev 2006;19:111−26.

[5] Walker J, Gongora R, Vasquez JJ, Drummelsmith J, Burchmore R, Roy G, et al. Discovery of factors linked to antimony resistance in Leishmania panamensis through differential proteome analysis. Mol Biochem Parasitol 2012;183:166−76.

[6] Passero LF, Bonfim-Melo A, Corbett CE, Laurenti MD, Toyama MH, de Toyama DO, et al. Anti-leishmanial effects of purified compounds from aerial parts of Baccharis uncinella C. DC. (Asteraceae). Parasitol Res 2011;108:529−36.

[7] Goodwin LG, Page JE. A study of the excretion of organic antimonials using a polarographic procedure. Biochem J 1943;37:198−209.

[8] Burguera JL, Burguera M, Petit de Pena Y, Lugo A, Anez N. Selective determination of antimony (III) and antimony (IV) in serum and urine and of total antimony in skin biopsies of patients with entaneous Leishmaniasis treated with meglumine antimoniate. J Trace Elem Med Biol 1993;10:66−70.

[9] Shaked-Mishan P, Ulrich N, Ephros M, Zilberstein D. Novel Intracellular SbV reducing activity correlates with antimony susceptibility in Leishmania donovani. J Biol Chem 2001;276:3971−6.

[10] Cunningham ML, Fairlamb AH. Trypanothione reductase from Leishmania donovani. Purification, characterisation and inhibition by trivalent antimonials. Eur J Biochem 1995;230:460−8.

[11] Wyllie S, Cunningham ML, Fairlamb AH. Dual action of antimonial drugs on thiol redox metabolism in the human pathogen Leishmania donovani. J Biol Chem 2004;279:39925−32.

[12] Lucumi A, Robledo S, Gama V, Saravia NG. Sensitivity of Leishmania viannia panamensis to pentavalent antimony is correlated with the formation of cleavable DNA-protein complexes. Antimicrob Agents Chemother 1998;42:1990−5.

[13] Sereno D, Holzmuller P, Mangot I, Cuny G, Ouaissi A, Lemesre JL. Antimonial-mediated DNA fragmentation in Leishmania infantum amastigotes. Antimicrob Agents Chemother 2001;45:2064−9.

[14] Sudhandiran G, Shaha C. Antimonial-induced increase in intracellular Ca2+ through non-selective cation channels in the host and the parasite is responsible for apoptosis of intracellular Leishmania donovani amastigotes. J Biol Chem 2003;278: 25120−32.

[15] Chakraborty AK, Majumder HK. Mode of action of pentavalent antimonials: specific inhibition of type I DNA topoisomerase of Leishmania donovani. Biochem Biophys Res Commun 1988;152:605−11.

[16] Walker J, Saravia NG. Inhibition of Leishmania donovani promastigote DNA topoisomerase I and human monocyte DNA topoisomerases I and II by antimonial drugs and classical antitopoisomerase agents. J Parasitol 2004;90:1155−62.

[17] Berman JD, Waddell D, Hanson BD. Biochemical mechanisms of the antileishmanial activity of sodium stibogluconate. Antimicrob Agents Chemother 1985;27:916−20.

[18] Berman JD, Gallalee JV, Best JM. Sodium stibogluconate (Pentostam) inhibition of glucose catabolism via the glycolytic pathway, and fatty acid beta-oxidation in Leishmania mexicana amastigotes. Biochem Pharmacol 1987;36:197−201.

[19] Muniz-Junqueira MI, de Paula-Coelho VN. Meglumine antimonate directly increases phagocytosis, superoxide anion and TNF-alpha production, but only via TNF-alpha it indirectly increases nitric oxide production by phagocytes of healthy individuals, in vitro. Int Immunopharmacol 2008;8:1633−8.

[20] Dutcher JD, Gold W, Pagano JF, Vandeputte J. Amphotericin B, its production and its salts. U.S. Pat 1959;2,908,611.

[21] Kleinberg M. What is the current and future status of conventional amphotericin B? Int J Antimicrob Agents 2006;1:12−6.

[22] Brittain HG. Circular dichroism studies of the self-association of Amphotericin B. Chirality 1994;6: 665−9.

[23] de Kruijff B, Gerritsen WJ, Oerlemans A, van Dijck PW, Demel RA, et al. Polyene antibiotic-sterol interactions in membranes of Acholesplasma laidlawii cells and lecithin liposomes. II. Temperature dependence of the polyene antibiotic-sterol complex formation. Biochim Biophys 1974;339:44−56.

[24] Brajtburg J, Elberg S, Bolard J, Kobayashi GS, Levy RA, Ostlund Jr RE, et al. Interaction of plasma proteins and lipoproteins with amphotericin B. J Infect Dis 1984;6:986−97.

[25] Murray HW. Treatment of visceral leishmaniasis. Am J Trop Med Hyg 2004;72:359.

[26] Singh S, Sivakumar R. Challenges and new discoveries in the treatment of leishmaniasis. J Infect Chemother 2004;10:307−15.

[27] Meyerhoff AUS. Food and Drug Administration approval of AmBisome (liposomal amphotericin B) for treatment of visceral leishmaniasis. Clin Infect Dis 1999;28:42−8.

[28] Murray HW. Leishmaniasis in the United States: treatment in 2012. Am J Trop Med Hyg 2012;86: 434−40.

[29] Chia JK, McManus EJI. *In vitro* tumor necrosis factor induction assay for analysis of febrile toxicity associated with amphotericin B preparations. Antimicrob Agents Chemother 1990;34:906−8.

[30] Wilson E, Thorson L, Speert DP. Enhancement of macrophage superoxide anion production by amphotericin B. Antimicrob Agents Chemother 1991; 35:796−800.

[31] Mozaffarian NB, Berman JW, Casadevall A. Enhancement of nitric oxide synthesis by macrophages represents an additional mechanism of action for amphotericin. Antimicrob Agents Chemother 1997;41:1825−9.

[32] Carvalho L, Luque-Ortega JR, Manzano JI, Castanys S, Rivas L, Gamarro F. Tafenoquine, an antiplasmodial 8-aminoquinoline, targets leishmania respiratory complex III and induces apoptosis. Antimicrob Agents Chemother 2010;54:5344−51.

[33] Lux H, Hart DT, Parker PJ, Klenner T. Ether lipid metabolism, GPI anchor biosynthesis, and signal transduction are putative targets for anti-leishmanial alkyl phospholipid analogues. Adv Exp Med Biol 1996;416:201−11.

[34] Rakotomanga M, Blanc S, Gaudin K, Chaminade P, Loiseau PM. Miltefosine affects lipid metabolism in Leishmania donovani promastigotes. Antimicrob Agents Chemother 2007;51:1425−30.

[35] Lux H, Heise N, Klenner T, Hart D, Opperdoes FR. Ether−lipid (alkyl-phospholipid) metabolism and the mechanism of action of ether−lipid analogues in Leishmania. Mol Biochem Parasitol 2000;111: 1−14.

[36] Murray HW. Treatment of visceral leishmaniasis in 2010: direction from Bihar State, India. Future Microbiol 2010;5:1301−3.

[37] Sundar S, Thakur BB, Tandon AK, Agrawal NR, Mishra CP, Mahapatra TM, et al. Clinicoepidemiological study of drug resistance in Indian kala-azar. BMJ 1994;308:307.

[38] Ibrahim ME, Hag-Ali M, el-Hassan AM, Theander TG, Kharazmi A. Leishmania resistant to sodium stibogluconate: drug-associated macrophage-dependent killing. Parasitol Res 1994;80:569−74.

[39] Grogl M, Thomason TN, Franke ED. Drug resistance in leishmaniasis: its implication in systemic chemotherapy of cutaneous and mucocutaneous disease. Am J Trop Med Hyg 1992;47:117−26.

[40] Roberts WL, Berman JD, Rainey PM. *In vitro* anti-leishmanial properties of tri- and pentavalent antimonial preparations. Antimicrob Agents Chemother 1995;39:1234−9.

[41] Leprohon P, Légaré D, Ouellette M. Intracellular localization of the ABCC proteins of Leishmania and their role in resistance to antimonials. Antimicrob Agents Chemother 2009;53:2646−9.

[42] Leprohon P, Légaré D, Raymond F, Madore E, Hardiman G, Corbeil J, et al. Gene expression modulation is associated with gene amplification, supernumerary chromosomes and chromosome loss in antimony-resistant Leishmania infantum. Nucleic Acids Res 2009;37:1387−99.

[43] Kumar D, Singh R, Bhandari V, Kulshrestha A, Negi NS, Salotra P. Biomarkers of antimony resistance: need for expression analysis of multiple genes to distinguish resistance phenotype in clinical isolates of Leishmania donovani. Parasitol Res. 2012;111(1): 223−30.

[44] El Fadili K, Messier N, Leprohon P, Roy G, Guimond C, Trudel N, et al. Role of the ABC transporter MRPA (PGPA) in antimony resistance in Leishmania infantum axenic and intracellular amastigotes. Antimicrob Agents Chemother 2005;49: 1988−93.

[45] Escobar P, Matu S, Marques C, Croft SL. Sensitivities of Leishmania species to hexadecylphosphocholine (miltefosine), ET-18-OCH(3) (edelfosine) and amphotericin B. Acta Trop 2002;81:151−7.

[46] Coelho AC, Boisvert S, Mukherjee A, Leprohon P, Corbeil J, Ouellette M. Multiple mutations in heterogeneous miltefosine-resistant Leishmania major population as determined by whole genome sequencing. PLoS Negl Trop Dis 2012;6:1512.

[47] Chan-Bacab MJ, Peña-Rodríguez LM. Plant natural products with leishmanicidal activity. Nat Prod Rep 2001;18:674−88.

[48] Rocha LG, Almeida JR, Macêdo RO, Barbosa-Filho JM. A review of natural products with anti-leishmanial activity. Phytomedicine 2005;12:514−35.

[49] Staerk D, Lemmich E, Christensen J, Kharazmi A, Olsen CE, Jaroszewski JW. Leishmanicidal, antiplasmodial and cytotoxic activity of indole alkaloids from Corynanthe pachyceras. Planta Med 2000;66: 531−6.

[50] Bringmann G, Hamm A, Günther C, Michel M, Brun R, Mudogo V. Ancistroealaines A and B, two new bioactive naphthylisoquinolines, and related naphthoic acids from Ancistrocladus ealaensis. J Nat Prod 2000;11:1465−70.

[51] Mahiou V, Roblot F, Fournet A, Hocquemiller R. Bis-benzylisoquinoline alkaloids from Guatteria boliviana (Annonaceae). Phytochemistry 2000;54:709−16.

[52] Muhammad I, Dunbar DC, Khan SI, Tekwani BL, Bedir E, Takamatsu S, et al. Antiparasitic alkaloids from Psychotria klugii. J Nat Prod 2003;66:962−7.

[53] Camacho MR, Mata R, Castaneda P, Kirby GC, Warhurst DC, et al. Bioactive compounds from

Celaenodendron mexicanum. Planta Med 2000;66: 463—8.

[54] Bravo BJA, Sauvain M, Gimenez TA, Balanza E, Serani L, Laprévote O, et al. Trypanocidal withanolides and withanolide glycosides from Dunalia brachyacantha. J Nat Prod 2001;64:720—5.

[55] Delmas F, Di Giorgio C, Elias R, Gasquet M, Azas N, Mshvildadze V, et al. Antileishmanial activity of three saponins isolated from ivy, alpha-hederin, beta-hederin and hederacolchiside A1, as compared to their action on mammalian cells cultured in vitro. Planta Méd 2000;66:343—7.

[56] Ridoux O, Di Giorgio C, Delmas F, Elias R, Mshvildadze V, Dekanosidze G, et al. In vitro antileishmanial activity of three saponins isolated from ivy, alpha-hederin, beta-hederin and hederacolchiside A(1), in association with pentamidine and amphotericin B. Phytother Res 2001;15:298—301.

[57] Germonprez N, Maes L, Van Puyvelde L, Van Tri M, Tuan DA, De Kimpe N. In vitro and in vivo antileishmanial activity of triterpenoid saponins isolated from Maesa balansae and some chemical derivatives. J Med Chem 2005;48:32—7.

[58] Fuchino H, Koide T, Takahashi M, Sekita S, Satake M. New sesquiterpene lactones from Elephantopus mollis and their leishmanicidal activities. Planta Med 2001;67:647—53.

[59] Villaescusa-Castillo L, Díaz-Lanza AM, Gasquet M, Delmas F, Ollivier E, Bernabé M, et al. Antiprotozoal activity of sesquiterpenes from Jasonia glutinosa. Pharm Biol 2000;38:176—80.

[60] Berger I, Passreiter CM, Cáceres A, Kubelka W. Antiprotozoal activity of Neuroláena lobata. Phytother Res 2001;15:327—30.

[61] Tiuman TS, Ueda-Nakamura T, Garcia Cortez DA, Dias Filho BP, Morgado-Díaz JA, de Souza W, et al. Antileishmanial activity of parthenolide, a sesquiterpene lactone isolated from Tanacetum parthenium. Antimicrob Agents Chemother 2005;49:176—82.

[62] Jullian V, Bonduelle C, Valentin A, Acebey L, Duigou AG, Prévost MF, et al. New clerodane diterpenoids from Laetia procera (Poepp.) Eichler (Flacourtiaceae), with antiplasmodial and antileishmanial activities. Bioorg Med Chem Lett 2005;15:5065—70.

[63 Tan N, Kaloga M, Radtke OA, Kiderlen AF, Oksüz S, Ulubelen A, et al. Abietane diterpenoids and triterpenoic acids from Salvia cilicica and their antileishmanial activities. Phytochemistry 2002;61:881—4.

[64] Sairafianpour M, Kayser O, Christensen J, Asfa M, Witt M, Staerk D, et al. Leishmanicidal and antiplasmodial activity of constituents of Smirnowia iranica. J Nat Prod 2002;65:1754—8.

[65] Hermoso A, Jiménez IA, Mamani ZA, Bazzocchi IL, Piñero JE, Ravelo AG, et al. Antileishmanial activities of dihydrochalcones from Piper elongatum and synthetic related compounds. Structural requirements for activity. Bioorg Med Chem 2003;11:3975—80.

[66] Pirani JR. A revision of Helietta and Balfourodendron (Rutaceae-Pteleinae). Britonia 1998;50:348—80.

[67] Ferreira ME, de Arias AR, Yaluff G, de Bilbao NV, et al. Antileishmanial activity of furoquinolines and coumarins from Helietta apiculata. Phytomedicine 2010;17:375—8.

[68] Lorenzi H, Matos FJA. Plantas medicinais no Brasil: Nativas e Exóticas. Nova Odessa: Instituto Plantarum 2008;2:223—4.

[69] Muzitano MF, Falcão CA, Cruz EA, Bergonzi MC, Bilia AR, Vincieri FF, et al. Oral metabolism and efficacy of Kalanchoe pinnata flavonoids in a murine model of cutaneous leishmaniasis. Planta Med 2009;75:307—11.

[70] Torres-Santos EC, Da Silva SA, Costa SS, Santos AP, Almeida AP, Rossi-Bergmann B. Toxicological analysis and effectiveness of oral Kalanchoe pinnata on a human case of cutaneous leishmaniasis. Phytother Res 2003;17,801—803.

[71] Gomes DC, Muzitano MF, Costa SS, Rossi-Bergmann B. Effectiveness of the immunomodulatory extract of Kalanchoe pinnata against murine visceral leishmaniasis. Parasitology 2010;137:613—8.

[72] Caris P, Ronse Decrane LP, Smets E, Clinklemaillie D. Floral development of three Maesa species, with special emphasis on the position of the genus within Primulales. Ann Bot 2000;86:87—97.

[73] Desta B. Ethiopian traditional herbal drugs. Part I: Studies on the toxicity and therapeutic activity of local taenicidal medication. J Ethnopharmacol 1995;45: 27—33.

[74] Maes L, Germonprez N, Quirijnen L, Van Puyvelde L, Cos P, Vanden Berghe D. Comparative activities of the triterpene saponin maesabalide III and liposomal amphotericin B (AmBisome) against Leishmania donovani in hamsters. Antimicrob Agents Chemother 2004;48:2056—60.

[75] Mols JB, Kessler PJA, Rogstad SH, Saunders RMK. Reassignment of Six Polyalthia Species to the New Genus Maasia (Annonaceae): Molecular and Morphological Congruence. Systematic Botany 2008; 33:490—4.

[76] Rastogi RP. Compendium of Indian Medicinal Plants. New Delhi: CSIR; 1997.

[77] Marthanda Murthy M, Subramanyam M, Hima Bindu M, Annapurna J. Antimicrobial activity of clerodane diterpenoids from Polyalthia longifolia seeds. Fitoterapia 2005;76:336—9.

[78] Misra P, Sashidhara KV, Singh SP, Kumar A, Gupta R, Chaudhaery SS, et al. 16alpha-Hydroxycleroda-3,13(14)Z-dien-15,16-olide from Polyalthia longifolia: a safe and orally active antileishmanial agent. Br J Pharmacol 2010;159:1143−50.

[79] Bennett BC, Alarcón R. Osteophloeum platyspermum and Virola duckei (Myristicaceae): Newly Reported as Hallucinogens from Amazonian Ecuador. Econ Bot 1994;48:152−8.

[80] Davis EW, Yost JA. The ethnomedicine of the Waorani of Amazonian Ecuador. J Ethnopharmacol 1983;9: 273−97.

[81] Roumy V, Garcia-Pizango G, Gutierrez-Choquevilca AL, Ruiz L, Jullian V, Winterton P, et al. Amazonian plants from Peru used by Quechua and Mestizo to treat malaria with evaluation of their activity. J Ethnopharmacol 2007;112:482−9.

[82] Barata LE, Santos LS, Ferri PH, Phillipson JD, Paine A, Croft SL. Anti-leishmanial activity of neolignans from Virola species and synthetic analogues. Phytochemistry 2000;55:589−95.

[83] Seidemann J. World Spice Plants: Economic Usage, Botany, Taxonomy. Springer-Verlag 2005:399−402.

[84] Chase MW, Morton CM, Kallunki JA. Phylogenetic Relationships of Rutaceae: a Cladistic Analysis of the Subfamilies Using Evidence from rbcL and atpB Sequence Variation. American J Bot 1999;86: 1191−9. I.

[85] Adesina SK. The Nigerian Zanthoxylum: chemical and biological values. Afr J Trad CAM 2005;2: 282−301.

[86] Jullian V, Bourdy G, Georges S, Maurel S, Sauvain M. Validation of use of a traditional antimalarial remedy from French Guiana, Zanthoxylum rhoifolium Lam. J Ethnopharmacol 2006;106:348−52.

[87] Bertani S, Bourdy G, Landau I, Robinson JC, Esterre P, Deharo E. Evaluation of French Guiana traditional antimalarial remedies. J Ethnopharmacol 2005;98: 45−54.

[88] Gómez Y, Gil K, González E, Farías. LM. Actividad antifúngica de extractos orgánicos del árbol Fagara monophylla (Rutaceae) en Venezuela. Rev Biol Trop 2007;55:767−75.

[89] Díaz W, Ortega F. Inventario de recursos botánicos útiles y potenciales de la cuenca del río Morón, estado Carabobo, Venezuela. Ernstia 2006;16:31−67.

[90] Diéguez R, Rivas Y, Prieto-Gonzáles S, Garrido G, Molina-Torres J. Potencialidad del Género Zanthoxylum como Fuente de Agentes con Actividad Biológica. Acta Farm Bonaerense 2004;23:243−51.

[91] Ferreira ME, Rojas de Arias A, Torres de Ortiz S, Inchausti A, Nakayama H, Thouvenel C, et al. Leishmanicidal activity of two canthin-6-one alkaloids, two major constituents of Zanthoxylum chiloperone var. angustifolium. J Ethnopharmacol 2002;80: 199−202.

[92] Ferreira ME, Nakayama H, de Arias AR, Schinini A, de Bilbao, et al. Effects of canthin-6-one alkaloids from Zanthoxylum chiloperone on Trypanosoma cruzi-infected mice. J Ethnopharmacol 2007;109: 258−63.

[93] Pereira IO, Marques MJ, Pavan AL, Codonho BS, Barbieri CL, Beijo LA, et al. Leishmanicidal activity of benzophenones and extracts from Garcinia brasiliensis Mart. fruits. Phytomedicine 2010;17:339−45

[94] Passero LF, Castro AA, Tomokane TY, Kato MJ, Paulinetti TF, Corbett CE, et al. Anti-leishmania activity of semi-purified fraction of Jacaranda puberula leaves. Parasitol Res 2007;101:677−80.

[95] Corrêa DS, Tempone AG, Reimão JQ, Taniwaki NN, Romoff P, Fávero OA, et al. Anti-leishmanial and anti-trypanosomal potential of polygodial isolated from stem barks of Drimys brasiliensis Miers (Winteraceae). Parasitol Res 2011;109:231−6.

[96] Sartorelli P, Carvalho CS, Reimão JQ, Ferreira MJ, Tempone AG. Antiparasitic activity of biochanin A, an isolated isoflavone from fruits of Cassia fistula (Leguminosae). Parasitol Res 2009;104:311−4.

Natural Products with Activity Against *Schistosoma* Species

Clarissa Campos Barbosa de Castro[1], Mirna Meana Dias[1], Túlio Pessoa de Rezende[1], Lizandra Guidi Magalhães[2], Ademar A. Da Silva Filho[1]

[1]Department of Pharmaceutical Sciences, Faculty of Pharmacy, Federal University of Juiz de Fora, Juiz de Fora, Minas Gerais State, Brazil and [2]Núcleo de Pesquisas em Ciências Exatas e Tecnológicas, University of Franca, Franca, São Paulo State, Brazil

INTRODUCTION: SCHISTOSOMIASIS

Schistosomiasis, also called bilharziasis, is a chronic disease caused by parasitic worms of the genus *Schistosoma* and one the most significant neglected tropical diseases [1,2]. According to World Health Organization (WHO), schistosomiasis affects more than 240 million people in tropical and subtropical areas, mainly in poor communities without potable water and adequate sanitation conditions [3,4]. In addition, it is estimated that approximately 700 million people are at risk of infection in 74 endemic countries [3,4]. The main species of *Schistosoma* that cause human schistosomiasis are *Schistosoma haematobium*, *Schistosoma intercalatum*, *Schistosoma japonicum*, *Schistosoma mansoni*, and *Schistosoma mekongi*. *S. mansoni* is transmitted by *Biomphalaria* snail species in Africa, the eastern Mediterranean, the Caribbean, and South America, and causes intestinal and hepatic schistosomiasis. On the other hand, *S. japonicum* and *S. mekongi* are transmitted by *Oncomelania* or *Neotricula* snail species, respectively, in China, the Philippines, Southeast Asia, and the Western Pacific region, and cause intestinal and hepatic schistosomiasis. *S. haematobium* and *S. intercalatum* are both transmitted by *Bulinus* snail species and cause urinary and intestinal schistosomiasis, respectively, mainly in countries in Africa and the eastern Mediterranean [1].

The schistosome parasites are digenetic trematodes with complex life cycles that undergo extensive body remodeling during development to sustain metabolic adaptations within their invertebrate and vertebrate hosts, as well as during their aquatic phase [5]. Briefly, eggs produced by adult worms are released by feces or urine into water, where the eggs reach

Fighting Multidrug Resistance with Herbal Extracts, Essential Oils and Their Components
http://dx.doi.org/10.1016/B978-0-12-398539-2.00008-2

maturation and release miracidia, which are the infective forms for the snails. Inside the snails, miracidia are transformed into sporocysts by asexual reproduction and form cercariae that are released into the water and, after contact with the skin, penetrate into the definitive host (humans or another mammalian host). Inside the definitive host, the cercariae transform into schistosomula, which migrate to the hepatoportal circulation, where they mature into male and female adult worms. After coupling, pairs of adult worms migrate to their final niche in the mesenteric circulation, where they begin egg production, which is responsible for the resulting immunopathologic lesions [6,7].

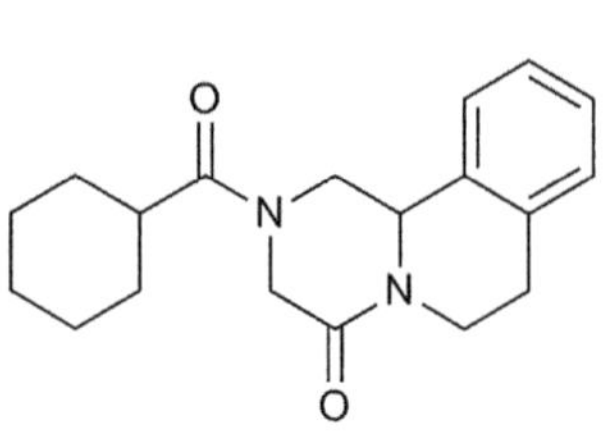

SCHISTOSOMICIDAL DRUGS AND RESISTANCE

Although some promising results of schistosomiasis vaccines have been reported in animal models, at the moment there is no available vaccine against *Schistosoma* spp. [8]. Therefore, the current strategy, as recommended by WHO Expert Committee for the control of schistosomiasis, is the use of an effective drug for reducing morbidity [9,10]. Historically, the first drug used against *Schistosoma* was tartar emetic but its use has been suspended due to its severe toxic effects, which are common in antimony compounds [7,11]. With advances in drug development leading to the introduction of safer schistosomicidal drugs that can be orally administered in a single dose, it has been possible to initiate control programs in various endemic areas [8,12]. Nowadays, praziquantel (compound 1, Fig. 8-1) and oxamniquine (compound 2, Fig. 8-2) are included in the WHO list as essential drugs for schistosomiasis control [13].

Praziquantel, a pyrazino-isoquinoline derivative, is a racemic compound containing equal amounts of its optical isomers [14], although

Praziquantel (1)

FIGURE 8-1 Chemical structure of praziquantel.

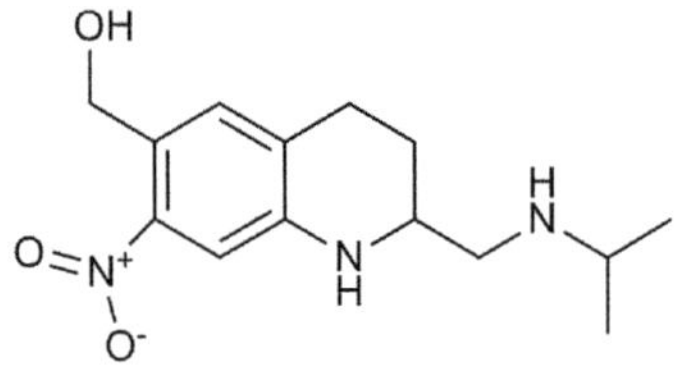

Oxamniquine (2)

FIGURE 8-2 Chemical structure of oxaminiquine.

in vitro [15,16] and *in vivo* [15] studies have identified the (*levo*)-isomer as the active compound. This drug is effective against all *Schistosoma* species that infect humans, showing *in vivo* activity against cercariae, young schistosomula, and adult worms. However, juvenile parasites, between 7 and 28 days old, appear to be less susceptible to this drug [17,18]. Although praziquantel has been used since the 1980s, its mechanism of action has yet to be fully elucidated [11]. *In vitro* studies have demonstrated that praziquantel causes intense muscular paralysis in adult worms, accompanied by a rapid influx of calcium ions, a slower influx of sodium ions, and a reduction in potassium ion uptake [19,20]. An additional suggested mechanism is that praziquantel disrupts the *Schistosoma* tegument, causing the exposure of antigens on the parasite surface that, together with the immune response, is responsible to the death of parasites [8]. Recently, following microarrays analysis, Aragon et al. [20] observed changes in gene expression after exposure of mature *S. mansoni* to a sublethal dose of praziquantel. It was suggested that schistosomes may undergo a transcriptomic response after praziquantel exposure, similar to that observed during oxidative stress.

Oxamniquine, a tetrahydroquinoline derivative, is also effective against cercariae, schistosomula, and adult worms of *Schistosoma* spp. Despite oxamniquine being as effective as praziquantel, it is more expensive and may cause more adverse effects, such as drowsiness, sleepiness, and epileptic seizures [1]. As for praziquantel, it was previously believed that the mechanism of action for oxamniquine was related to its anticholinergic effects. Currently, its mechanism is thought to involve irreversible inhibition of nucleic acid synthesis in the parasites [8].

Regarding the effectiveness of praziquantel and oxamniquine for current schistosomiasis treatment, several reports have been published demonstrating a significant reduction in their efficacy and suggesting that multidrug resistance may be present [8,21].

Some studies have shown specific resistance against praziquantel [22–26]. Ismail et al. [24] demonstrated that mice infected with *S. mansoni* and treated with subcurative doses of praziquantel over several generations, produced parasites that were less sensitive to the drug. In addition, possible parasite resistance to praziquantel was reported in Senegal in 1994. For a dose of 40 mg/kg, the cure rates against *S. mansoni*, ranged from 18% to 39%; the rates normally expected are between 60% and 90% [22,23,26]. Moreover, when the isolated strain was evaluated in laboratory, it was found to be less susceptible to praziquantel treatment than laboratory isolates [23]. In addition, when oxamniquine was used in the same region, the cure rate was approximately 79% [22]. There have been reports that the parasite cannot be killed by praziquantel in Egyptian patients, with treatment failing after the repeated administration of high doses [24,25]. In a school-based treatment program

in Kenya, King et al. [27] evaluated the long-term efficacy of praziquantel against *S. haematobium*, reporting variation in the response to this drug that suggested a possible resistance to praziquantel.

Similarly, some research groups have investigated parasite resistance against oxamniquine. In 1973, Katz et al. [28] conducted clinical studies in Brazil and reported oxamniquine resistance in isolated *Schistosoma* strains. Five years later, resistant strains to praziquantel and oxamniquine were found in Brazil, Egypt, Kenya, and Senegal [29]. In 1993, Pica-Mattoccia et al. [30] reported that resistant schistosomes survived oxamniquine treatment with doses 1000-fold higher than the dose effective against sensitive parasites. In addition, it was noted that Central and East African strains of *S. mansoni* are significantly less susceptible to oxamniquine than are South American and West African strains [8].

Currently, the mechanism of drug resistance is unclear. However, schistosomes are becoming less sensitive to the major schistosomicidal drugs, since a complete cure after praziquantel treatment is now rarely achieved [21]. Therefore, there is an urgent need to develop new safe and effective schistosomicidal drugs [21].

NATURAL PRODUCTS AS SCHISTOSOMICIDAL AGENTS

Natural products are compounds that are produced by living systems [31–33], including extracts (e.g., a crude plant extract) and pure compounds (e.g., secondary metabolites) isolated from plants, animals, or microorganisms [34,35]. Many of the biologically active constituents of medicinal and poisonous plants are secondary metabolites; these have attracted interest over the years because of their biologic activities, mainly against microorganisms and parasites [36,37].

Herbal extracts have been used as anthelmintics for the treatment of human and veterinary parasitic diseases in traditional medicines since ancient times [38–41]. Subsequently, many plants have been chemically and biologically investigated in an attempt to discovery useful herbal preparations or natural active constituents that may be used as lead compounds for the development of new antiparasitic drugs [33,42]. Ascaridol (compound 3, Fig. 8-3), a peroxide monoterpene, is the active ascaricide constituent of *Chenopodium ambrosioides* (Chenopodiaceae), a traditional antiparasitic herb used for treating intestinal helminths. Thymol (compound 4, Fig. 8-3), a phenol derived from *Thymus vulgaris* (Lamiaceae), was also used as antiparasitic at the end of 19th century [43].

Since the last few decades of the 20th century, natural products have attracted renewed interest as source of biologically and pharmacologically active antiparasitic compounds [33,44]. Artemisinin (compound 5, Fig. 8-3), quinine (compound 6, Fig. 8-3), and licochalcone A (compound 7, Fig. 8-3) are examples of plant-derived compounds that possess antiparasitic activity, while ivermectin (compound 8, Fig. 8-3) and amphotericin B (compound 9, Fig. 8-3) are antiparasitic molecules isolated from microorganisms [33,45–47].

On the other hand, there has been little incentive to invest in the discovery and development of schistosomicidal drugs. Nevertheless, since the beginning of the 21st century, a large number of samples have been evaluated against *Schistosoma* spp. as part of worldwide attempts to find new effective and safe drugs to overcome this serious disease. To illustrate the importance of natural products to the drug discovery process, several studies showing the effects of herbal extracts, essential oils (EOs) and isolated natural compounds on *Schistosoma* spp. are presented in the following paragraphs.

Crude Extracts and Herbal Preparations with Activity Against Schistosomes

Most of the studies on the schistosomicidal activity of crude herbal preparations are focused on *in vitro* screening against different schistosome life-stages, mainly cercariae, schistosomula, and adult worms. Several crude plants extracts reported to display activity against *Schistosoma* spp., are shown in Table 8-1. Numerous extracts and herbal preparations, from different plant species have been evaluated. For example, Yousif et al. [48] screened 346 methanol extracts from different plant species against adult worms of *S. mansoni*. They have shown strong *in vitro* schistosomicidal activity in a number of plants, including extracts of *Pinus canariensis* (LC50 of 12.8 ppm), *Agave lophantha* (LC50 of 8.2 ppm), *Furcraea selloa* (LC50 of 7.10 ppm), and *Solanum elaeagnifolium* (LC50 of 6.0 ppm). Ferreira et al. [45] tested some crude extracts against adult *S. mansoni* and showed that hydroalcohol extracts of *Artemisia absinthium* and *Artemisia annua*, both of which contain artemisinin, are able to kill all adult schistosomes. Also, Koné et al. [49] screened 65 extracts (at 2.0 mg/mL) from 50 medicinal plants against schistosomula and adult forms of *S. mansoni*: the hydroalcohol extract of *Eriosema griseum* leaves was one of the most active.

In contrast, just a few plant extracts have been evaluated in *in vivo* studies. As summarized in Table 8-1, most of the crude extracts tested exhibited weak *in vivo* activity. For example, the *in vivo* treatment of *S. mansoni* with alcoholic extracts of *Ambrosia maritima* (Asteraceae) [53] and *Cleome droserifolia* (Asteraceae) [59] resulted in a weak reduction in the worm burden. On the other hand, *in vivo* treatment with aqueous extracts of *Sida pilosa* (Malvaceae) or *Clerodendrum umbellatum* (Verbenaceae) resulted in significant reductions in parasite burden [63,71]. Other crude extracts were also evaluated *in vivo*, such as *Citrus*

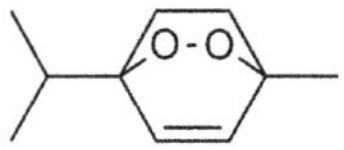

Ascaridol (3) 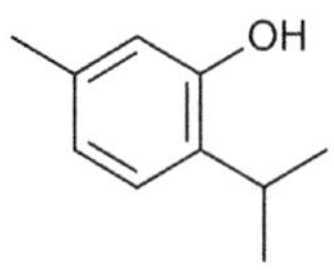   Thymol (4) 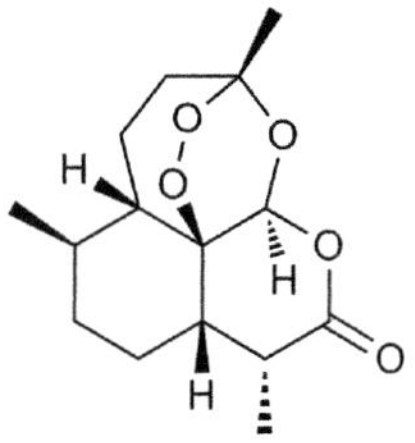   Artemisinin (5) 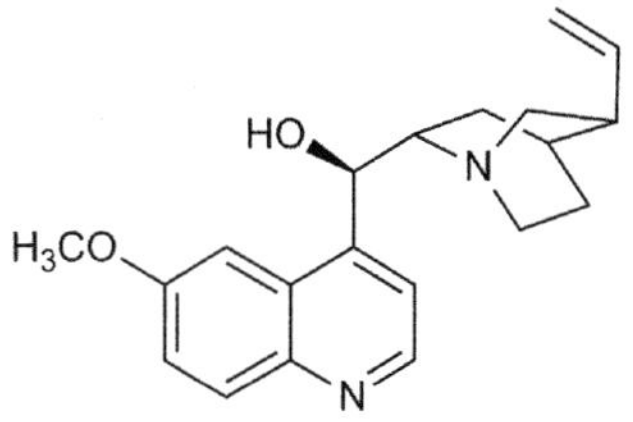   Quinine (6)

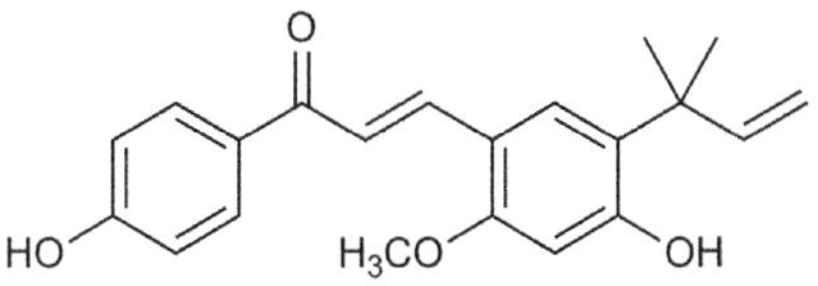

Licochalcone-A (7) 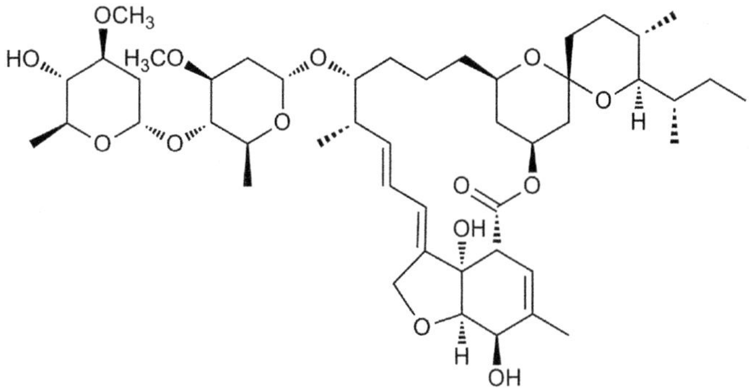   Ivermectin (8)

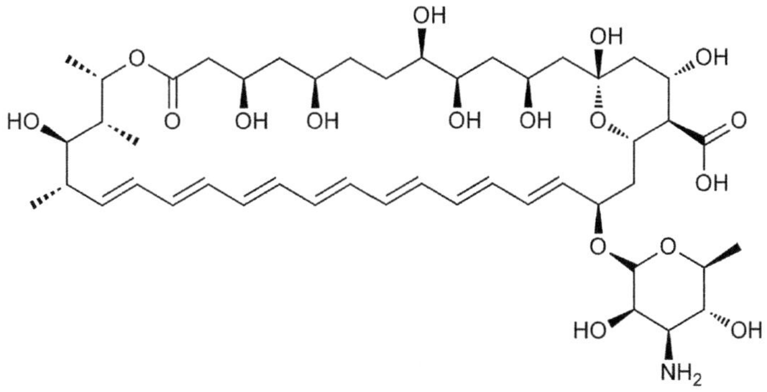

Amphotericin B (9)

FIGURE 8-3 Plant-derived compounds with antiparasitic activity.

reticulata (Rutaceae) and *Pavetta owariensis* (Rubiaceae), which led to a reduction in both the worm burden and egg count [65,66]. Different methods of plant extraction may produce different herbal preparations, which may be chemically distinct and, therefore, have different biologic activities. For example, *in vivo* studies with *Zingiber officinalis* (Zingiberaceae) demonstrated that its ethyl acetate extract was inactive, while its aqueous extract induced

TABLE 8-1   Crude Extracts and Herbal Preparations with Activity Against *Schistosoma* Species

Family	Species	Extract/Part	Activity	Reference(s)
Agavaceae	*Agave americana*	Methanol/leaves	LC50: 11.45 ppm; LC90: 18.40 ppm[a]	[48]
	Agave lophantha	Methanol/leaves	LC50: 8.20 ppm; LC90: 11.70 ppm[a]	
	Furcraea selloa	Methanol/leaves	LC50: 7.10 ppm; LC90: 9.40 ppm[a]	
	Furcraea selloa	Methanol/leaves	Death of all worms at 50 mg/mL after 24 h[a]	[50]
Anacardiaceae	*Lannea barteri*	Hydroalcohol/leaves	MLC: 200 μg/mL and NTS-MLC: 10 μg/mL[a]	[49]
Annonaceae	*Asimina triloba*	Hydroalcohol/stems	Death of all worms at 2.0 mg/mL within 17 h[a]	[45]
	Xylopia aethiopica	Hydroalcohol/bark	MLC: 160 μg/mL and NTS-MLC: 10 μg/mL[a]	[49]
Apocynaceae	*Nerium oleander*	Methanol	Death of all worms at 100 μg/mL after 24 h[a]	[51]
Asclepiadaceae	*Calotropis procera*	Methanol/herb	LC50: 11.50 ppm; LC90: 19 ppm[a]	[48]
	Pergularia tomentosa	Methanol/herb	LC50: 9.40 ppm; LC90: 12 ppm[a]	
	Asclepias sinaica	Methanol/herb	LC50: 11.70 ppm; LC90: 16.20 ppm[a]	
Asparagaceae	*Scilla natalensis*	Aqueous	Inhibitory concentration: 0.4 mg/mL[d]	[52]
Asteraceae	*Ambrosia maritima*	Alcohol/leaves	Inactive against *S. mansoni* adult worms *in vivo*	[53]
	Artemisia absinthium	Hydroalcohol/leaves	Death of all worms at 2.0 mg/mL within 17 h[a]	[45]
	Artemisia annua	Hydroalcohol/leaves	Death of all worms at 2.0 mg/mL within 17 h[a]	
Boraginaceae	*Alkanna orientalis*	Methanol/herb	LC50: 7 ppm; LC90: 13 ppm[a]	[48]
Burseraceae	*Commiphora molmol*	Oleo-gum resin/stem (Myrrh)	Cure rate: 91.7–100% at 10 mg/kg[c]	[54,55]
			Cure rate: 97.4% for *S. haematobium* and 96.2% for *S. mansoni*[c]	[56]
			Inactive against *S. mansoni* adult worms *in vivo*	[57,58]
Capparaceae	*Cleome droserifolia*	Alcohol/leaves	Weak reduction in worm burden[b]	[59]

TABLE 8-1 Crude Extracts and Herbal Preparations with Activity Against *Schistosoma* Species (*cont'd*)

Family	Species	Extract/Part	Activity	Reference(s)
Combretaceae	*Anogeissus leiocarpus*	Hydroalcohol/stem bark	MLC: 160 μg/mL and NTS-MLC: 10 μg/mL[a]; reduction: 57% of worms at 400 mg/kg[b]	[49]
	Combretum mucronatum	Hydroalcohol/leaves	MLC: 200 μg/mL and NTS-MLC: 20 μg/mL[a]	
Dipterocarpaceae	*Monotes kerstingii*	Hydroalcohol/stem bark	MLC: 160 μg/mL and NTS-MLC: 40 μg/mL[a]	[49]
		Hydroalcohol/roots	MLC: 200 μg/mL and NTS-MLC: 20 μg/mL[a]	
Euphorbiaceae	*Euphorbia royleana*	Lyophilized latex	LC90: 11 ppm[a]	[60]
	Euphorbia mauritanica	Natural latex	LC90: 60 ppm[a]	
	Jatropha curcas	Acetonitrile, chloroform	LC90: 6 ppm[a]; LC90: 55 ppm[a]	
	Milbraedia paniculata	Hydroalcohol/leaves	MLC: 200 μg/mL and NTS-MLC: 40 μg/mL[a]	[49]
Fabaceae	*Eriosema griseum*	Hydroalcohol/leaves	MLC: 40 μg/mL and NTS-MLC: 20 μg/mL[a]; reduction: 49.5% of worms at 400 mg/kg[b]	[49]
	Erythrina senegalensis	Hydroalcohol/roots	MLC: 160 μg/mL and NTS-MLC: 10 μg/mL[a]	
Fumariaceae	*Fumaria officinalis*	Methanol/leaves	Inactive against *S. mansoni* adult worms *in vitro*	[45]
Humiriaceae	*Sacoglottis gabonensis*	Hydroalcohol/bark	MLC: 80 μg/mL; NTS-MLC: 5 μg/mL[a]	[49]
Hyacinthaceae	*Ledeboria ovatifolia*	Aqueous	Death of all worms at 1.6 mg/mL[d]	[52]
Lamiaceae	*Plectranthus tenuiflorus*	Methanol/herb	Cercariae, IC50: 123 μg/mL; schistosomula, IC50: 174 μg/mL and miracidia, IC50: 244 μg/ts}mL[a]	[61]
Liliaceae	*Allium sativum*	Aqueous	NTS-MLC[b]: dose of 125 mg/kg	[62]
Malpighiaceae	*Flabellaria paniculata*	Hydroalcohol/leaves	MLC: 200 μg/mL and NTS-MLC: 20 μg/mL[a]	[49]
Malvaceae	*Sida pilosa*	Aqueous/whole plant	Significant reduction in egg-laying[b]	[63]

(*Continued*)

TABLE 8-1   Crude Extracts and Herbal Preparations with Activity Against *Schistosoma* Species (*cont'd*)

Family	Species	Extract/Part	Activity	Reference(s)
Meliaceae	*Khaya grandifoliola*	Methanol/branches	LC50: 8.70 ppm; LC90: 24.40 ppm[a]	[48]
	Swietenia mahogani	Methanol/branches	LC50: 7.40 ppm; LC90: 13.30 ppm[a]	
Myrtaceae	*Pimenta racemosa*	Methanol/bark	LC50: 14.20 ppm; LC90: 23.70 ppm[a]	[48]
Olacaceae	*Olax subscorpioidea*	Hydroalcohol/roots	MLC: 160 μg/mL and NTS-MLC: 40 μg/mL[a]; reduction: 60.2% of worms at 400 mg/kg[b]	[49]
Pinaceae	*Pinus canariensis*	Methanol/fruit	LC50: 12.80 ppm; LC90: 22 ppm[a]	[48]
Polygalaceae	*Securidaca longipedunculata*	Hydroalcohol/roots	MLC: 80 μg/mL and NTS-MLC: 40 μg/mL[a]	[49]
Ranunculaceae	*Nigella sativa*	Seeds	LC50: about 50 ppm[a]	[64]
Rubiaceae	*Craterispermum caudatum*	Hydroalcohol/leaves	MLC: 200 μg/mL and NTS-MLC: 10 μg/mL[a]	[49]
	Crossopteryx febrifuga	Hydroalcohol/roots	MLC: 10 μg/mL and NTS-MLC: 40 μg/mL[a]; reduction: 46.3% of worms at 400 mg/kg[b]	[49]
	Pavetta owariensis	Alcohol/barks	Modulation of the granulomatous reaction[b]	[65]
Rutaceae	*Citrus reticulata*	Alcohol/roots	Reduction of enzyme activities and decrease egg production by 59.45%[b]	[66]
	Zanthoxylum naranjillo	Ethanol/leaves	Decrease *S. mansoni* egg production by 16.3%	[67]
Sapotaceae	*Mimusops kummel*	Hydroalcohol/stem bark	MLC: 200 μg/mL and NTS-MLC: 40 μg/mL[a]	[49]
Scrophulariaceae	*Verbascum sinuatum*	Methanol/fruit	LC50: 8.0 ppm; LC90: 14.20 ppm[a]	[48]
Solanaceae	*Solanum elaeagnifolium*	Methanol/fruit	LC50: 6.0 ppm; LC90: 8.41 ppm[a]	
	Solanum lycocarpum	Alkaloid/fruit	Death of all worms at 20 μg/mL after 24 h[a]	[68]
	Solanum nigrum	Aqueous/leaves	LC100: 30 mg/L within 30 min of exposure[e]	[69]
		Methanol/herb	LC50: 13.30 ppm; LC90: 28.80 ppm[a]	[48]
	Solanum tuberosum	Methanol/callus	IC50: 104 ppm after 24 h[a]	[70]

TABLE 8-1 Crude Extracts and Herbal Preparations with Activity Against *Schistosoma* Species (*cont'd*)

Family	Species	Extract/Part	Activity	Reference(s)
Sterculiaceae	*Brachychiton rupestris*	Methanol/branches and leaves	LC50: 11.60 ppm; LC90: 14.53 ppm[a]	[48]
Verbenaceae	*Clerodendrum umbellatum*	Aqueous/leaves	Dose $\geq$ 80 mg/kg b.w.[b]	[71]
	Stachytarpheta cayennensis	Hydroalcohol/whole plant	MLC: 160 μg/mL. NTS-MLC: 40 μg/mL[a]	[49]
Zingiberaceae	*Curcuma longa*	Methanol/roots	Death of all worms at 100 μg/mL after 24 h[a]	[51]
		Chloroform/roots	EC50: 28.92 μg/mL for male and 31.58 μg/mL for female worms[a]	
	Zingiber officinale	Aqueous/rhizome	Dose of 500 mg/kg b.w.[b]	[72]
	Zingiber officinale	Ethyl acetate/rhizome	Inactive against *S. mansoni* adult worms *in vivo*. Death of almost all at 200 mg/L after 24 h[a]	[73]

[a] *In vitro* against *S. mansoni* adult worms.
[b] *In vivo* against *S. mansoni* adult worms.
[c] *In vivo* against *S. mansoni* and *S. haematobium* adult worms.
[d] *In vitro* against *S. haematobium* adult worms.
[e] *In vitro* against *S. mansoni* and *S. haematobium* adult worms.
b.w., body weight; EC50, effective concentration 50; IC50, half maximal inhibitory concentration; LC50/90/100, lethal concentration 50/90; MLC, minimal lethal concentration; NTS, newly transformed schistosomula.

a significant reduction in worm recovery and egg density [72,73]. Recently, Egyptian researchers have reported the *in vivo* effects of Mirazid, an oleo-resin derived from myrrh trees (*Commiphora molmol*, Burseraceae; Mirazid). Several *in vivo* experiments in mice were conducted, as well as clinical trials in Egyptian patients [54,74]. Initial studies using oral administration of Mirazid reported a very high cure rate (91.7%) in *S. mansoni*-infected patients [54,74], while other trials reported significant reductions in the numbers of eggs [55,56,74]. However, more recent studies have shown that Mirazid is ineffective against schistosome infection [57,58].

The use of whole plant preparations may be more therapeutically effective than using isolated compounds [33]. However, since crude plant extracts contain a wide range of constituents for which the chemical compositions may vary as a result of seasonal variation, it is important to use a chemically defined standard of a particular type of extract for biologic evaluation [75,76]. Discovering unexploited natural sources of new schistosomicidal compounds from crude extracts remains a challenge. However, given the large number of plant species worldwide that have not yet been evaluated against schistosomes, there is still much scope for investigations.

Essential Oils with Activity Against Schistosomes

Scientific investigations have revealed many biologic activities of EOs, including antimicrobial and anthelmintic effects [37,77,78]. Very few EOs have been screened for schistosomicidal activity and most researches have focused on *in vitro* assays against adult *S. mansoni* worms (shown in Table 8-2). Several EOs displayed good *in vitro* activity against schistosomes. For example, *Baccharis dracunculifolia* (Asteraceae) EO at a concentration of 10 µg/mL had *in vitro* activity against adult *S. mansoni*, while EOs from *Bidens pilosa* (Asteraceae) and *Tagetes erecta* (Asteraceae) were able to kill all adult worms of *S. mansoni* at concentrations of around 100 µg/mL. In contrast, (*E*)-nerolidol, the major compound identified in *B. dracunculifolia* EO, was inactive against adult schistosomes at the concentrations tested [37]. In addition, EOs of *Piper cubeba* (Piperaceae) and *Ageratum conyzoides* (Apiaceae) were also active *in vitro* against adult schistosomes (Table 8-2). Cercaricidal *in vitro* effects against *S. mansoni* were also observed for other EOs, such as those of *Apium graveolens* (Apiaceae), *P. cubeba*, and *Piper marginatum* (Piperaceae), as well as *Capsicum annum* (Solanaceae) and *Eucalyptus* spp. (Myrtaceae). Moreover, *Zingiber officinale* (Zingiberaceae) EO was active against cercariae of *S. japonicum* [79–84].

Similar to crude plant extracts, the schistosomicidal activities of only a few EOs have been evaluated *in vivo*. Oral treatment of *S. mansoni*-infected mice with 2.5 and 5.0 mL/kg of *Nigella sativa* (Ranunculaceae) EO produced a significant reduction in the number of adult worms recovered, as well as a significant decrease in the total number of eggs. Treatment with a combination of praziquantel and *N. sativa* EO led to the greatest reduction in the total number of eggs produced. According to one study, the schistosomicidal effects of *N. sativa* EO on *S. mansoni*-infected mice may correlate with its antioxidant effect and ability to improve the host immune system [89].

EOs are complex mixtures of a large number of chemical compounds, mainly comprising monoterpenes, sesquiterpenes, and phenylpropanoids [77,78]. Similar to crude plant extracts, the chemical composition of EOs may vary considerably, which reinforces the need to also test the major isolated compounds in schistosomicidal assays.

NATURAL PLANT COMPOUNDS WITH SCHISTOSOMICIDAL ACTIVITY

Many reports have demonstrated *in vivo* or *in vitro* schistosomicidal activity in herbal extracts and some of the active molecules have purified and identified by fractionation and isolation of the major compounds [63]. After isolating and identifying the major compounds present in an active herbal extract, the next step is to evaluate the purified compounds in schistosomicidal assays. A number of natural molecules have been shown to have activity against schistosomes. The following discussion will focus on recently reported examples.

Alkaloids and Alkamides

Glycoalkaloids from *Solanum* spp. (Solanaceae) are some of the isolated natural products reported to have schistosomicidal activity. It was shown that an alkaloid extract of *S. lycocarpum* fruit kills *S. mansoni* adult worms within 24 h of incubation. Solasodine (compound 10, Fig. 8-4) and solamargine (compound 11, Fig. 8-4), isolated from the active alkaloid extract, showed promising *in vitro* schistosomicidal activity in concentrations ranging from 20 to 50 µM, causing the separation of coupled worms, extensive disruption of worm teguments, and the death of all *S. mansoni* adult worms [74].

Moreover, epiisopiloturine (compound 13, Fig. 8-4), an imidazole alkaloid isolated from the leaves of *Pilocarpus microphyllus* (Rutaceae), showed an *in vitro* effect on the survival time of *S. mansoni* at different stages (schistosomula and adult worms). Epiisopiloturine (compound

TABLE 8-2 Essential Oils with Activity Against *Schistosoma*

Family	Species	Extraction/Parts	Major compounds	Oil activity	Reference
Apiaceae	*Apium graveolens*	Distillation/aerial parts	Limonene; myrcene and β-selinene	Death of 96% at 50 ppm within 15 min[a]	[81]
Asteraceae	*Ageratum conyzoides*	Hydrodistillation/leaves	Precocene I and (E)-caryophyllene	LC50: 198.8 μg/mL in 24 h[b]	[80]
	Baccharis dracunculifolia	Hydrodistillation/leaves	(E)-nerolidol and spathulenol	Death of all worms at 10 μg/mL (E)-nerolidol was inactive[b]	[37]
	Bidens sulfurea	Hydrodistillation/flowers	2,6-di-tert-butyl-4-methylphenol; germacrene D- and β-caryophyllene	Death of all worms at 100 μg/mL within 48 h[b]	[85]
	Tagetes erecta	Hydrodistillation/leaves	α-terpinolene; (E)-ocimenone and dihydrotagetone	LC50: 81.47 μg/mL in 24 h[b]	[86]
Cupressaceae	*Thujopsis dolabrata*	Wood oil extract	Hinokitiol	Hinokitiol prevented cercariae penetration in host In skin *vivo*	[87]
Lamiaceae	*Plectranthus neochilus*	Hydrodistillation/leaves	β-Caryophyllene; α-thujene and α-pinene	LC50: 89.65 μg/mL in 24 h[b]	[88]
Myrtaceae	*Eucalyptus species*	Hydrolates and essential oils obtained by vapor dragging	ND	*In vivo* hydrolates and essential oils presented activity on cercariae[a]	[84]
Piperaceae	*Piper cubeba*	Essential oil/fruits	Sabinene; eucalyptol and 4-terpineol	*In vitro* against cercariae, schistosomula and adult worms of *S. mansoni*	[79]
	Piper marginatum	Hydrodistillation/leaves	Water soluble, unsaturated	Death of 90–96% cercariae within 15 min[a]	[82]
Ranunculaceae	*Nigella sativa*	Hydrodistillation/seeds	ND	Reduction: 32% of worms at 5 mL/kg[c]	[89]
Solanaceae	*Capsicum annuum*	Hydrodistillation/fruits	Water soluble, unsaturated	Death of 90–96% cercariae within 15 min[a]	[82]
Zingiberaceae	*Zingiber officinale*	Distillation/rhizome	ND	Death of all cercariae at 100 μL within 20 s[d]	[83]

[a] *In vitro* against cercariae of *S. mansoni*
[b] *In vitro* against *S. mansoni* adult worms
[c] *In vivo* against *S. mansoni* adult worms
[d] *In vitro* against cercariae of *S. japonicum*
ND, not determined.

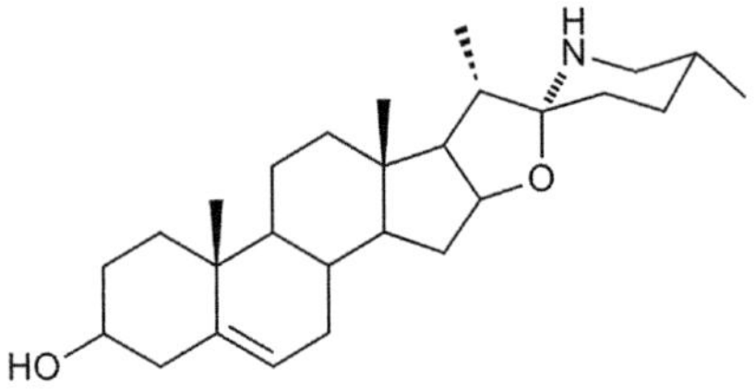

Solasodine (**10**)

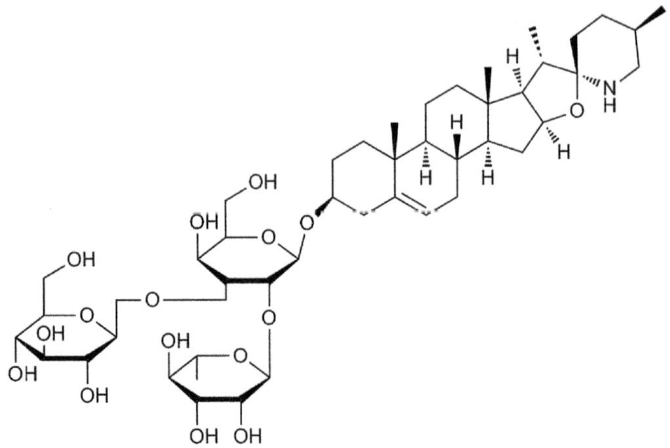

Solamargine (**11**)

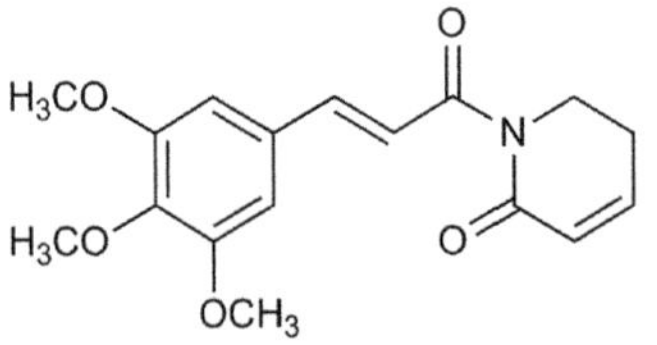

Piplartine (**12**)

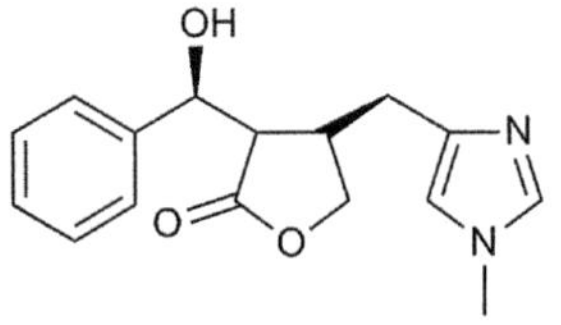

Epiisopiloturine (**13**)

FIGURE 8-4 Alkaloids and alkamides with schistosomicidal activity.

13, Fig. 8-4) at a concentration of 300 μg/mL killed all schistosomula and led to extensive tegumental alterations, as well as causing death, in adult worms but exhibited no cytotoxicity toward mammalian cells [90].

Similar to alkaloids, some alkamides from *Piper* spp. are reported to exert *in vitro* schistosomicidal activity. Piplartine (compound 12, Fig. 8-4), an amide found in several *Piper* spp., such as *Piper tuberculatum*, showed *in vitro* activity against adult *S. mansoni* worms. It was reported that treatment with 15.8 μM piplartine reduced motor activity and egg production, as well as killing all adult worms. Piplartine also induced morphologic changes in the tegument, causing extensive tegumental destruction and damage to the parasite tubercles. This damage was dose dependent in the range of 15.8–630.2 μM. More schistosomicidal and phytochemical studies should be conducted with *Piper* spp. since most have shown to be active against schistosomes [91].

Lignans and Neolignans

Neolignans and lignans are classes of natural compounds that have fascinated researchers over the years because of their variety of chemical structures and broad range of biologic activities, such as antileishmanial [92], anti-inflammatory [93,94], and trypanocidal [95,96] activities. Recently, our research group demonstrated that dibenzylbutyrolactone and tetrahydrofuran lignans are potent groups of natural products with both *in vitro* and *in vivo* schistosomicidal activity against *S. mansoni*. According to

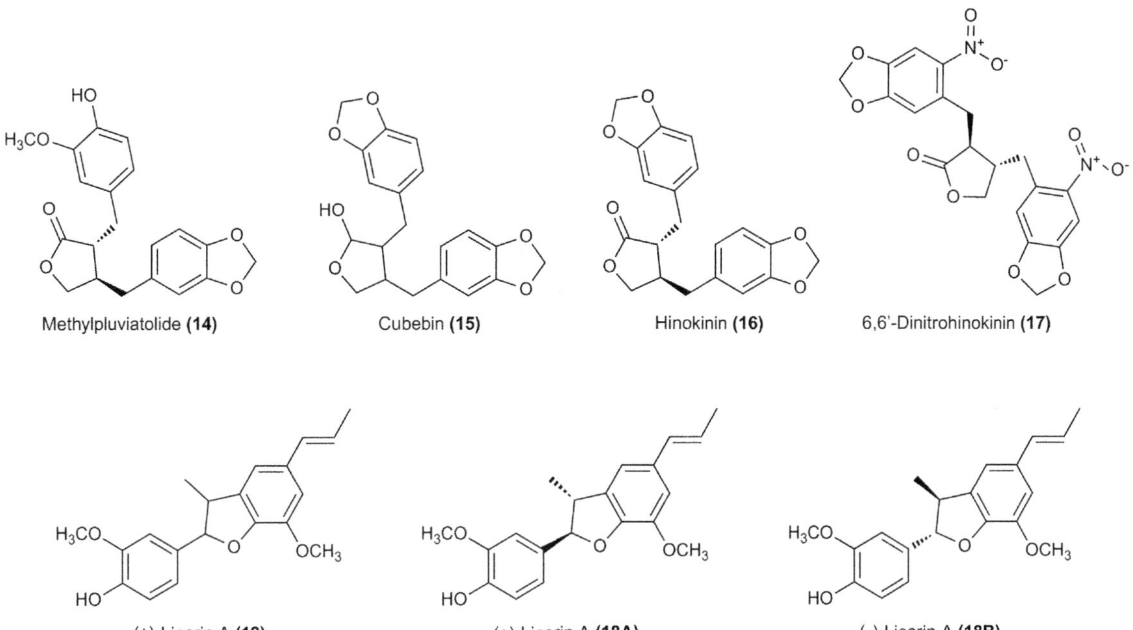

Methylpluviatolide (14) Cubebin (15) Hinokinin (16) 6,6'-Dinitrohinokinin (17)

(±)-Licarin A (18) (+)-Licarin A (18A) (-)-Licarin A (18B)

FIGURE 8-5 Lignans and neolignans with schistosomicidal activity.

a patent [97], some dibenzylbutyrolactone lignans, such as methylpluviatolide (compound 14, Fig. 8-5) and cubebin (compound 15, Fig. 8-5), as well as their semisynthetic lignan derivatives hinokinin (compound 16, Fig. 8-5) and 6,6'-dinitrohinokinin (compound 17, Fig. 8-5), possess high *in vitro* and *in vivo* schistosomicidal activity that is not associated with toxicity to host animals.

Continuing our study of schistosomicidal activity in lignoids, we showed that licarin A (compound 18, Fig. 8-5), a benzofuran neolignan active against multidrug-resistant mycobacteria [98] found in plants of the genus *Licaria* (Lauraceae), exhibits high *in vitro* schistosomicidal activity against adult *S. mansoni* worms. A racemic mixture of (±)-licarin A was obtained by oxidative coupling of isoeugenol to horseradish peroxidase and enantiomers were isolated by chiral high performance liquid chromatography. It was observed that the racemic mixture of licarin A was more active than the individual (+) (compound 18A, Fig. 8-5) and (−) (compound 18B, Fig. 8-5) enantiomers. In addition, treatment with 100 μM (±)-licarin A killed all coupled adult worms and was associated with extensive tegumental alterations without separation of the worms (Fig. 8-6C and D), in contrast to untreated (Fig. 8-6A) and vehicle controls (Fig. 8-6B). As a consequence of their schistosomicidal potential, further schistosomicidal studies with lignans and neolignans are now in progress.

Diarylheptanoids

Curcumin (compound 19, Fig. 8-7), a major diarylheptanoid found in the rhizomes of *Curcuma longa* (Zingiberaceae), has been tested for schistosomicidal activity. *In vitro* studies revealed that 50 μM curcumin is capable of

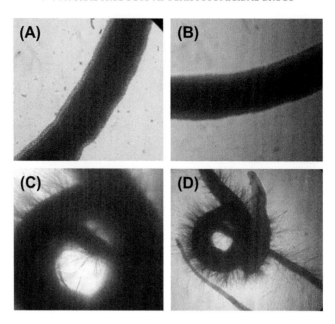

FIGURE 8-6 *In vitro* **tegumental alterations in** *Schistosoma mansoni* **induced by (±)-licarin A.** (A) Coupled adult worms in RPMI 1640 medium. (B) Coupled adult worms in RPMI 1640 medium containing 0.4% DMSO (vehicle control). (C) Coupled adult worms treated with 100μM (±)-licarin A for 120 h. 10× magnification. (D) Coupled adult worms treated with 100μM (±)-licarin A for 120 h. 4× magnification.

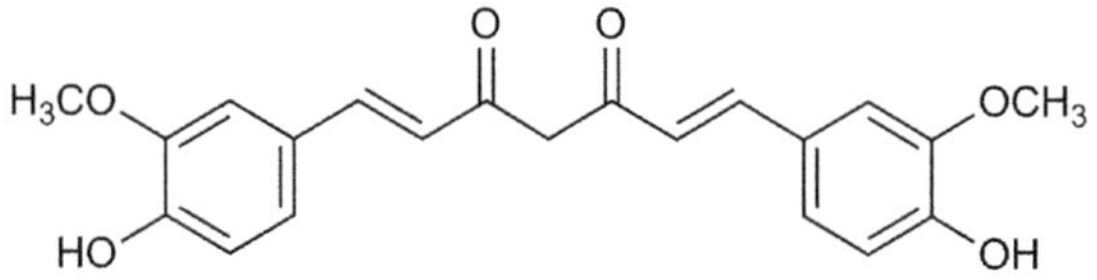

Curcumin (19)

FIGURE 8-7   Chemical structure of curcumin.

killing all coupled adult worms and separating them into individual males and females. Additionally, it was observed that the incorporation of curcumin into poly(lactic-coglycolic)acid (PLGA) nanospheres enhances its activity [99]. As curcumin was considered a promising schistosomicidal compound, *in vivo* studies were performed in Swiss albino mice infected with *S. mansoni* and demonstrated that treatment with this compound induced a significant reduction (67.3%) in worm burden compared

with controls [100]. Also, serum levels of interferon gamma (IFN-γ) and interleukin-2 (IL-2) were significantly decreased in the group treated with curcumin [100]. Additional *in vivo* studies carried out in mice demonstrated that intraperitoneal administration of curcumin at 400 mg/kg is effective in reducing not only the worm burden (44.4%) but also the tissue egg burden (30.9%) and hepatic granuloma volume (79.0%). Moreover, curcumin was able to modulate cellular and humoral immune responses in infected mice, leading to a significant reduction in parasite burden in acute murine schistosomiasis mansoni [101]. Recently, it was shown that curcumin is effective against liver fibrosis caused by schistosomes, which closely correlates with a decrease in type I/III collagen synthesis [102]. These results suggest that curcumin represents a new inexpensive and effective schistosomicidal drug.

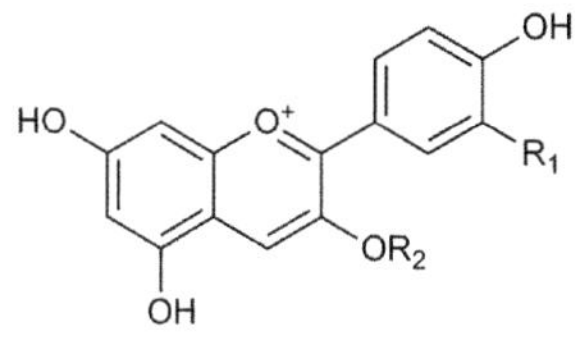

Cyanidin (R_1 = OH; R_2 = H) **(20)**
Pelargonidin (R_1 = H; R_2 = H) **(21)**

Kuromanin (R_1 = H); R_2 = 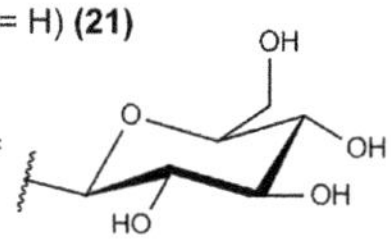 **(22)**

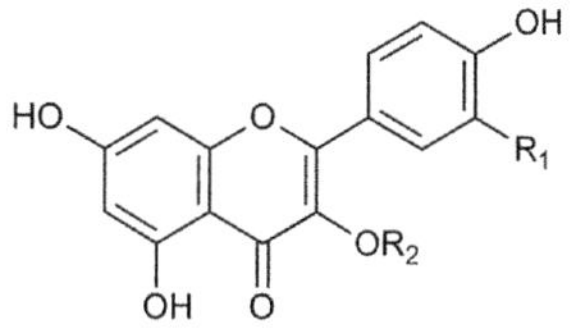

Kaempferol (R_1 = H; R_2 = H) **(23)**
Quercetin (R_1 = OH; R_2 = H) **(26)**

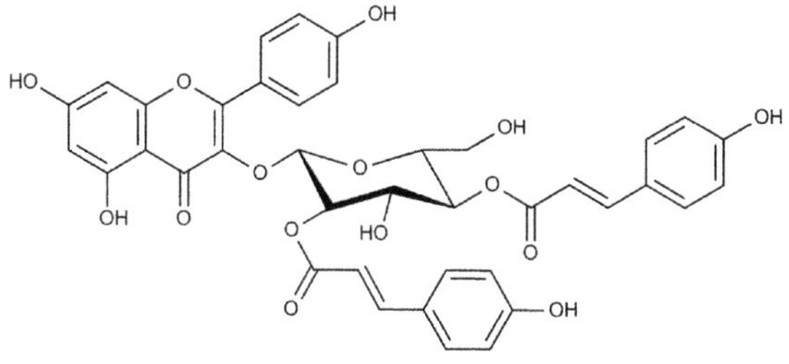

Kaempferol-3-O-(2",4"-di-O-(E)-p-Coumaroyl)-β-D-Glucopyranoside **(24)**

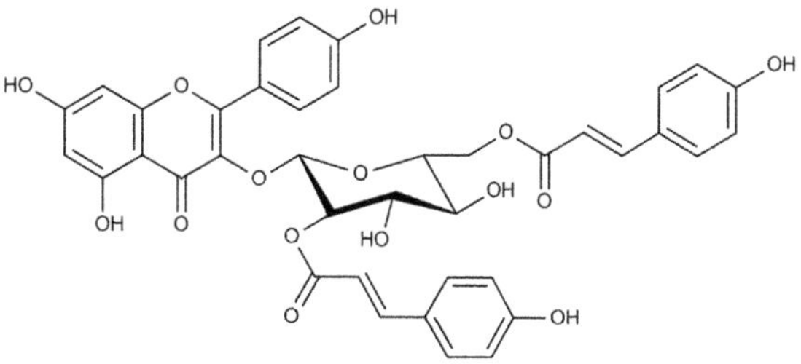

Kaempferol-3-O-(2",6"-di-O-(E)-p-Coumaroyl)-β-D-Glucopyranoside **(25)**

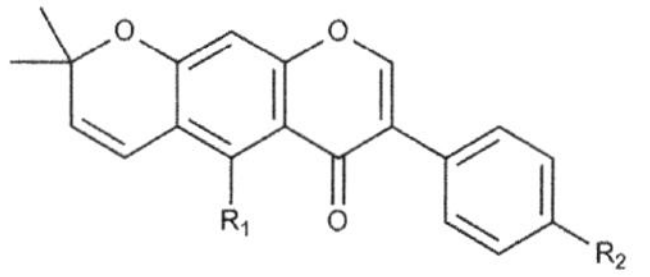

Alpinumisoflavone (R_1 = OH and R_2 = OH) **(27)**
Dimethylalpinumisoflavone (R_1 = OCH$_3$ and R_2 = OCH$_3$) **(28)**

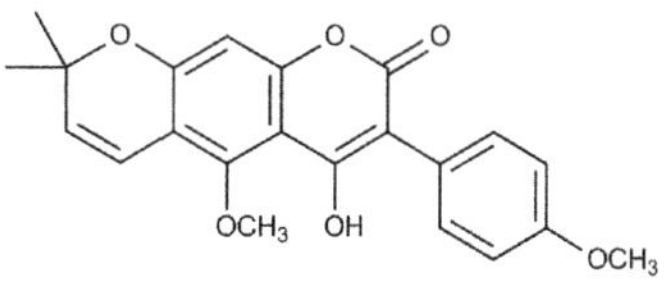

Robustic acid **(29)**

FIGURE 8-8 Flavonoids with schistosomicidal activity.

Flavonoids

Several isoflavonoids were isolated from a dichloromethane extract of *Millettia thonningii* (Fabaceae) in a bioguided fractionation study. The isolation process led to the identification of alpinumisoflavone (compound 27, Fig. 8-8), dimethylalpinumisoflavone (compound 28, Fig. 8-8), and robustic acid (compound 29, Fig. 8-8) from active fractions of the dichloromethane extract. It was shown that isoflavones (compounds 27–29, Fig. 8-8) at a concentration

of 50 mg/mL were highly active *in vitro* against miracidia, cercariae and adult worms of *S. mansoni*. According to the report's authors, the schistosomicidal activity of the crude extract and isolated isoflavones may correlate with their ability to interfere with the energy metabolism of parasites by inhibiting mitochondrial electron transport at complex I (NADH dehydrogenase) [103].

Similarly, flavonoids isolated from two species of *Styrax* (Styracaceae) were evaluated *in vitro* for activity against *S. mansoni* adult

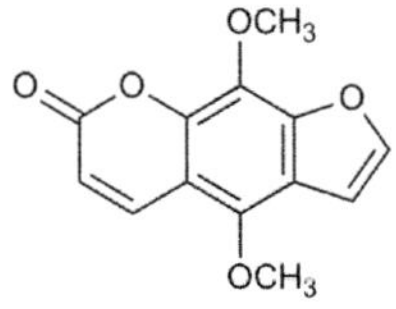

Isopimpinellin **(30)**

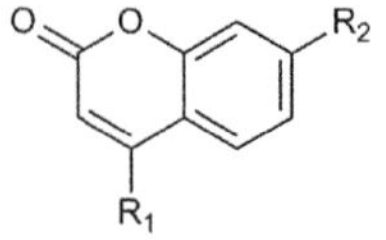

7-Acetoxycoumarin (R_1 = H and R_2 = OCOCH$_3$) **(31)**
4-Methyl-7-Hydroxycoumarin (R_1 = CH$_3$ and R_2 = OH) **(32)**
7-Hydroxycoumarin (R_1 = H and R_2 = OH) **(33)**
Coumarin (R_1 = H and R_2 = H) **(34)**

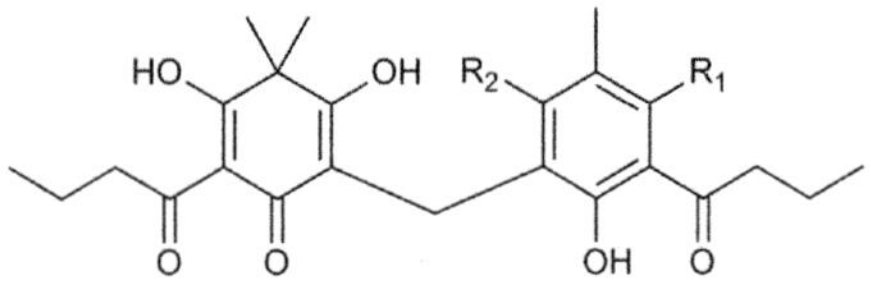

Aspidin (R_1 = OCH$_3$; R_2 = OH) **(35)**
Desaspidin (R_1 = OH; R_2 = OCH$_3$) **(36)**
Flavaspidic acid (R_1 = OH; R_2 = OH) **(37)**

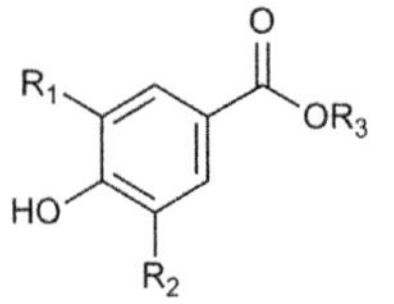

Protocatechuic acid (R_1 = OH; R_2 = H; R_3 = H) **(38)**
Gallic acid (R_1 = OH; R_2 = OH; R_3 = H) **(39)**
p-Hydroxybenzoic acid (R_1 = H; R_2 = H; R_3 = H) **(40)**

5-O-Caffeoylshikimic acid (R_1 = OH; R_2 = H); R_3 = 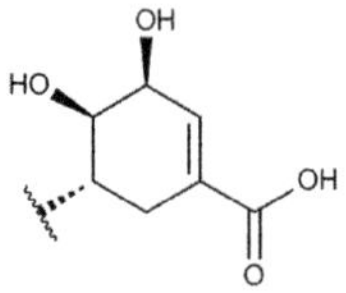 **(41)**

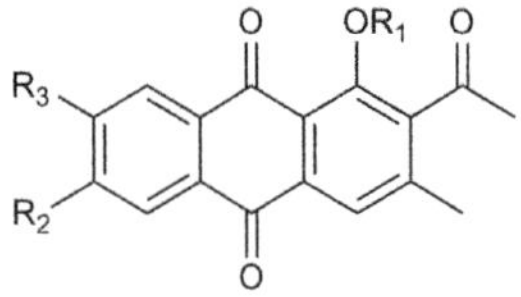

Kwanzoquinone A (R_1 = H, R_2 = CH$_3$ and R_3 = H) **(43A)**
Kwanzoquinone B (R_1 = H, R_2 = H and R_3 = CH$_3$) **(43B)**

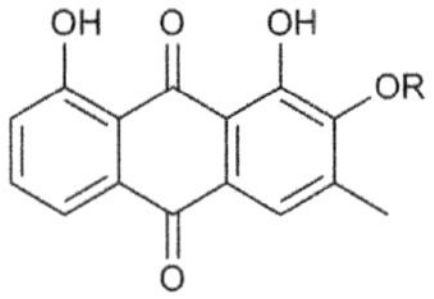

2-Hydroxychrysophanol (R = H) **(42)**
Kwanzoquinone C (R = β-D-Glucopyranoside) **(43C)**
Kwanzoquinone D (R = Malonyl-(1→6)-β-D-Glucopyranoside) **(43D)**

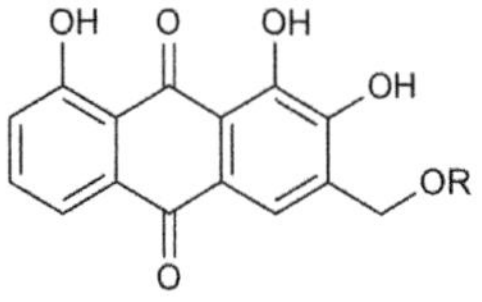

Kwanzoquinone E (R = H) **(43E)**
Kwanzoquinone F (R = β-D-Glucopyranoside) **(43F)**

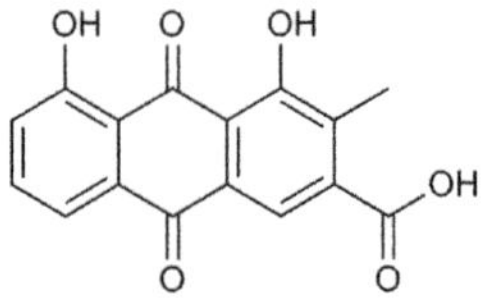

Kwanzoquinone G **(43G)**

FIGURE 8-9 Coumarins, phloroglucinols, benzoic acid derivatives, and quinones with schistosomicidal activity.

worms. It was observed that 100 μM kaempferol (compound 23, Fig. 8-8) had schistosomicidal active, while kaempferol-3-O-(2″,4″-di-O-(E)-p-coumaroyl)-β-D-glucopyranoside (compound 24, Fig. 8-8), kaempferol-3-O-(2″,6″-di-O-(E)-p-coumaroyl)-β-D-glucopyranoside (compound 25, Fig. 8-8), and quercetin (compound 26, Fig. 8-8) were inactive at the same concentration [104].

Moreover, studies using high-throughput screening assays revealed that several flavonoids are natural inhibitors of SmNACE, a key NAD(P)+ catabolizing enzyme recently identified on the surface of *S. mansoni* worms [105]. Of all the tested flavonoids, cyanidin (compound 20, Fig. 8-8), pelargonidin (compound 21, Fig. 8-8), and kuromanin (compound 22, Fig. 8-8) were found to be the most active. These results indicate that further investigations are necessary to investigate the full potential of flavonoids as schistosomicidal compounds [105].

Other Phenolic Compounds

Coumarins isolated from the leaves of *Citrus aurantifolia* and *Citrus limonia* (Rutaceae) have also been evaluated for their schistosomicidal activity. Among the compounds isolated from *Citrus* leaves, isopimpinellin (compound 30, Fig. 8-9) is the most active [106]. It was also observed that several other natural and semisynthetic coumarins possess immunostimulatory effects, at both the humoral and cellular levels, in mice infected with *S. mansoni* [107]. Therefore, in order to find chemoprophylactic agents for treating schistosomiasis, a group of natural and semisynthetic coumarins were evaluated for their capacity to inhibit penetration by *S. mansoni* cercariae. It was observed that 7-acetoxycoumarin (compound 31, Fig. 8-9), 4-methyl-7-hydroxycoumarin (compound 32, Fig. 8-9), 7-hydroxycoumarin (compound 33, Fig. 8-9), and coumarin (compound 34, Fig. 8-9) inhibited cercarial penetration by more than 70% [108].

Similarly, phloroglucinols isolated from the rhizomes of *Dryopteris* spp. have shown *in vitro* schistosomicidal activity against adult *S. mansoni* worms. Among the compounds tested, the acylphloroglucinols aspidin (compound 35, Fig. 8-9), desaspidin (compound 36, Fig. 8-9), and flavaspidic acid (compound 37, Fig. 8-9) were able to kill all worms when used at concentrations > 50 μM, as well as decreasing both motor activity and egg production in coupled adult worms [109].

Some benzoic acid derivatives have also been tested against adult *S. mansoni* worms. Protocatechuic acid (compound 38, Fig. 8-9), gallic acid (compound 39, Fig. 8-9), p-hydroxybenzoic acid (compound 40, Fig. 8-9), and 5-O-caffeoylshikimic acid (compound 41, Fig. 8-9), isolated from the leaves of *Zanthoxylum naranjillo* (Rutaceae), were all able to separate adult worm pairs into males and females, as well as decreasing egg production, but without causing the death of parasites [67].

Similarly, anthraquinones have been studied as molluscicidal agents against the snails of the *Biomphalaria* genus, the intermediate host of schistosomes [110]. Several potentially schistosomicidal anthraquinones were isolated from the roots of *Hemerocallis fulva* (Hemerocallidaceae) and evaluated *in vitro* against multiple life-stages (cercariae, schistosomula, and adult) of *S. mansoni* worms. Anthraquinones 2-hydroxychrysophanol (compound 42, Fig. 8-9) and kwanzoquinones A−G (compounds 43A−G, Fig. 8-9) induced significant mortality in cercariae and exhibited a high activity against adult worms, but most compounds were inactive against schistosomula. Considering their potential as topical cercaricidal agents, further studies should be conducted in order to determine their mode of action [111−113].

Saponins

There are many reports of molluscicidal activity for saponins in the literature, mainly against *Biomphalaria* snail species. However,

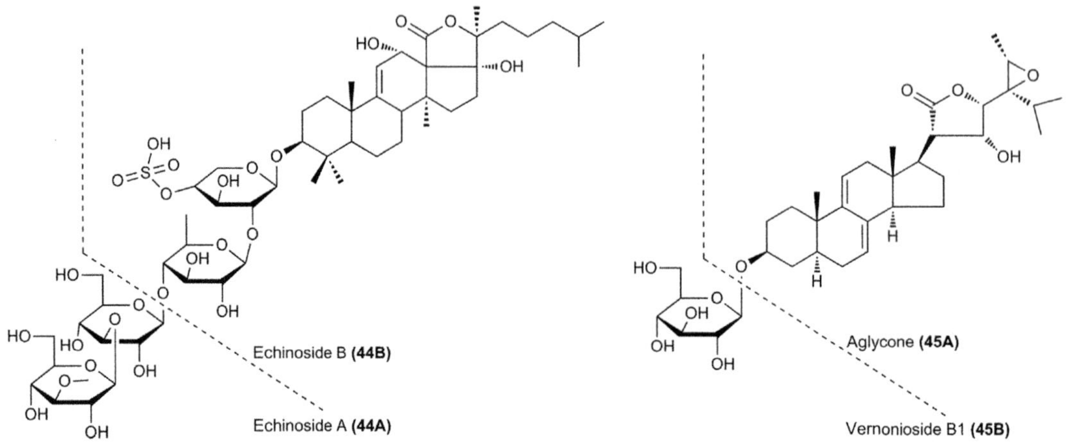

FIGURE 8-10 Saponins and glycosides with schistosomicidal activity.

a few studies have reported the *in vitro* activity of saponins against adult worms and *in vivo* activity in infected mice. *In vitro* schistosomicidal screening of methanol extracts from 79 marine organisms led to the identification of several marine extracts with activity against adult *S. mansoni* worms. Bioguided fractionation of the active extracts enabled the isolation of echinosides A (compound 44A, Fig. 8-10) and B (compound 44B, Fig. 8-10) from the sea cucumbers *Actinopyga echinites* and *Holothuria polii*, respectively. Both of these saponins exhibit high levels of *in vitro* antischistosomal activity and are promising lead compounds for the development of new schistosomicidal agents [114].

In addition, steroidal saponins and their aglycones, isolated from *Vernonia amygdalina* (Asteraceae) in a zoopharmacologic study in chimpanzees, were examined for *in vitro* and *in vivo* activities against *S. japonicum*. Vernonioside B1 (compound 45B, Fig. 8-10) exhibited schistosomicidal activity *in vitro* and its aglycone (compound 45A, Fig. 8-10) showed a high level of activity in the same test. Interestingly, the results obtained in this study suggest that sick chimpanzees may use steroidal saponins in *V. amygdalina* as a remedy against common

parasitosis, such as schistosomiasis. Furthermore, active steroidal saponins, such as vernonioside B1 may be metabolized into the more active aglycone form during digestion [115].

Artemisinin and its Derivatives

Artemisinin (compound 5, Fig. 8-3), also called *qinghaosu*, is a sesquiterpene lactone with an endoperoxide bridge that is the main active antimalarial compound in the leaves of *Artemisia annua* (Asteraceae), a plant with a long history of use as febrifuge in China [33,39,116]. The limited stability and poor solubility of artemisinin stimulated researchers to develop semisynthetic derivatives with improved solubility and stability, such as artemether (compound 46, Fig. 8-11) and artesunate (compound 47, Fig. 8-11).

Artemether has been effectively used in the treatment of human schistosomes; its major activity is against the juvenile stages, which makes this drug useful as a prophylactic agent against schistosomiasis, mainly against *S. japonicum* [116–118]. Reports of the therapeutic efficacy of artemether against *S. mansoni*-infected mice suggest that artemether is efficacious against the Egyptian strain of *S. mansoni*, mainly affecting female worms [119]. Artesunate is also

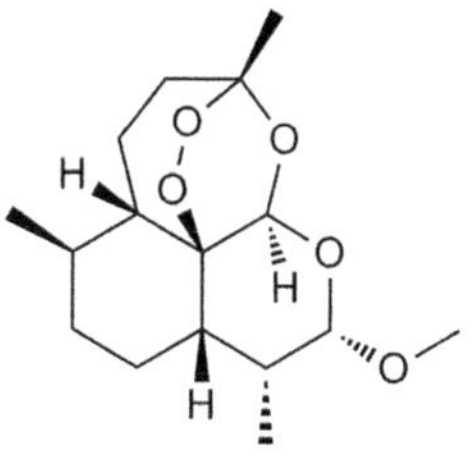

FIGURE 8-11 Artemisin derivatives artemether and artesunate have schistosomicidal activity.

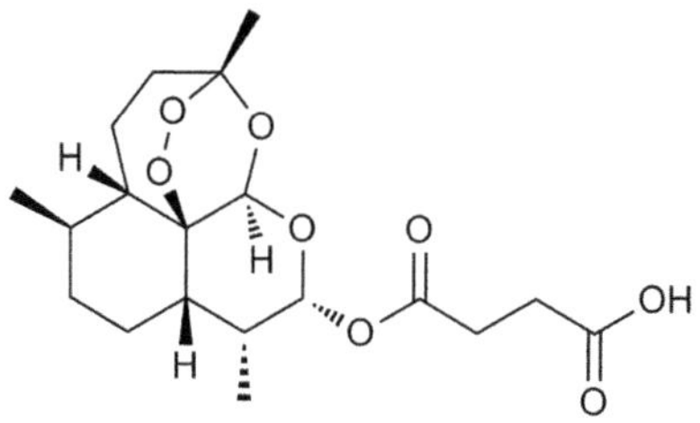

Artemether (46)   Artesunate (47)

a good chemoprophylactic drug used for preventing *S. japonicum* infection [120]. Several published reports and trials have demonstrated the protective effects of artesunate and artemether against schistosome infection. However, there are conflicting reports with regard to cure rates (ranging from 23% to 100%) and egg reduction rates (ranging from 55% to 100%) [58,116−118,121−124]. As different experimental protocols and treatments were employed in those studies, it is difficult to compare their results. However, it was recently reported that the sensitivity of *S. japonicum* to artesunate and artemether has decreased in China following years of use [120].

Other Terpenoids

Terpenoids were initially shown to be able to inhibit penetration by cercariae; several reports showed the chemoprophylactic potential of terpenes in protecting against infection by *S. mansoni* cercariae [125−127]. In the original assays, 14,15-epoxygeranylgeraniol (compound 49, Fig. 8-12) was one of the most active compounds, showing efficacy at low concentrations. Sodium abietate (compound 50, Fig. 8-12), a diterpene sodium salt, was also evaluated for activity against adult *S. japonicum* worms *in vitro*; its activity was found to be mainly against male worms through effects on their intestinal canals. Compared with negative controls, sodium abietate has a negative effect on protein metabolism in worms [128]. In addition, thapsigargin (compound 51, Fig. 8-12), a sesquiterpene

lactone isolated from *Thapsia garganica* (Apiaceae), has been reported to inhibit *S. mansoni* ATP diphosphohydrolases, which are ubiquitous enzymes present on the parasite surface and capable of hydrolyzing nucleoside di- and triphosphates. As specific inhibitors of these enzymes would be of significant therapeutic value, further studies should be conducted to identify other natural *S. mansoni* ATP diphosphohydrolase inhibitors that may be used for the development of schistosomicidal compounds [129].

The sesquiterpene lactones vernodalin (compound 52, Fig. 8-12), vernolide (compound 53, Fig. 8-12), hydrovernolide (compound 54, Fig. 8-12), and vernodalol (compound 55, Fig. 8-12) were isolated from *Vernonia amygdalina* (Asteraceae), a tropical African plant used by wild chimpanzees to treat parasite-related diseases. After *in vitro* evaluation against *S. japonicum*, vernodalin was found to be the most active sesquiterpene lactone and evaluated *in vivo* against *S. japonicum*. When orally administered to mice infected with *S. japonicum*, 120 mg/kg vernodalin observed to be highly toxic and lethal to parasites. However, at 60 mg/kg, vernodalin had no effect on parasites [115,130]. In a similar way, the sesquiterpene lactones eremanthine (compound 56, Fig. 8-12), costunolide (compound 57, Fig. 8-12), and α-cyclocostunolide (compound 58, Fig. 8-12) isolated from the heartwood oil of *Eremanthus elaeagnus* (Asteraceae) were assayed for their ability to inhibit penetration by *S. mansoni*

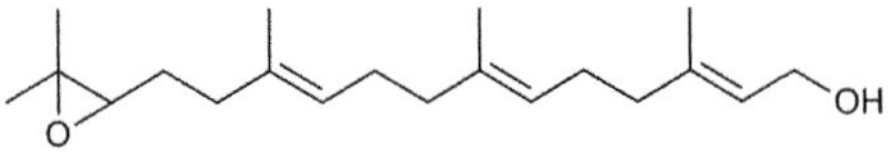

(-)-14,15-Epoxygeranylgeraniol (49) 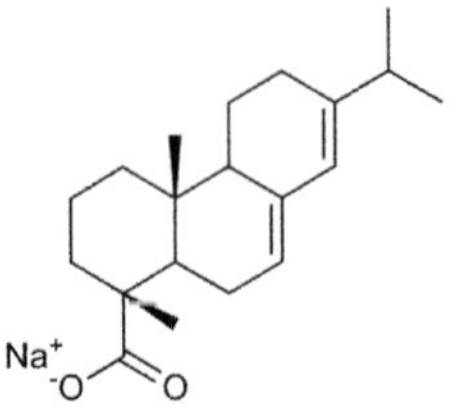   Sodium abietate (50) 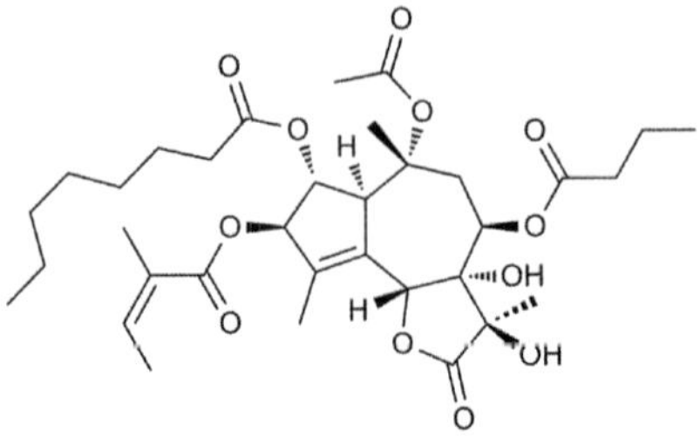   Thapsigargin (51)

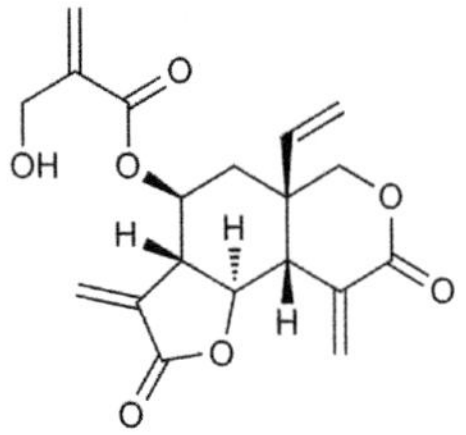

Vernodalin (52)

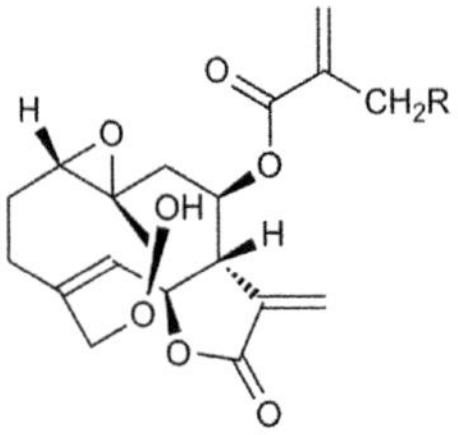

Vernolide (R = H) (53)
Hydrovernolide (R = OH) (54)

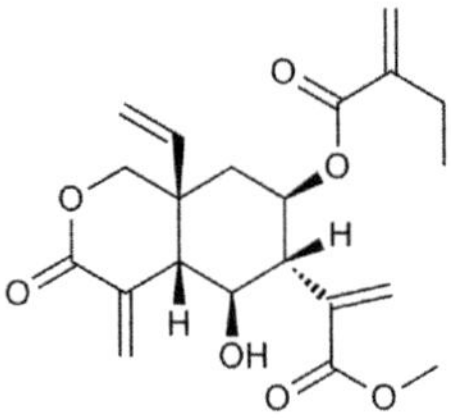

Vernodalol (55)

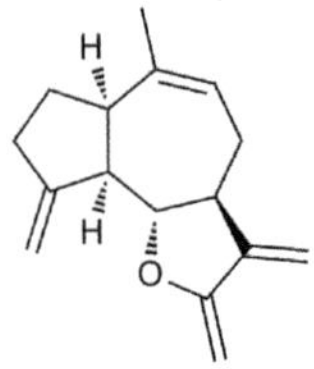

Eremanthine (56)

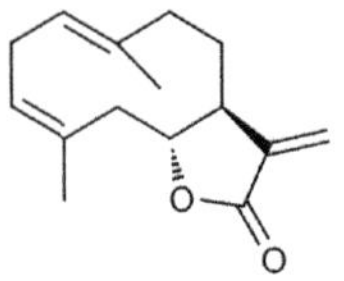

Costunolide (57) 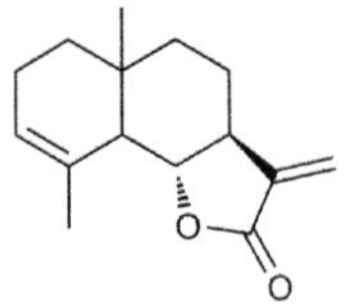   α–Cyclocostunolide (58)

FIGURE 8-12   Terpenes with schistosomicidal activity.

cercariae. Of these, eremanthine was shown to be the compound responsible for the major prophylactic activity of this plant against *S. mansoni* [131,132].

CONCLUSIONS

Natural products remain the main sources of lead compounds, especially for the development of chemotherapeutic drugs [133]. In this chapter, the *in vitro* and *in vivo* schistosomicidal evaluation of several classes of natural compounds has been described, including crude plant extracts, EOs, and purified natural compounds. A number of these possess schistosomicidal activity *in vitro*, but only a few studies investigating their *in vivo* activities have been conducted so far. Most of these reports did not present and/or discuss the cytotoxic potential of the evaluated samples, which is especially important for drug discovery. In addition, some plant preparations evaluated against *Schistosoma* spp. have a long history of use as natural antiparasitic agents. Therefore, most of these plant extracts should be evaluated in cytotoxicity assays, including plants that are used in the traditional folk medicine or have been used therapeutically for centuries. Among natural products, curcumin and some of the lignans discussed in this chapter seem to have the greatest potential as natural lead compounds for developing new schistosomicidal compounds. Nonetheless, an ideal natural schistosomicidal lead candidate has yet to be identified. Much research is required to overcome the limitations of current schistosomicidal drugs and the appearance of resistant schistosome strains. Schistosomiasis is a neglected disease that mainly affects people in developing countries that cannot afford to pay the high costs of research and drug development. In view of this alarming situation, efforts should be made to identify new lead structures, as well as effective and safe drugs. Nature has been demonstrated to be a good source of lead compounds for many diseases, but only a small proportion of existing natural compounds have been evaluated against *Schistosoma* spp. Therefore, further studies are required to identify natural lead molecules that may be used to develop new safe and effective drugs against schistosomiasis.

Acknowledgments

The authors thank the Fundação de Amparo à Pesquisa do Estado de Minas Gerais (FAPEMIG) and Fundação de Apoio a Pesquisa do Estado de São Paulo (FAPESP) for financial support. The authors are also grateful to the Programa Institucional de Bolsas de Iniciação Científica da Universidade Federal de Juiz de Fora (PIBIC/CNPq/UFJF), Coordenação de Aperfeiçoamento de Pessoal de Nível Superior (CAPES), and the CNPq for funding fellowships.

References

[1] Gryseels B, Polman K, Clerinx J, Kestens L. Human schistosomiasis. Lancet 2006;368:1106–18.

[2] Hotez PJ, Molyneux DH, Fenwick A, Kumaresan J, Sachs SE, Sachs JD, et al. Control of neglected tropical diseases. N Engl J Med 2007;357:1018–27.

[3] van der Werf MJ, de Vlas SJ, Brooker S, Looman CWN, Nagelkerke NJD, Habbema JDF, et al. Quantification of clinical morbidity associated with schistosome infection in sub-Saharan Africa. Acta Trop 2003;86:125–39.

[4] Steinmann P, Keiser J, Bos R, Tanner M, Utzinger J. Schistosomiasis and water resources development: systematic review, meta-analysis, and estimates of people at risk. Lancet Infect Dis 2006;6:411–25.

[5] Berriman M, Haas BJ, LoVerde PT, Wilson RA, Dillon GP, Cerqueira GC, et al. The genome of the blood fluke Schistosoma mansoni. Nature 2009;460:352–8.

[6] Kunz W. Schistosome male–female interaction: induction of germ-cell differentiation. Trends Parasitol 2001;17:227–31.

[7] Ribeiro-dos-Santos G, Verjovski-Almeida S, Leite L. Schistosomiasis—a century searching for chemotherapeutic drugs. Parasitol Res 2006;99:505–21.

[8] Abdul-Ghani R, Loutfy N, Sahn A, Hassan A. Current chemotherapy arsenal for schistosomiasis mansoni: alternatives and challenges. Parasitol Res 2009;104:955–65.

[9] World Health Organization. WHO technical report series 912: prevention and control of schistosomiasis and soil-transmitted helminthiasis; 2002. Geneva.

[10] World Health Organization. Strategic emphases for tropical diseases research: a TDR perspective. TRENDS in Parasitology 2002;18(No.10):421–6.

[11] Cioli D, Pica-Mattoccia L, Archer S. Antischistosomal drugs: Past, present.and future? Pharmacol and Ther 1995;68:35–85.

[12] Katz N, Coelho PMZ. Clinical therapy of schistosomiasis mansoni: The Brazilian contribution. Acta Trop 2008;108:72–8.

[13] Abdul-Ghani RA, Loutfy N, Hassan A. Experimentally promising antischistosomal drugs: a review of some drug candidates not reaching the clinical use. Parasitol Res 2009;105:899–906.

[14] Xiao S, Chollet J, Booth M, Weiss NA, Tanner M. Therapeutic effect of praziquantel enantiomers in mice infected with Schistosoma mansoni. Trans R Soc Trop Med Hyg 1999;93:324–5.

[15] Xiao SH, Catto BA. Comparative *in vitro* and *in vivo* activity of racemic praziquantel and its levorotated isomer on Schistosoma mansoni. J Infect Dis 1989;159:589–92.

[16] Staudt U, Schmahl G, Blaschke G, Mehlhorn H. Light and scanning electron microscopy studies on the effects of the enantiomers of praziquantel and its main metabolite on Schistosoma mansoni *in vitro*. Parasitol Res 1992;78:392–7.

[17] Gönnert R, Andrews P. Praziquantel, a new broad-spectrum antischistosomal agent. Parasitol Res 1977;52:129–50.

[18] Utzinger J, Keiser J, Shuhua X, Tanner M, Singer BH. Combination Chemotherapy of Schistosomiasis in

Laboratory Studies and Clinical Trials. Antimicrob. Agents Chemother 2003;47:1487—95.

[19] Pax R, Bennett JL, Fetterer R. A benzodiazepine derivative and praziquantel: effects on musculature of Schistosoma mansoni and Schistosoma japonicum. Naunyn Schmiedebergs Arch Pharmacol 1978;304:309—15.

[20] Aragon AD, Imani RA, Blackburn VR, Cupit PM, Melman SD, Goronga T, et al. Towards an understanding of the mechanism of action of praziquantel. Mol Biochem Parasitol 2009;164:57—65.

[21] Botros SS, Bennett JL. Praziquantel resistance. Expert Opin Drug Discovery 2007;2:35—S40.

[22] Stelma FF, Talla I, Sow S, Kongs A, Niang M, Polman K, et al. Efficacy and side effects of praziquantel in an epidemic focus of Schistosoma mansoni. Am J Trop Med Hyg 1995;53: 167—70.

[23] Fallon PG, Sturrock RF, Niang AC, Doenhoff MJ. Short report: diminished susceptibility to praziquantel in a Senegal isolate of Schistosoma mansoni. Am J Trop Med Hyg 1995;53:61—2.

[24] Ismail MM, Taha SA, Farghaly AM, el-Azony AS. Laboratory induced resistance to praziquantel in experimental schistosomiasis. J Egypt Soc Parasitol 1994;24:685—95.

[25] Ismail M, Metwally A, Farghaly A, Bruce J, Tao LF, Bennett JL. Characterization of isolates of Schistosoma mansoni from Egyptian villagers that tolerate high doses of praziquantel. Am J Trop Med Hyg 1996;55:214—8.

[26] Picquet M, Vercruysse J, Shaw DJ, Diop M, Ly A. Efficacy of praziquantel against Schistosoma mansoni in northern Senegal. Trans R Soc Trop Med Hyg 1998;92:90—3.

[27] King CH, Muchiri EM, Ouma JH. Evidence against rapid emergence of praziquantel resistance in Schistosoma haematobium, Kenya. Emerging Infect Dis 2000;6:585—94.

[28] Katz N, Filho Santos, dos D, Sarti SJ, Mendes NM, Rocha Filho PA, et al. Estudo de uma cepa humana de Schistosoma mansoni resistente a agentes esquistossomicidas. Rev Soc Bras Med Trop 1973;7: 381—7.

[29] Rokni MB, Allegretti SM, de Oliveira CNF, de Oliveira RN, Frezza TF. and Rehder VLG. The use of Brazilian medicinal plants to combat Schistosoma mansoni. Schistosomiasis. InTechOpen; 2012. p. 27—70. http://www.intechopen.com/books/schistosomiasis/the-use-of-brazilian-medicinal-plants-to-combat-schistosoma-mansoni [chapter 3]. [accessed 21.12.12].

[30] Pica-Mattoccia L, Dias LCD, Moroni R, Cioli D. Schistosoma mansoni: Genetic complementation

analysis shows that two independent hycanthone/oxamniquine-resistant strains are mutated in the same gene. Exp Parasitol 1993;77:445—9.

[31] Verpoorte R. Exploration of nature's chemodiversity: The role of secondary metabolites as leads in drug development. Drug Discov Today 1998;3:232—8.

[32] Fischbach MA, Clardy J. One pathway, many products. Nature Chemical Biology 2007;3:353—5.

[33] Kayser O, Kiderlen AF, Croft SL. Natural products as antiparasitic drugs. Parasitol Res 2003;90:55—62.

[34] Hanson JR. (Royal Society of Chemistry: London, United Kingdom, 2003). Natural Products: The Secondary Metabolites.

[35] Sarker SD, Latif Z. and Gray AI. (Humana Press Inc.: Totowa, New Jersey, 2005). Natural Products Isolation.

[36] Andrade SF, Da Silva Filho AA, Resende DDO, Silva MLA, Cunha WR, Nanayakkara NPD, et al. Antileishmanial, antimalarial and antimicrobial activities of the extract and isolated compounds from Austroplenckia populnea (Celastraceae). Z Naturforsch C 2008;63:497—502.

[37] Parreira NA, Magalhães LG, Morais DR, Caixeta SC, de Sousa JPB, Bastos JK, et al. Antiprotozoal, schistosomicidal, and antimicrobial activities of the essential oil from the leaves of Baccharis dracunculifolia. Chem Biodivers 2010;7:993—1001.

[38] Geary TG, Chibale K, Abegaz B, Andrae-Marobela K, Ubalijoro E. A new approach for anthelmintic discovery for humans. Trends Parasitol 2012;28:176—81.

[39] Klayman DL. Qinghaosu (artemisinin): An antimalarial drug from China. Science 1985;228:1049—55.

[40] Neto AG, da Silva Filho AA, Costa JMLC, Vinholis AHC, Souza GHB, Cunha WR, et al. Evaluation of the trypanocidal and leishmanicidal in vitro activity of the crude hydroalcoholic extract of Pfaffia glomerata (Amarathanceae) roots. Phytomedicine 2004;11:662—5.

[41] Ndamba J, Nyazema N, Makaza N, Anderson C, Kaondera KC. Traditional herbal remedies used for the treatment of urinary schistosomiasis in Zimbabwe. J Ethnopharmacol 1994;42:125—32.

[42] Kron M, Yousif F, Ramirez B. Capacity building in anthelmintic drug discovery. Exp Opin Drug Discovery 2007;2:75—82.

[43] Sharma S. and Anand N. (Elsevier: Amsterdam, Netherlands, 1997). Approaches to design and synthesis of antiparasitic drugs. Pharmaco Chemistry Library 25.

[44] Anthony J-P, Fyfe L, Smith H. Plant active components—a resource for antiparasitic agents? Trends Parasitol 2005;21:462—8.

[45] Ferreira J, Peaden P, Keiser J. *In vitro* trematocidal effects of crude alcoholic extracts of Artemisia annua, A. absinthium, Asimina triloba, and Fumaria officinalis. Parasitol Res 2011;109:1585−92.

[46] Pontin K, Da Silva Filho AA, Santos FF, Silva MLAE, Cunha WR, Nanayakkara NPD, et al. *In vitro* and *in vivo* antileishmanial activities of a Brazilian green propolis extract. Parasitol Res 2008;103:487−92.

[47] Abdin MZ, Israr M, Rehman RU, Jain SK. Artemisinin, a novel antimalarial drug: biochemical and molecular approaches for enhanced production. Planta Med 2003;69:289−99.

[48] Yousif F, Hifnawy MS, Soliman G, Boulos L, Labib T, Mahmoud S, et al. Large-scale *in vitro* screening of Egyptian native and cultivated plants for schistosomicidal activity. Pharm Biol 2007;45:501−10.

[49] Koné WM, Vargas M, Keiser J. Anthelmintic activity of medicinal plants used in Côte d'Ivoire for treating parasitic diseases. Parasitol Res 2011;110(6):2351−62.

[50] El-Nahas HA, Abdel-Hameed ES, Sabra AA, El-Wakil EA. Steroidal glycosides of Furcraea selloa and their biological properties against different Schistosoma mansoni stages. Bull Pharm Sci 2005;28: 169−83.

[51] Abdel-Hameed ES, El-Nahas HA, Abo-Sedera SA. Antischistosomal and antimicrobial activities of some Egyptian plant species. Pharm Biol 2008;46: 626−33.

[52] Sparg SG, van Staden J, Jäger AK. Pharmacological and phytochemical screening of two Hyacinthaceae species: Scilla natalensis and Ledebouria ovatifolia. J Ethnopharmacol 2002;80:95−101.

[53] Abadome F, Geerts S, Kumar V. Evaluation of the activity of Ambrosia maritima L. against Schistosoma mansoni infection in mice. J Ethnopharmacol 1994;44:195−8.

[54] Sheir Z, Nasr AA, Massoud A, Salama O, Badra GA, El-Shennawy H, et al. A safe, effective, herbal antischistosomal therapy derived from myrrh. Am J Trop Med Hyg 2001;65:700−4.

[55] Soliman OE, El-Arman M, Abdul-Samie ER, El-Nemr HI, Massoud A. Evaluation of myrrh (Mirazid) therapy in fascioliasis and intestinal schistosomiasis in children: immunological and parasitological study. J Egypt Soc Parasitol 2004;34: 941−66.

[56] Abo-Madyan AA, Morsy TA, Motawea SM. Efficacy of Myrrh in the treatment of schistosomiasis (haematobium and mansoni) in Ezbet El-Bakly, Tamyia Center, El-Fayoum Governorate. Egypt J Egypt Soc Parasitol 2004;34:423−46.

[57] Barakat R, Elmorshedy H, Fenwick A. Efficacy of myrrh in the treatment of human Schistosomiasis mansoni. Am J Trop Med Hyg 2005;73:365−7.

[58] Botros S, Sayed H, El-Dusoki H, Sabry H, Rabie I, El-Ghannam M, et al. Efficacy of mirazid in comparison with praziquantel in Egyptian Schistosoma mansoni -infected school children and households. Am J Trop Med Hyg 2005;72:119−23.

[59] El-Shenawy NS, Soliman MFM, Abdel-Nabi IM. Does Cleome droserifolia have anti-schistosomiasis mansoni activity? Rev Inst Med Trop São Paulo 2006;48:223−8.

[60] Abdel-Hamid HF. Molluscicidal and in-vitro schistosomicidal activities of the latex and some extracts of some plants belonging to Euphorbiaceae. J Egypt Soc Parasitol 2003;33:947−54.

[61] Abdel-Aziz IZ, El-Bady AA, El-Gayed SH. *In vitro* anti-schistosomal activity of "Plectranthus tenuiflorus" on miracidium, cercaria and schistosomula stages of Schistosoma mansoni. Res J Parasitol 2011;6:74−82.

[62] EL-Shenawy NS, Soliman MFM, Reyad SI. The effect of antioxidant properties of aqueous garlic extract and Nigella sativa as anti-schistosomiasis agents in mice. Rev Inst Med Trop São Paulo 2008;50:29−36.

[63] Jatsa HB, Endougou DRA, Kemeta DRA, Kenfack CM, Tchuem Tchuente LA, Kamtchouing P. *In vivo* antischistosomal and toxicological evaluation of Sida pilosa retz on mice Balb/c. Pharmacologyonline 2009;3:531−8.

[64] Mohamed AM, Metwally NM, Mahmoud SS. Sativa seeds against Schistosoma mansoni different stages. Mem Inst Oswaldo Cruz 2005;100:205−11.

[65] Baldé AM, Van Marck E, Vanhaelen M. In vivo activity of an extract of Pavetta owariensis bark on experimental Schistosoma mansoni infection in mice. J Ethnopharmacol 1986;18:187−92.

[66] Hamed MA, Hetta MH. Efficacy of Citrus reticulata and mirazid in treatment of Schistosoma mansoni. Mem Inst Oswaldo Cruz 2005;100:771−8.

[67] Braguine CG, Costa ES, Magalhães LG, Rodrigues V, da Silva Filho AA, Bastos JK, et al. Schistosomicidal evaluation of Zanthoxylum naranjillo and its isolated compounds against Schistosoma mansoni adult worms. Z Naturforsch C 2010;64:793−7.

[68] Miranda M, Magalhães L, Tiossi R, Kuehn C, Oliveira L, Rodrigues V, et al. Evaluation of the schistosomicidal activity of the steroidal alkaloids from Solanum lycocarpum fruits. Parasitol Res 2012;111(1):257−62.

[69] Ahmed AH, Ramzy RM. Laboratory assessment of the molluscicidal and cercaricidal activities of the

Egyptian weed, Solanum nigrum L. Ann Trop Med Parasitol 1997;91:931−7.

[70] Al-Ashaal HAA-H. Regeneration, in vitro glyco-alkaloids production and evaluation of bioactivity of callus methanolic extract of Solanum tuberosum L. Fitoterapia 2010;81:600−6.

[71] Jatsa HB, Ngo Sock ET, Tchuem Tchuente LA, Kamtchouing P. Evaluation of the in vivo activity of different concentrations of Clerodendrum umbellatum poir against Schistosoma mansoni infection in mice. Afr J Tradit Complement Altern Med 2009;6: 216−21.

[72] Mostafa O, Eid R, Adly M. Antischistosomal activity of ginger (Zingiber officinale) against Schistosoma mansoni harbored in C57 mice. Parasitol Res 2011;109:395−403.

[73] Sanderson L, Bartlett A, Whitfield PJ. In vitro and in vivo studies on the bioactivity of a ginger (Zingiber officinale) extract towards adult schistosomes and their egg production. J Helminthol 2002;76:241−7.

[74] Keiser J, Utzinger J. Advances in the discovery and development of trematocidal drugs. Exp Opin Drug Discovery 2007;2:S9−S23.

[75] Chitwood DJ. Phytochemical based strategies for nematode control. Annu Rev Phytopathol 2002;40: 221−49.

[76] Cos P, Vlietinck AJ, Berghe DV, Maes L. Anti-infective potential of natural products: How to develop a stronger in vitro "proof-of-concept."J. Ethnopharmacol 2006;106:290−302.

[77] Edris AE. Pharmaceutical and therapeutic potentials of essential oils and their individual volatile constituents: a review. Phytother Res 2007;21: 308−23.

[78] Bakkali F, Averbeck S, Averbeck D, Idaomar M. Biological effects of essential oils—A review. Food Chem Toxicol 2008;46:446−75.

[79] Magalhães L, de Souza J, Wakabayashi K, da S, Laurentiz R, Vinhólis A, et al. In vitro efficacy of the essential oil of Piper cubeba L. (Piperaceae) against Schistosoma mansoni. Parasitol Res 2012;110: 1747−54.

[80] de Melo NI, Magalhaes LG, de Carvalho CE, Wakabayashi KAL, de P, Aguiar G, et al. Schistosomicidal activity of the essential oil of Ageratum conyzoides L. (Asteraceae) against adult Schistosoma mansoni worms. Molecules 2011;16:762−73.

[81] Saleh MM, Zwaving JH, Malingré TM, Bos R. The essential oil of Apium graveolens var. secalinum and its cercaricidal activity. Pharm World Sci 1985;7: 277−9.

[82] Frischkorn CGB, Frischkorn HE, Carrazzoni E. Cercaricidal activity of some essential oils of plants from Brazil. Naturwissenschaften 1978;65:480−3.

[83] Bian Z, Huang B, Ding Y, Xia P. The experimental study of the ginger oil on killing Schistosoma japonicum cercariae. Lishizhen Medicine and Materia Medica Res 2007-05.

[84] Mendes NM, Araújo N, de Souza CP, Pereira JP, Katz N. Molluscacide and cercariacide activity of different species of Eucalyptus. Rev Soc Bras de Med Trop 1990;23:197−9.

[85] Aguiar GP, Melo NI, Wakabayashi KAL, Lopes MHS, Mantovani ALL, Dias HJ, et al. Chemical composition and in vitro schistosomicidal activity of the essential oil from the flowers of Bidens sulphurea (Asteraceae). Nat Prod Res 2012 [Epub ahead of print].

[86] Tonuci LRS, Melo NI, de, Dias HJ, Wakabayashi KAL, Aguiar GP, Aguiar DP, et al. In vitro schistosomicidal effects of the essential oil of Tagetes erecta. Rev Bras Farmacogn 2012;22:88−93.

[87] Nargis M, Sato H, Ozaki T, Inaba T, Chisty M, Kamiya H. Cercaricidal effects of hinokitiol on Schistosoma mansoni. Hirosaki Med 1 J 1997;49: 49−57.

[88] Caixeta SC, Magalhães LG, de Melo NI, Wakabayashi KAL, de P, Aguiar G, et al. Chemical composition and in vitro schistosomicidal activity of the essential oil of Plectranthus neochilus grown in southeast Brazil. Chem Biodivers 2011;8:2149−57.

[89] Mahmoud MR, El-Abhar HS, Saleh S. The effect of Nigella sativa oil against the liver damage induced by Schistosoma mansoni infection in mice. J Ethnopharmacol 2002;79:1−11.

[90] Veras LM, Guimaraes MA, Campelo YD, Vieira MM, Nascimento C, Lima DF, et al. Activity of epi-isopiloturine against Schistosoma mansoni. Curr Med Chem 2012;19:2051−8.

[91] Rapado LN, Nakano E, Ohlweiler FP, Kato MJ, Yamaguchi LF, Pereira CAB, et al. Molluscicidal and ovicidal activities of plant extracts of the Piperaceae on Biomphalaria glabrata (Say, 1818). J Helminthol 2011;85:66−72.

[92] da Silva Filho AA, Costa ES, Cunha WR, Silva MLA, Nanayakkara NPD, Bastos JK. In vitro anti-leishmanial and antimalarial activities of tetrahydrofuran lignans isolated from Nectandra megapotamica (Lauraceae). Phytother Res 2008;22: 1307−10.

[93] Souza GHB, da Silva Filho AA, de Souza VA, Pereira AC, Royo V, de A, et al. Analgesic and anti-inflammatory activities evaluation of (−)-O-acetyl, (−)-O-methyl, (−)-O-dimethylethylamine

cubebin and their preparation from (−)-cubebin. Farmaco 2004;59:55−61.

[94] da Silva R, de Souza GHB, da Silva AA, de Souza VA, Pereira AC, Royo V, et al. Synthesis and biological activity evaluation of lignan lactones derived from (−)-cubebin. Bioorg Med Chem Lett 2005;15:1033−7.

[95] da Silva Filho AA, Albuquerque S, Silva MLAE, Eberlin MN, Tomazela DM, Bastos JK. Tetrahydrofuran lignans from Nectandra megapotamica with trypanocidal activity. J Nat Prod 2004;67:42−5.

[96] de Souza VA, da Silva R, Pereira AC, Royo VDA, Saraiva J, Montanheiro M, et al. Trypanocidal activity of (−)-cubebin derivatives against free amastigote forms of Trypanosoma cruzi. Bioorg Med Chem Lett 2005;15:303−7.

[97] Silva MLA, da Silva R, Rodrigues V, Pereira OS, da Silva Filho AA, Donate PM, et al. Process to obtain synthetic and semi-synthetic lignan derivatives, their synthetic and semi-synthetic derivatives, their analgesic and anti-inflammatory activities, topical and/or systemic formulations containing said lignans and their respective therapeutic method. WO 2007/009201 A2; 2007.

[98] León-Díaz R, Meckes M, Said-Fernández S, Molina-Salinas GM, Vargas-Villarreal J, Torres J, et al. Antimycobacterial neolignans isolated from Aristolochia taliscana. Mem Inst Oswaldo Cruz 2010;105:45−51.

[99] Luz PP, Magalhães LG, Pereira AC, Cunha WR, Rodrigues V, Andrade E Silva ML. Curcumin-loaded into PLGA nanoparticles: preparation and in vitro schistosomicidal activity. Parasitol Res 2012;110:593−8.

[100] EL-Sherbiny M, Abdel-Aziz MM, Elbakry KA, Toson EA, Abbas AT. Schistosomicidal efffect of curcumin. Trends Appl Sci Res 2006;1:627−33.

[101] Allam G. Immunomodulatory effects of curcumin treatment on murine schistosomiasis mansoni. Immunobiology 2009;214:712−27.

[102] El-Agamy DS, Shebl AM, Said SA. Prevention and treatment of Schistosoma mansoni-induced liver fibrosis in mice. Inflammopharmacology 2011;19: 307−16.

[103] Lyddiard JRA, Whitfield PJ, Bartlett A. Antischistosomal bioactivity of isoflavonoids from Millettia thonningii (Leguminosae). J Parasitol 2002;88: 163−70.

[104] Braguine CG, Bertanha CS, Gonçalves UO, Magalhães LG, Rodrigues V, Melleiro Gimenez VM, et al. Schistosomicidal evaluation of flavonoids from two species of Styrax against Schistosoma mansoni adult worms. Pharm Biol 2012;50(7):925−9.

[105] Kuhn I, Kellenberger E, Said-Hassane F, Villa P, Rognan D, Lobstein A, et al. Identification by high-throughput screening of inhibitors of Schistosoma mansoni NAD+ catabolizing enzyme. Bioorg Med Chem 2010;18:7900−10.

[106] Haggag EG, Abdel Wahab SM, El-Zalabany SM, Moustafa EAA, El-Kherasy EM, Mabry TJ. Coumarin content and anti-bilharzial activity of extracts of leaves of Citrus aurantifolia (lime) and Citrus limonia (lemon). Asian J Chem 1998;10:583−6.

[107] Maghraby A, Bahgat M. Immunostimulatory effect of coumarin derivatives before and after infection of mice with the parasite Schistosoma mansoni. Arzneim Forsch 2004;54:545−50.

[108] Lopes JL, Lopes JN, Mello RT, Pellegrino J, Vieira PC. Chemoprophylactic study of schistosomiasis. IV: Inhibition of the penetration of cercariae by coumarin derivatives. Rev Bras Biol 1980;40:283−5.

[109] Magalhães L, Kapadia G, da Silva Tonuci L, Caixeta S, Parreira N, Rodrigues V, et al. In vitro schistosomicidal effects of some phloroglucinol derivatives from Dryopteris species against Schistosoma mansoni adult worms. Parasitol Res 2010;106: 395−401.

[110] Liu SY, Sporer F, Wink M, Jourdane J, Henning R, Li YL, et al. Anthraquinones in Rheum palmatum and Rumex dentatus (Polygonaceae), and phorbol esters in Jatropha curcas (Euphorbiaceae) with molluscicidal activity against the schistosome vector snails Oncomelania, Biomphalaria and Bulinus. Trop Med Int Health 1997;2:179−88.

[111] Cichewicz RH, Lim K-C, McKerrow JH, Nair MG. Kwanzoquinones A−G and other constituents of Hemerocallis fulva "Kwanzo" roots and their activity against the human pathogenic trematode Schistosoma mansoni. Tetrahedron 2002;58:8597−606.

[112] Cichewicz RH, Nair MG, McKerrow JH. Antihelminthic anthraquinones and method of use thereof US Patent Number 7132403 B2; 2006.

[113] Austin FG. Schistosoma mansoni chemoprophylaxis with dietary lapachol. Am J Trop Med Hyg 1974;23: 412−9.

[114] Melek FR, Tadros MM, Yousif F, Selim MA, Hassan MH. Screening of marine extracts for schistosomicidal activity in vitro. Isolation of the triterpene glycosides echinosides A and B with potential activity from the Sea Cucumbers Actinopyga echinites and Holothuria polii. Pharm Biol 2012;50:490−6.

[115] Jisaka M, Kawanaka M, Sugiyama H, Takegawa K, Huffman MA, Ohigashi H, et al. Antischistosomal activities of sesquiterpene lactones and steroid glucosides from Vernonia amygdalina, possibly used by wild chimpanzees against parasite-related diseases. Biosci Biotechnol Biochem 1992;56:845−6.

[116] Utzinger J, Shuhua X, N'Goran EK, Bergquist R, Tanner M. The potential of artemether for the control of schistosomiasis. Int J Parasitol 2001;31:1549–62.

[117] Utzinger J, N'Goran EK, N'Dri A, Lengeler C, Xiao S, Tanner M. Oral artemether for prevention of Schistosoma mansoni infection: randomised controlled trial. Lancet 2000;355:1320–5.

[118] N'Goran EK, Utzinger J, Gnaka HN, Yapi A, N'Guessan NA, Kigbafori SD, et al. Randomized, double-blind, placebo-controlled trial of oral artemether for the prevention of patent Schistosoma haematobium infections. Am J Trop Med Hyg 2003;68:24–32.

[119] Abdul-Ghani R, Loutfy N, Sheta M, Hassan A. Artemether shows promising female schistosomicidal and ovicidal effects on the Egyptian strain of Schistosoma mansoni after maturity of infection. Parasitol Res 2011;108:1199–205.

[120] Liu R, Dong H-F, Jiang M-S. The sensitivity of artesunate against Schistosoma japonicum decreased after 10 years of use in China? Parasitol Res 2012;110:1563–4.

[121] Borrmann S, Szlezák N, Faucher JF, Matsiegui PB, Neubauer R, Binder RK, et al. Artesunate and praziquantel for the treatment of Schistosoma haematobium infections: a double-blind, randomized, placebo-controlled study. J Infect Dis 2001;184:1363–6.

[122] De Clercq D, Vercruysse J, Kongs A, Verlé P, Dompnier JP, Faye PC. Efficacy of artesunate and praziquantel in Schistosoma haematobium infected schoolchildren. Acta Trop 2002;82:61–6.

[123] Inyang-Etoh PC, Ejezie GC, Useh MF, Inyang-Etoh EC. Efficacy of artesunate in the treatment of urinary schistosomiasis, in an endemic community in Nigeria. Ann Trop Med Parasitol 2004;98:491–9.

[124] Boulanger D, Dieng Y, Cisse B, Remoue F, Capuano F, Dieme J-L, et al. Antischistosomal efficacy of artesunate combination therapies administered as curative treatments for malaria attacks. Trans R Soc Trop Med Hyg 2007;101:113–6.

[125] Mors WB, dos Santos Filho MF, Monteiro HJ, Gilbert B, Pellegrino J. Chemoprophylactic agent in schistosomiasis: 14,15-epoxygeranylgeraniol. Science 1967;157:950–1.

[126] Gilbert B, de Souza JP, Fortes CC, Santos D, Seabra A, Kitagawa M, et al. Chemoprophylactic agents in schistosomiasis: active and inactive terpenes. J Parasitol 1970;56:397–8.

[127] Gilbert B, Mors WB, Baker PM, Tomassini TCB, Coulart EG, De Holanda JC, et al. Anthelminthic activity of essential oils and their chemical components. An Acad Bras Cienc 1972;44:423–8.

[128] Wang W-B, Liu H-J, Wang B-J, Zhou X, Zhang J, Liu C-C, et al. Killing effect of sodium abietate on adult worms of Schistosoma japonicum in vitro. Chinese J. Schistosomiasis Control 2011; 23:268–72.

[129] Martins SM, Torres CR, Ferreira ST. Inhibition of the ecto-ATPdiphosphohydrolase of Schistosoma mansoni by thapsigargin. Biosci Rep 2000;20:369–81.

[130] Ohigashi H, Huffman MA, Izutsu D, Koshimizu K, Kawanaka M, Sugiyama H, et al. Toward the chemical ecology of medicinal plant use in chimpanzees: The case of Vernonia amygdalina, a plant used by wild chimpanzees possibly for parasite-related diseases. J Chem Ecol 1994;20:541–53.

[131] Vichnewski W, Gilbert B. Schistosomicidal sesquiterpene lactone from Eremanthus elaeagnus. Phytochem 1972;11:2563–6.

[132] Baker PM, Fortes CC, Fortes EG, Gazzinelli G, Gilbert B, Lopes JNC, et al. Chemoprophylactic agents in schistosomiasis: eremanthine, costunolide, α-cyclocostunolide and bisabolol. J Pharm Pharmacol 1972;24:853–7.

[133] Newman DJ, Cragg GM. Natural products as sources of new drugs over the 30 years from 1981 to 2010. J Nat Prod 2012;75:311–35.

Botanicals for Adjunct Therapy and the Treatment of Multidrug-Resistant Staphylococcal Infections

T.O. Lawal[1], K.K. Soni[2], B.A. Adeniyi[1], B.J. Doyle[3], G.B. Mahady[2]

[1]Department of Pharmaceutical Microbiology, University of Ibadan, Ibadan, Nigeria,
[2]Department of Pharmacy Practice and Department of Medicinal Chemistry and Pharmacognosy,
PAHO/WHO Collaborating Centre for Traditional Medicine, College of Pharmacy,
University of Illinois at Chicago, Chicago, Illinois 60612, USA and
[3]Department of Biology and Department of Biochemistry, Alma College, Alma, Michigan, USA

INTRODUCTION

Infectious diseases account for approximately 2 % of deaths in wealthy nations, but are a significant (21%) cause of morbidity and mortality in low- to middle-income countries [1]. Infectious and parasitic diseases remain the primary killers of children under the age of 5 years in developing countries, partly as a result of the human immunodeficiency virus (HIV)/ acquired immunodeficiency syndrome (AIDS) epidemic [2]. Malaria, tuberculosis, and HIV/ AIDS represent approximately 10% of the deaths in low- and middle-income nations. Approximately 8.1 million children less than 5 years of age died in 2009, of which 98% lived in developing countries [1]. Mortality rates caused by infectious diseases have also been increasing in Western countries such as the United States, where death from infectious disease has increased by approximately 58% in the past decade [3]. One of the primary contributing factors to the increase in mortality from infectious diseases is an increase in antibiotic resistance in both nosocomial and community acquired pathogens [3]. Furthermore, although much progress has been made in the understanding and control of bacteria, outbreaks of drug-resistant bacteria, along with the emergence of previously unknown disease-causing microbes, are a threat to public health and safety. These negative health trends require immediate attention by scientific and pharmaceutical researchers and institutions to establish initiatives for the development of novel strategies for the prevention and treatment of

Fighting Multidrug Resistance with Herbal Extracts,
Essential Oils and Their Components
http://dx.doi.org/10.1016/B978-0-12-398539-2.00009-4

infectious diseases [4]. Proposed solutions have been coordinated by the US Center for Disease Control into a multipronged approach that includes prevention (such as vaccination), improved monitoring, and the development of new treatments [5]. It is this last solution that would encompass the development of new antimicrobial therapies [4–6].

USE OF TRADITIONAL BOTANICAL MEDICINES AS ANTIBACTERIAL AGENTS

Written history has shown that humans have used plants in systems of traditional medicine for thousands of years to treat all types of diseases. Due to these records we know a great deal about the traditional systems of medicine used in most countries worldwide [4,7–13]. Records dating back 5000 years describe the use of plant-based medicines for the treatment of almost all diseases known to humans, including infectious diseases [4,11]. Thus, it can be said that for thousands of years prior to the development of antibiotics in the twentieth century, humans used plant-based medicines almost exclusively for the treatment of infectious diseases. From many historical texts, we know much about the origin, development, and use of herbal medicines from traditional Chinese medicine (TCM), Ayurveda, Siddha, Unani, and the medicines of the ancient Greeks [12–13]. While some traditional systems of medicine were never written down, the information from many of these countries has since been described by anthropologists, missionaries, and ethnobotanists [9,10]. In fact, medicinal plant use in North America began with the Native Americans, or First Nations, and these plant-based medicines were incorporated into the medical practices of the European settlers of North America from about 1620 AD onward [11]. Until the 1970s, approximately 25% of all prescription drugs dispensed in the

United States were plant-based compounds or derivatives thereof, and in 2000, 11% of the 252 drugs on the World Health Organization's Essential Drugs List were still exclusively obtained from flowering plants [14,15]. Thus, the historical use of plants as the basis of treating the diseases of mankind, including infectious diseases, is long and continues to this day.

In terms of the antimicrobial activities of botanical medicines and their use in the treatment of infectious disease, there are volumes of published scientific information describing the plant materials, extraction types, and details of the bacterial strains used for testing; there are enormous amounts of published scientific data that describe the antimicrobial activities of plant extracts [4,23]. A search of the PubMed database (from 1975 to 2012) shows that there are at least 1200 publications in the scientific and medical literature describing the antimicrobial activities of various plant species and their chemical constituents. Searches of the NAPRALERT database, a natural products database housed within the University of Illinois at Chicago, shows that of the 60,000 plant species listed, at least 6255 species of plants have *in vitro* or *in vivo* antibacterial activities, of which there is traditional knowledge indicating that approximately 3859 of these plant species have been used in one or more countries to treat infectious diseases [16]. Many of the plant species tested in these assays were native to tropical countries as a result of the extraordinary biodiversity within these countries.

There is an increasingly need to research and develop plant-based medicines as antibacterial agents, as many treatable bacteria have become resistant to traditional antibiotics [17]. In addition, the alarming rate of plant species extinction has been a definite driving force for the renewed interest in investigating plant antimicrobials since the late 1900s. The possibility that plant species that may prove useful for treating disease and novel phytochemicals with the potential for treating infectious diseases may be lost forever should further stimulate research in this field. Furthermore, the research and development of new antibiotics since the late 1900s has dramatically decreased due to a lack of interest on the part of pharmaceutical companies and the large costs of developing these agents relative to the perceived payback. However, if novel antimicrobial therapies can be developed from medicinal plants or compounds from plants can be used in combination with common antibiotics to reduce resistance and increase the lifespan of these antibiotics, then the payback in improving human life could be substantial. This review will focus on one specific bacterium, *Staphylococcus aureus*, and how medicinal plants may be used in drug discovery for the treatment of multidrug-resistant *S. aureus* (MRSA).

METHICILLIN- AND VANCOMYCIN-RESISTANT *STAPHYLOCOCCUS AUREUS*

S. aureus is a facultative anaerobic Gram-positive coccus that normally colonizes the epithelial surfaces of humans, such as the skin and nasal passages [18,19]. The bacterium is named for the yellow pigmentation of its colonies when grown in culture. The bacteria can grow under conditions of high osmotic pressure and low moisture, which may explain why the cocci can survive in nasal secretions, on skin, and even in foods such as cured meats [18,19]. Under normal conditions, *S. aureus* colonization usually does not lead to infection. However, the bacteria produce many toxins that allow it to invade the epithelial layer and damage tissues, thereby causing serious infections [20–21]. One major problem is the infection of surgical wounds by *S. aureus*, which is a common and serious issue in hospitals. *S. aureus* also has numerous cell surface virulence factors (such as protein A

and clumping factor) and secrets exotoxin, as well as enterotoxin that can cause nausea and vomiting associated with food poisoning. Its production of exotoxin has been associated with infections that range from furuncles to scalded skin syndrome, sepsis, necrotizing pneumonia, and toxic shock syndrome [20–21].

The ability of *S. aureus* to develop resistance to antibiotics such as penicillin, a β-lactam, makes it dangerous to patients, particularly those in hospital for surgical procedures. There are now many published reports worldwide concerning the emergence of multidrug-resistant (MDR) staphylococci against all classes of β-lactam antibiotics. Antibiotic resistance results from the expression of PC1 β-lactamase and the acquisition by the bacillus of the *mecA* gene encoding a penicillin-binding protein, PBP2A [20–21]. *S. aureus* can enter the host through the epidermal and mucosal epithelia, and no single cell-surface virulence factor appears to be required for mucous membrane attachment. Once *S. aureus* colonization occurs, exotoxins (including the pyrogenic toxin superantigens and exfoliative toxins) cause serious and sometimes life-threatening infections. Numerous strains of MRSA contain genes that encode the Panton-Valentine leukocidin (PVL) exotoxin. These bicomponent leukocidal toxins are cytotoxins that can lyse leukocytes, and thus PVL-positive infections are life-threatening infections of soft tissues and bones [20–21].

The presence of MDR infections, including those of *S. aureus*, in both hospitals and communities is disturbing to healthcare professionals due to the difficulty in treating these infections. Compared with methicillin-sensitive *S. aureus* (MSSA), MDR bacterial infections result in longer hospital stays and increased patient morbidity and mortality [20]. Recent estimates suggest that approximately 20% of *S. aureus* isolates in Europe and approximately 33–55% of the *S. aureus* isolates in the United States are multidrug resistant [21]. Furthermore, it is estimated that 60% of nosocomial infections are due to MDR bacteria, most of which are vancomycin-resistant enterococci (VRE), methicillin-resistant *S. aureus* (MRSA) and, more recently, vancomycin-resistant *S. aureus* (VRSA) [22]. The case reports of vancomycin resistance present an incredibly serious problem, since vancomycin is one of the most commonly used antibiotics for MRSA infections in the United States [22]. Vancomycin, one of the antibiotics to which MRSA was susceptible, is a drug associated with very serious adverse events, but is usually considered a drug of last resort for resistant strains [19–22]. The first reports of vancomycin resistance were published in 1997 and 12 cases of VRSA infection have been reported in the United States since [22]. Resistance in VRSA appears to be due to specific modifications of the peptidoglycan cell wall target in *S. aureus*; this alteration is mediated by the *VanA* operon [22]. There is little in the way of therapeutic options for the treatment of VRSA infections. Thus, there needs to be a concerted effort to research and develop new preventative and treatment strategies for the management of both MRSA and VRSA.

IN VITRO STUDIES ON THE USE OF BOTANICAL EXTRACTS FOR TREATING METHICILLIN-RESISTANT *STAPHYLOCOCCUS AUREUS* AND VANCOMYCIN-RESISTANT *STAPHYLOCOCCUS AUREUS* INFECTIONS

As previously mentioned, numerous scientific publications from the worldwide literature describe the *in vitro* susceptibility of *S. aureus* to botanical extracts [4,23]. Additionally, since the threat of MRSA and VRSA is now well recognized, there has been an increase in the amount of data suggesting that specific plant extracts may also be of benefit for the

treatment of these infections [4,23–41]. These recent data (2008–2012) have been generated using plant species from countries around the world, indicating that the global potential antimicrobial drug development from plants is still high.

In Thailand, bacterial infections have long been a serious problem due to the hot, humid climate and the lack of access for many patients to Western medicine. Thai traditional plant medicines have been used for thousands of years for the treatment of many bacterial diseases [24]. In a study of 2011, ethanol extracts of Thai medicinal plants, including *Curcuma longa* L., *Rhinacanthus nasutus* L., *Garcinia mangostana* L., *Caesalpinia sappan* L., and *Centella asiatica* L., were tested in the disk diffusion and broth dilution assays against *Acinetobacter baumannii*, *Escherichia coli*, and *Klebsiella pneumoniae*, *Pseudomonas aeruginosa*, *S. aureus*, and MRSA, [24]. An ethanol extract of *C. sappan* L. was active against *S. aureus* and MRSA with a minimum inhibitory concentration (MIC) of 0.625 mg/mL [24]. Another 2011 study from Thailand [25] reported that an alcohol extract of *Mangifera indica* L. cv. Fahlun (Thai mango; Anacardiaceae) seed kernel extract (MSKE), and its phenolic principles (gallic acid, methyl gallate, and 1,2,3,4,6-pentagalloyl-β-D-glucopyranose; Fig. 9-1) had antibacterial activity against *S. aureus* and 19 clinical MRSA isolates. Scanning electron microscopy and transmission electron microscopy of bacteria treated with the MSKE showed a concentration-dependent reduction in cell division and ultrastructural changes in bacterial morphology, including the thickening of cell walls [25]. MSKE and its phenolic compounds were synergistic with penicillin G against 19 clinical MRSA isolates and lowered the MIC by at least five-fold. The active chemical constituent was isolated and identified as 1,2,3,4,6-pentagalloyl-β-D-glucopyranose. These results suggest that MSKE may be useful as an alternative therapeutic agent or an adjunctive therapy combined

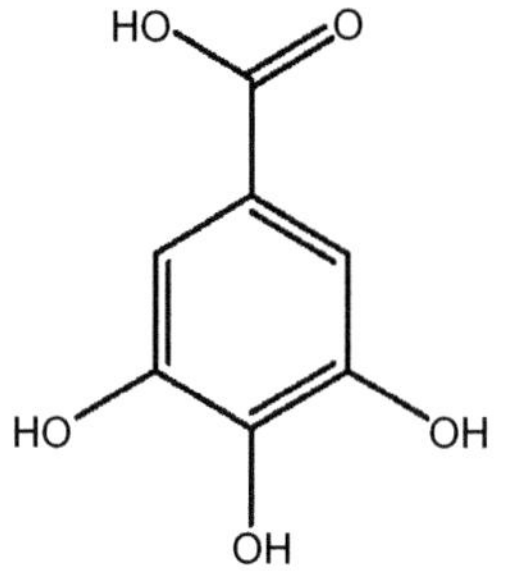

Gallic acid

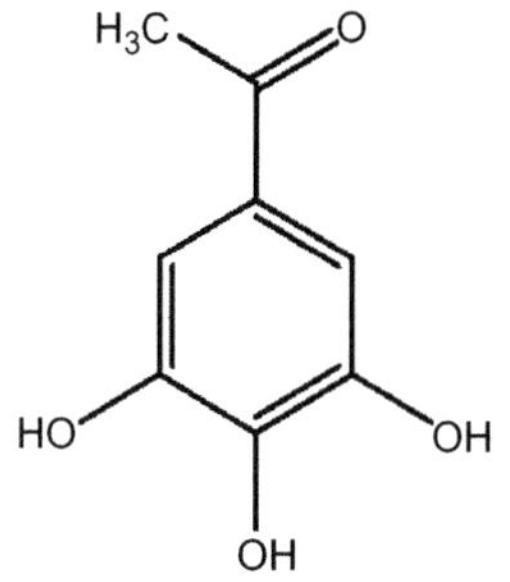

Methyl gallate

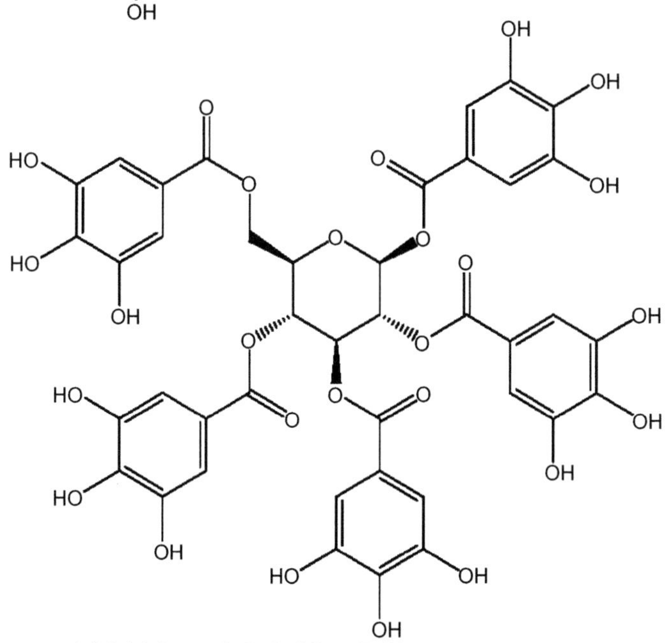

1,2,3,4,6-Penta-O-Galloyl-Beta-D-Glucopyranose

FIGURE 9-1 Structures of the major components of an alcohol extract of *Mangifera indica* L. cv. Fahlun (Thai mango) seed kernel. An alcohol extract of *Mangifera indica* L. cv. Fahlun (Thai mango) seed kernel contained the phenolic constituents gallic acid, methyl gallate and 1,2,3,4,6-pentagalloyl-β-D-glucopyranose), which have antibacterial activity against *S. aureus* and 19 clinical MRSA isolates.

with penicillin G in the treatment of MRSA [25]. An extract of another Thai plant, *G. mangostana* L. (purple mangosteen; Clusiaceae) inhibited the growth of MRSA *in vitro* [25] and the active constituent, α-mangostin (a prenylated xanthone; Fig. 9-2) was the most active compound, with an MIC and minimum bactericidal concentration (MBC) values of 1.95 and 3.91 μg/mL, respectively. If one compares this compound to chloramphenicol (MIC of 62.5 μg/mL), the MIC of α-mangostin is 20–30 times lower [25]. Earlier studies on α-mangostin showed synergism between α-mangostin and vancomycin hydrochloride against MRSA, suggesting that the combination may be useful for the treatment of both MRSA and VRSA [26]. Another group of researchers from Thailand isolated a series of seven pterocarpans, erybraedin A and B, erycristagallin, erythrabyssin II, erythrabissin-1, erystagallin A, and phaseollin, along with two flavanones, 5-hydroxysophoranone and glabrol, and one isoflavone, erysubin F, derived from the stems of *Erythrina subumbrans* (Leguminosae) [27]. The compound erycristagallin was very active against *Staphylococcus* spp. *in vitro*, including MRSA and VRSA strains, with an MIC range of 0.39–1.56 μg/mL. Erycristagallin showed the highest level of activity against all VRSA strains tested, with an MIC range of 0.39–1.56 μg/mL [27], thereby demonstrating the potential of naturally occurring compounds in the development of novel VRSA treatments.

In 2010, Yang and coworkers [28] from Taiwan identified four new antimicrobial compounds from the Chinese herbal medicine *Illicium verum* Hook. f. (star anise; Schisandraceae). Supercritical carbon dioxide and ethanol extracts of the plant inhibited the growth of 67 clinical drug-resistant bacterial isolates, including 20 MRSA strains. The diethyl ether and ethyl acetate fractions had MICs of 0.15–0.70 μg/mL and 0.11 μg/mL, respectively [28]. The chemical components of the extracts included (*E*)-anethole, anisylacetone, anisyl alcohol, and anisyl aldehyde (Fig. 9-3) [28].

In 2010, Turkish researchers Polatoglu and colleagues [29] published a report concerning the activity of a water-distilled essential oil (EO) from *Tanacetum parthenium* (L.) Sch. Bip. (Feverfew; Asteraceae) collected from two different localities in Turkey against *S. aureus* and MRSA. The antibacterial effects of the EO were evaluated against five Gram-positive and five Gram-negative bacteria using a broth microdilution assay [29]. The strongest activity was observed on *Bacillus subtilis*, *S. aureus*, and MRSA. The oil showed highest activity against *B. subtilis* (125 μg/mL) and MRSA (125 μg/mL), which was only two fold higher than the MIC of the positive control, chloramphenicol

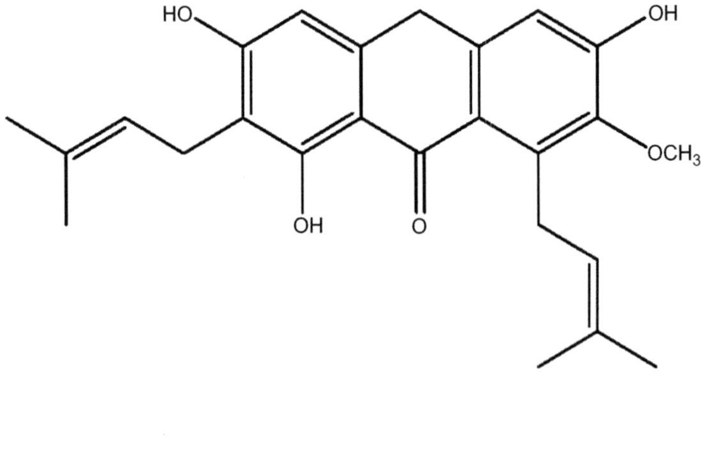

FIGURE 9-2 Structure of α-mangostin. α-Mangostin, isolated from a *Garcinia mangostana* (mangosteen) extract from Thailand, inhibits the growth of MRSA *in vitro*.

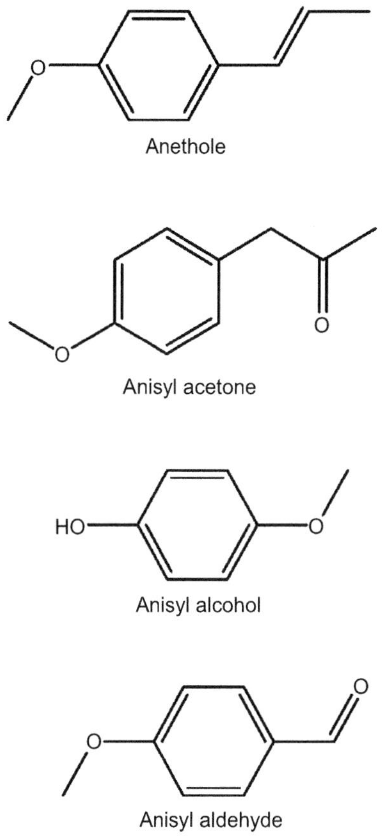

Anethole

Anisyl acetone

Anisyl alcohol

Anisyl aldehyde

FIGURE 9-3 Structures of (E)-anethole, anisyl-acetone, anisyl alcohol, and anisyl aldehyde. These antimicrobial compounds were present in a diethyl ether partition of a supercritical carbon dioxide extract of Illicium verum.

Camphor

Camphene

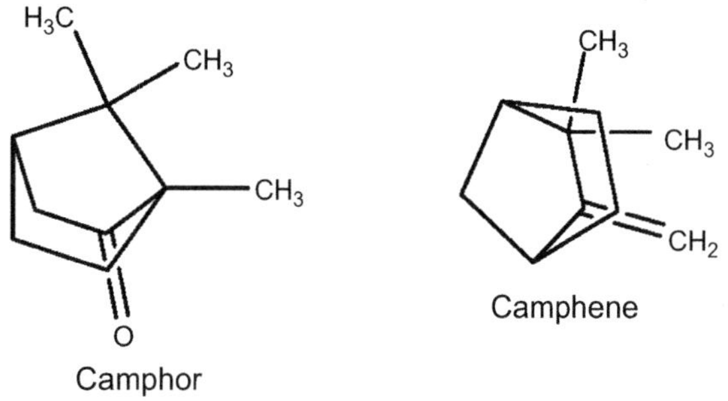

FIGURE 9-4 Structures of camphor and camphene. Camphor and camphene isolated from the essential oil of Tanacetum parthenium (Feverfew; Asteraceae) show activity against methicillin-resistant S. aureus.

processed using silica gel and reverse phase chromatography to purify kaempferol and aloe-emodin (Fig. 9-5), which exhibited MICs of 13.0 µg/mL and 12.0 µg/mL, respectively. The authors suggested that there was a structure-activity relationship for the anti-MRSA activity, in that a free hydroxyl group at C-3 of the flavonol was a structural requirement for the inhibition of MRSA [30].

In 2005, a 50% ethanol extract from the dried fruit of the traditional medicinal plant Terminalia chebula Retz. (Combretaceae), was found to inhibit the growth of MRSA, with an MIC of 31.3 µg/mL [31]. The primary active constituents of the fruit were gallic acid and its methyl ester [31]; the same compounds as those purified from M. indica L. cv. Fahlun (Thai mango) seed kernel extract that were also found to be active against MRSA [25]. In 2010, Mahaboubi and coworkers from Iran reported on the anti-staphylococcal effects of the EO of Zataria multiflora Boiss. distilled from the aerial parts of the plant [32]. Thymol (38.7%), carvacrol (15.3%), and rho-cymene (10.2%) were the primary chemical constituents of the oil. The oil was effective against both MRSA and MSSA strains of S. aureus, with MIC and MBC values in the range of 0.25−1 and 0.5−2 µL/mL, respectively. The EO also exhibited synergism with vancomycin [32].

(62.5 µg/mL; 29). The active EO contained 49% camphor, 22.1% trans-chrysanthenyl acetate, and 9.4% camphene (Fig. 9-4).

In 2008, Hazni and coworkers [30] from Malaysia sequentially fractionated an extract of the leaves of Cassia alata L. (candle bush; Fabaceae) to obtain hexane, chloroform, butanol, and residual extracts. All of these extracts were evaluated against MRSA using an agar well diffusion assay [30]. The butanol and chloroform extracts of the leaves inhibited the growth of MRSA strains at a concentration of 50 mg/mL. The butanol extract was further

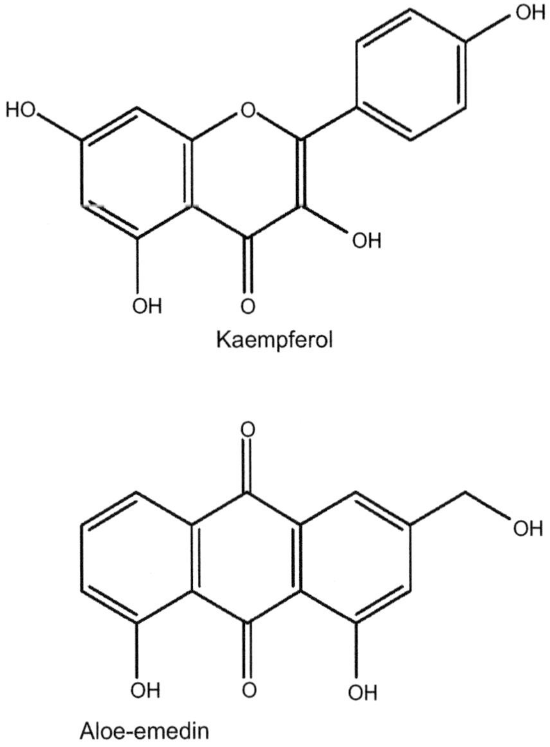

Kaempferol

Aloe-emedin

FIGURE 9-5 Structures of kaempferol and aloe-emodin. Active compounds, kaempferol and aloe-emodin, isolated from the butanol extract of the leaves of *Cassia alata* from Malaysia. The compounds have minimum inhibitory concentrations of 13.0 and 20.0 μg/mL, respectively.

HUMAN STUDIES ON THE USE OF BOTANICALS FOR TREATING METHICILLIN-RESISTANT *STAPHYLOCOCCUS AUREUS* INFECTIONS

The global scientific data from studies published since the start of the twenty-first century indicate that various extracts from medicinal plants used in traditional medicine have at least *in vitro* activity against MRSA and VRSA. However, while such data are important, very few medicinal plants have been tested in animal models of MRSA or in humans. Only one medicinal plant, *Melaleuca alternifolia* (Maiden & Betche)

Cheel (Myrtaceae), commonly referred to as tea tree oil, has been extensively scientifically investigated in humans [4,33–39]. Tea tree is indigenous to Australia, and the EO has been used by the native peoples of Australia for centuries to treat a wide range of infectious diseases and wounds. Historically, the Bundjalung Aborigines of northern New South Wales used tea tree leaves as a traditional medicine for the treatment of upper respiratory infections [33]. The crushed leaves of *tea trees* were used as an inhalant to treat coughs and colds, and were also sprinkled on wounds or used as an infusion to treat sore throats or other skin ailments [33]. The use of tea tree oil was not popularized until Penfold published the first reports of its antimicrobial activity in a series of papers in the 1920s and 1930s [33]. The EO of the leaves is obtained by steam distillation of the leaves and terminal branches.

In terms of *in vitro* studies, tea tree oil inhibits the growth of MRSA with an MIC and an MBC of 0.25 μg/mL and 0.50 μg/mL, respectively [33]. The major chemical constituents responsible for this activity in tea tree EO include 1,8-cineol (4.5–16.5%), terpinen-4-ol (29–45%), γ-terpinene (10–28%), and α-terpineol (2.7–13.0%; Fig. 9-6). Electron microscopy of MRSA cell morphology suggests that tea tree oil and its chemical constituents may compromise the bacterial cytoplasmic membrane [33]. Concerns regarding the development of bacterial resistance to tea tree oil appear to be controversial, as screening of 100 MRSA isolates revealed no resistance at concentrations up to 2.5% (v/v) [34]. However, the appearance of subpopulations of resistant mutants in isolates unexposed to tea tree oil suggests that exposure may eventually lead to the selection of resistant strains. Stepwise induction of resistance has been reported up to a concentration of 16% (v/v), which is greater than the 5–10% (v/v) concentration recommended in topical formulations to treat MRSA [34].

There are currently at least three published clinical studies or case reports that have

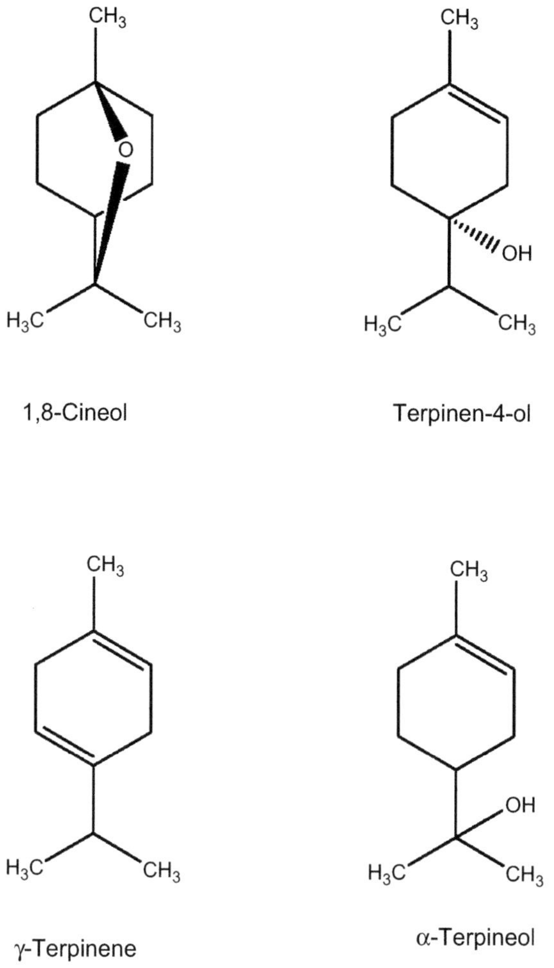

CH₃ → CH3 (1,8-Cineol)

1,8-Cineol

Terpinen-4-ol

γ-Terpinene

α-Terpineol

FIGURE 9-6 Structures of the major chemical constituents of tea tree oil. The major chemical constituents of tea tree oil, 1,8-cineol, terpinen-4-ol, γ-terpinene, and α-terpineol were shown to reduce the growth of *S. aureus* by altering the bacterial cytoplasmic membrane.

evaluated the efficacy of tea tree oil products for the treatment of MRSA infections. In addition, there are two clinical studies that describe the efficacy of a Japanese Kampo medicine for the treatment of MRSA, as well as a randomized Phase I study of *Atuna racemosa* Raf. (atun tree; Chrysobalanaceae) for the treatment of MRSA [35–41].

One of the first reported clinical trials of tea tree oil was published in 2000 by Caelli and coworkers [35]. This study compared the use of a combination of 4% tea tree nasal ointment and 5% tea tree body wash with that of a 2% mupirocin nasal ointment and triclosan body wash, which is considered the normal standard of care for the eradication of MRSA [35]. In this clinical trial, 30 inpatients infected with MRSA were randomly assigned to the tea tree oil treatment or standard of care treatment for a minimum of 3 days. The patients were also treated with intravenous vancomycin and followed up to determine the MRSA carriage at 48 and 96 h after treatment. The results of this study showed that of the 18 patients that completed the trial, two of the 10 patients in the standard of care group versus five of the eight patients in the tea tree oil group were cleared of MRSA. This study concluded that tea tree oil was an effective topical treatment for reducing MRSA carriage [35]. Adverse effects reported by the group receiving tea tree nasal ointment included swelling of the nasal mucosa and a burning sensation. No adverse events were reported from use of tea tree body wash or mupirocin ointment [35].

In a second clinical study published in 2004, the efficacy of tea tree oil on MRSA decolonization was compared with that of mupirocin, chlorhexidine gluconate, and silver sulfadiazine [36]. In this two-armed clinical trial, 224 patients were randomized to receive either (1) 10% tea tree oil cream applied to wounds three times daily for 5 days with 5% tea tree body wash used daily for 5 days and 10% tea tree oil cream applied to wounds or ulcers; or (2) 2% mupirocin nasal ointment applied three times daily for 5 days with 4% chlorhexidine gluconate body wash used daily for 5 days and 1% silver sulfadiazine cream (positive control) applied to wounds or ulcers daily for 5 days. All patients were tested for MRSA prior to starting treatment and 14 days after completion of treatment. Of the 110 subjects in the tea tree oil group, 64

(59%) remained positive for MRSA after treatment. Of the 114 patients who received the positive control treatments, 58 (51%) subjects were still positive for MRSA after completing the treatment. The results were not statistically different and the study demonstrated that tea tree oil may be considered as an option for MRSA decolonization but that the optimum dose to be used in topical products should be further evaluated for MRSA eradication [36].

In 2011, an uncontrolled, open-label, pilot study of tea tree oil solution investigating the decolonization of MRSA-positive wounds and its influence on wound healing was published [37]. Nineteen participants with acute and chronic wounds suspected of being colonized with MRSA were enrolled. Seven of the enrolled patients were shown not to have MRSA and were subsequently withdrawn from the study. Eleven of the remaining 12 participants were treated with a water-miscible tea tree oil (3.3% v/v) solution that was applied as part of the wound-cleansing routine during each dressing change. The study nurse, following assessment of each patient, determined whether wound-dressing changes were to be performed three times per week or daily. One participant withdrew from the study prior to treatment. The results of this study showed that no patients were MRSA negative after treatment; however, 8 of the 11 treated wounds had begun to heal and were reduced in size as measured by computer planimetry. This study concluded that tea tree oil had no effect on reducing MRSA colonization [37].

Currently, there is another ongoing multicenter, phase II/III prospective open-label randomized controlled clinical trial to evaluate the efficacy of a 5% tea tree oil product in the prevention of MRSA colonization as compared with a standard body wash [38]. This study began in 2008 and will evaluate the cost-effectiveness of tea tree oil body wash and determine its effectiveness using polymerase chain reaction assays to detect MRSA in critically ill patients [38]. The results of this study are yet to be published.

In addition to tea tree oil, other medicinal plants are being testing in human studies to determine whether they have efficacy against MRSA. In 2007, Buenz and coworkers reported that a seed extract of the atun tree that they identified by digital bioprospecting of a 400-year-old historic Dutch herbal text was potentially useful for treating MRSA [39]. Topical application of a kernel extract of seeds collected in Samoa inhibited the growth of MRSA, with an MIC of $16-32\,\mu g/mL$. In addition, a maximum tolerated topical application of the extract was assessed in a randomized, double-blind, placebo-controlled Phase I pilot trial [39]. The extract was well tolerated in short-term topical applications of 10 times the MIC and may therefore prove to be clinically useful in the treatment of MRSA [39].

Recently, from Kampo medicine (traditional Japanese medicine) Hochu-ekki-to, a multiple herbal formula consisting of 10 medicinal plants (*Angelicae radix, Astragali radix, Atractylodis lanceae rhizoma, Aurantii nobilis pericarpium, Bupleuri radix, Cimicifugae rhizoma, Ginseng radix, Glycyrrhizae radix, Zingiberis rhizoma*, and *Zizyphi fructus*) is reported to have anti-MRSA effects in a clinical setting [40,41]. Hochu-ekki-to was administered at a dose of 5 g/day to five MRSA-infected bedridden patients with cerebrovascular disorder, dementia, and, in two cases, bedsores, who were infected with MRSA and had resistance to several antibiotics. After treatment with Hochu-ekki-to, the patients appeared to be free of the MRSA infection, showed an improvement in their general condition, and reported no side effects [40]. In the second study, Hochu-ekki-to was administered at a dose of 7.5 g daily for at least 24 weeks to 38 patients with asymptomatic MRSA bacteriuria [41]. The prognostic nutritional index (PNI), albumin levels, and lymphocytes were measured in the peripheral blood before treatment and every 2 weeks after the start of treatment and urine was cultured for MRSA. The results were compared with 12 untreated control patients with asymptomatic MRSA bacteriuria. Four of the 38 patients treated with Hochu-ekki-to received antibacterial drugs during the treatment period and were subsequently excluded from the study. Of the 34 eligible patients treated with Hochu-ekki-to, urinary MRSA was eradicated in 12 patients and the bacterial counts in urine culture were reduced to ≤ 100 colony-forming units/mL in 10 patients. In patients treated with Hochu-ekki-to, numbers of urinary bacteria were significantly decreased from 10 weeks after treatment, as compared with the control group ($P < 0.05$). The PNI improved in all patients, compared with the baseline. The report concluded that Hochu-ekki-to reduced MRSA bacteriuria by enhancing the immune defenses of the patients [41].

THE FUTURE OF NATURAL PRODUCTS FOR TREATING MULTIDRUG-RESISTANT *STAPHYLOCOCCUS AUREUS* INFECTIONS

MDR bacteria are a growing threat to human health and welfare. MRSA, in particular, has become increasingly community acquired and is no longer a specifically nosocomial infection. Considering that both MRSA and VRSA infections are increasing and that production of new antibiotics to control these infections is lagging, there is an urgent need to identify new safe and effective treatments for MDR bacteria. Currently, there is an extensive body of published scientific literature from around the globe demonstrating that natural products are a good starting point for the research and development of new and adjunct treatments for both MRSA and VRSA. The abundance of supporting *in vitro* data suggests that plants and other natural products can be useful sources of new and novel treatments for MDR

organisms, if properly researched and developed. However, while there is an abundance of *in vitro* studies demonstrating the potential use of natural products as antimicrobials, data from animal models and clinical trials are still limited and few studies have been published. However, where there are published clinical data, such as for tea tree oil, the results look very promising and suggest that some botanicals may be useful in promoting MRSA decolonization. Data from larger randomized, controlled studies are needed, especially those including evaluations of efficacy in MRSA eradication. Recent docking studies have shown that computer-aided modeling is a novel approach to developing new MRSA agents from natural products [19]. The screening of lead compounds with good pharmacologic properties and drug safety via this method is an excellent platform for developing compounds with a favorable absorption, distribution, metabolism, excretion, and toxicity profiles [19]. A recent study using this method screened herbal extracts as therapeutic drugs for MRSA using ADMET (i.e., absorption-distribution-metabolism-excretion-toxicity) and docking studies and successfully compared this method with in vitro studies [19]. These methods are useful in determining the priority of botanical compounds for testing; however, there is still a need for *in vivo* animal studies and clinical trials to appropriately evaluate natural products for the treatment of MRSA.

References

[1] World Health Statics. pp. 1–50. Geneva, Switzerland: World Health Organization; 2011.

[2] World Health Report. pp. 1–10. Geneva, Switzerland: World Health Organization; 2008.

[3] Pinner R, Teutsch S, Simonsen L, Klug L, Graber J, Clarke M, et al. Trends in infectious diseases mortality in the United States. J Am Med Assoc 1996;275:189–93.

[4] Mahady GB. Medicinal plants for the prevention and treatment of bacterial infections. Curr Pharmaceut Des 2005;11:2405–27.

[5] Fauci A. The global challenge of infectious diseases: the evolving role of the National Institutes of Health in basic and clinical research. Nature Immunol 2005;6:743–5.

[6] Ymele-Leki P, Cao S, Sharp J, Lambert KG, McAdam AJ, Husson RN, et al. A high-throughput screen identifies a new natural product with broad-spectrum antibacterial activity. PLoS One 2012;7:e31307.

[7] Rosenbloom RA, Chaudhary J, Castro-Eschenbach D. Traditional botanical medicine: an introduction. Am J Ther 2011;18(2):158–61.

[8] Chan K, Shaw D, Simmonds MS, Leon CJ, Xu Q, Lu A, et al. Good practice in reviewing and publishing studies on herbal medicine, with special emphasis on traditional Chinese medicine and Chinese materia medica. J Ethnopharmacol 2012;140(3):469–75.

[9] Jarić S, Mitrović M, Djurdjević L, Kostić O, Gajić G, Pavlović D, et al. Phytotherapy in medieval Serbian medicine according to the pharmacological manuscripts of the Chilandar Medical Codex (15–16th centuries). J Ethnopharmacol 2011;137(1):601–19.

[10] Butler A, Keating R. Old herbal remedies and modern combination therapy. Scott Med J 2011;56(3):170–3.

[11] Mahady GB. Global harmonization of herbal health claims. J Nutr 2001;131(3s):1120S–3S.

[12] Hu J, Zhang J, Zhao W, Zhang Y, Zhang L, Shang H. Cochrane systematic reviews of Chinese herbal medicines: an overview. PLoS One 2011;6(12):e28696.

[13] Robinson N. Integrative medicine—traditional Chinese medicine, a model? Chin J Integr Med 2011; 17(1):21–5.

[14] Farnsworth NR, Morris RW. Higher plants—the sleeping giant of drug development. Amer J Pharm Educ 1976;148:46–52.

[15] Fabricant DS, Farnsworth NR. The value of plants used in traditional medicine for drug discovery. Environ Health Perspect 2001;109(Suppl. 1):69–75.

[16] Farnsworth NR. NAPRALERT database. IL, 2012 production. Chicago: University of Illinois at Chicago, www.napralert.org; 2012 [accessed 22.12.12].

[17] Martin K, Ernst E. Herbal medicines for treatment of bacterial infections: a review of controlled clinical trials. J Antimicrob Chemother 2003;51:241–6.

[18] Tenover FC, Gaynes RP. In: Fischetti VA, Novick RP, Ferretti JJ, Portnoy DA, Rood JI, editors. Gram-positive pathogens. Washington, DC, USA: ASM Press; 2000. p. 414–21.

[19] Skariyachan S, Krishnan RS, Siddap a SB, Salian C, Bora P, Sebastian D. Computer aided screening and evaluation of herbal therapeutics against MRSA infections. Bioinformation 2011;7:222–33.

[20] Ridenour GA, Wong ES, Call MA, Climo MW. Duration of colonization with methicillin-resistant Staphylococcus aureus among patients in the intensive care unit: implications for intervention. Infect Control Hosp Epidemiol 2006;27:271–8.

[21] Appelbaum PC. MRSA—the tip of the iceberg. Clin Microbiol Infect 2006;12(Suppl. 2):3–10.

[22] Kobayashi SD, Musser JM, Deleo FR. Genomic Analysis of the Emergence of Vancomycin-Resistant Staphylococcus aureus. MBio 26 2012;3(4):e00170.

[23] Gibbons S. Anti-staphylococcal plant natural products. Nat Prod Rep 2004;21:263–77.

[24] Temrangsee P, Kondo S, Itharat A. Antibacterial activity of extracts from five medicinal plants and their formula against bacteria that cause chronic wound infection. J. Med. Assoc. Thai 2011;94(Suppl. 7):S166–171.

[25] Jiamboonsri P, Pithayanukul P, Bavovada R, Chomnawang MT. The inhibitory potential of Thai mango seed kernel extract against methicillin-resistant Staphylococcus aureus. Molecules 2011;16:6255–70.

[26] Sakagami Y, Iinuma M, Piyasena KG, Dharmaratne HR. Antibacterial activity of alpha-mangostin against vancomycin resistant Enterococci (VRE) and synergism with antibiotics. Phytomedicine 2005;12:203–8.

[27] Rukachaisirikul T, Innok P, Aroonrerk N, Boonamnuaylap W, Limrangsun S, Boonyon C, et al. Antibacterial pterocarpans from Erythrina subumbrans. J Ethnopharmacol 2007;110(1):171–5.

[28] Yang JF, Yang CH, Chang HW, Yang CS, Wang SM, Hsieh MC, et al. Chemical composition and antibacterial activities of Illicium verum against antibiotic-resistant pathogens. J Med Food 2010;13: 1254–62.

[29] Polatoglu K, Demirci F, Demirci B, Gören N, Başer KH. Antibacterial activity and the variation of Tanacetum parthenium (L.) Schultz Bip. essential oils from Turkey. J Oleo Sci 2010;59:177–84.

[30] Hazni H, Ahmad N, Hitotsuyanagi Y, Takeya K, Choo CY. Phytochemical constituents from Cassia alata with inhibition against methicillin-resistant Staphylococcus aureus (MRSA). Planta Med 2008; 74:1802–5.

[31] Sato Y, Oketani H, Singyouchi K, Ohtsubo T, Kihara M, Shibata H, et al. Extraction and purification of effective antimicrobial constituents of Terminalia chebula Rets. against methicillin-resistant Staphylococcus aureus. Biol Pharm Bull 1997;20:401–4.

[32] Mahboubi M, Bidgoli FG. Antistaphylococcal activity of Zataria multiflora essential oil and its synergy with vancomycin. Phytomedicine 2010;17(7):548–50.

[33] Carson CF, Hammer KA, Riley TV. Melaleuca alternifolia (Tea Tree) Oil: a Review of Antimicrobial and Other Medicinal Properties. Clin Microbiol Rev 2006;19:50–62.

[34] Czech E, Kneifel W, Kopp B. Tea tree oil as an alternative topical decolonization agent for methicillin-resistant Staphylococcus aureus. Planta Medica 2001;67:263—9.

[35] Caelli M, Porteous J, Carson CF, Heller R, Riley TV. Tea tree oil as an alternative topical decolonization agent for methicillin-resistant Staphylococcus aureus. J Hosp Infect 2000;46:236—7.

[36] Dryden MS, Dailly S, Crouch M. A randomized, controlled trial of tea tree topical preparations versus a standard topical regimen for the clearance of MRSA colonization. J Hosp Infect 2004;56:283—6.

[37] Edmondson M, Newall N, Carville K, Smith J, Riley TV, Carson CF. Uncontrolled, open-label, pilot study of tea tree (Melaleuca alternifolia) oil solution in the decolonisation of methicillin-resistant Staphylococcus aureus positive wounds and its influence on wound healing. Int Wound J 2011;8:375—84.

[38] Thompson G, Blackwood B, McMullan R, Alderdice FA, Trinder TJ, Lavery GG, et al. A randomized controlled trial of tea tree oil (5%) body wash versus standard body wash to prevent colonization with methicillin-resistant Staphylococcus aureus (MRSA) in critically ill adults: research protocol. BMC Infect Dis 2008;28:161.

[39] Buenz E, Bauer BA, Schnepple D, Wahner-Roedler DJ, Vandel AG, Howe C. A randomized Phase I study of Atuna racemosa: A potential new anti-MRSA natural product extract. J Ethnopharmacol 2007;114:371—6.

[40] Itoh T, Itoh H, Kikuchi T. Five cases of MRSA-infected patients with cerebrovascular disorder and in a bedridden condition, for whom bu-zhong-yi-qi-tang (hochu-ekki-to) was useful. J Chin Med 2000;28:401—8.

[41] Nishida S. Effect of Am Hochu-ekki-to on asymptomatic MRSA bacteriuria. J Infect Chemother 2003;9:58—61.

Combining Essential Oils with Antibiotics and other Antimicrobial Agents to Overcome Multidrug-Resistant Bacteria

Kateryna Volodymyrivna Kon[1], Mahendra Kumar Rai[2,3]

[1]Kharkiv National Medical University, Kharkiv, Ukraine,
[2]Sant Gadge Baba Amravati University, Amravati, Maharashtra, India and
[3]Department of Chemical Biology, Institute of Chemistry, University of Campinas, Campinas, São Paulo, Brazil

INTRODUCTION

Multidrug-resistant (MDR) bacteria have become more prevalent in recent times owing to the inappropriate and irrational use of antibiotics, which provides favorable conditions for the selection of antibiotic-resistant mutants [1]. Resistance against all classes of antibiotics has been described, which leads to a constant need for the development and production of new drugs. However, difficulties in the identification of new substances with both high effectiveness and low toxicity have resulted in only a few new antibiotic classes being discovered since the 1970s [2].

MDR bacteria have become abundant, particularly in nosocomial infections. Many hospital infections are now caused by methicillin-resistant *Staphylococcus aureus* (MRSA), vancomycin-resistant enterococci (VRE), *Escherichia coli* and *Pseudomonas aeruginosa* resistant to fluoroquinolones; *Klebsiella pneumoniae* resistant to ceftazidime; MDR *Acinetobacter baumannii*; and other MDR bacteria. The clinical outcomes of infections caused by MDR bacteria are deteriorating, owing to the reduced treatment options [3,4].

A high prevalence of nosocomial infections caused by MDR bacteria has been documented by different international surveillance programs. In 2010, the proportion of MRSA in *S. aureus* isolates was found to be > 25% in seven out of the 28 European countries (Cyprus, Greece, Hungary, Italy, Malta, Romania, Spain) and > 50% in Portugal [5]. Both the high prevalence and the rapid increase in levels of antibiotic resistance are alarming. An illustration of

Fighting Multidrug Resistance with Herbal Extracts,
Essential Oils and Their Components
http://dx.doi.org/10.1016/B978-0-12-398539-2.00010-0

this is the prevalence of MRSA (39%) among invasive isolates in Mediterranean countries [6] and the 2.5-fold increase in the number of infection-related hospitalizations associated with antibiotic resistance [7]. The prevention and control of MRSA have been identified as public health priorities in the European Union because, in addition to healthcare-associated infections, new MRSA strains have recently emerged as community-associated (i.e., CA-MRSA) and livestock-associated human pathogens [8].

In moderate or severe infections, antibiotic therapy is commonly initiated empirically, before the results of sensitivity tests are known. A high prevalence of MDR bacteria increases the probability of inappropriate initial antibiotic treatment, which has been shown to significantly affect the results of treatment by increasing both morbidity and mortality rates [9–11].

Understanding the main mechanisms of antibiotic resistance is necessary for the development of new methods for coping with this problem. Antibiotics are inactivated by enzymatic modification or degradation; antibiotic binding to the bacterial cell is reduced by alterations to antibiotic targets, which in turn decreases antibiotic activity. Another mechanism of resistance includes removing the antibiotic from the cell through the action of efflux pumps. The change in bacterial physiology that occurs during the conversion of plankton cells into biofilms also leads to a decrease in antibiotic activity [2,12].

However, both the destruction of antibiotics by enzymes and changes to antibiotic targets become irrelevant if other classes of antimicrobial agents are used. Plant essential oils (EOs) represent a promising source [13–17]. Their multicomponent chemical composition and complex mechanism of action provides EOs with advantages over traditional antibiotics. EOs can have simultaneous activity against bacteria and fungi [18], protozoans [19], and viruses [20], which is especially important for mixed infections. Moreover, in addition to their antimicrobial properties, EOs have anti-inflammatory [21], immune modulatory [22], antioxidant [21,23], and regenerative activities [24], which indicates them to be promising agents for the treatment of different types of infections.

Great potential has been shown by combinations of different antibiotics and of antibiotics and nonantibiotics [2]. The antimicrobial properties of EOs have been known since ancient times [25]; however, during the era of antibiotic discovery, natural antimicrobial products derived from plants were largely forgotten. Nowadays, the high prevalence of antibiotic resistance has led to renewed interest in the antimicrobial properties of plant materials, and the number of studies devoted to the antibacterial activity of EOs and plant extracts is increasing year by year.

At the moment, EOs cannot substitute for antibiotics in the treatment of severe systemic infections owing to the absence of clinically applicable pharmaceutical forms and the modest level of clinical effectiveness. However, they are being widely investigated in both *in vitro* and *in vivo* studies for the treatment of local infections. The best studied applications of EOs are for the treatment of wounds, wound infections, and dermatologic infections. Good healing properties for experimental wounds have been demonstrated by the EOs of *Ocimum gratissimum* [26]; *Juniperus oxycedrus* subsp. *oxycedrus*; *J. phoenicea* [27]; and a mixture of EOs of *Origanum majorana*, *Origanum minutiflorum* (*Origani aetheroleum*), and *Salvia triloba*, together with an olive oil extract of the flowering aerial parts of *Hypericum perforatum* and olive oil [28]. A 5% tea tree (*Melaleuca alternifolia*) oil gel showed efficacy in the treatment of mild-to-moderate acne vulgaris in 60 patients in a randomized, double-blind clinical trial [29].

MECHANISM OF ACTION OF ESSENTIAL OILS ON MULTIDRUG-RESISTANT BACTERIA

EOs exhibit simultaneous activities toward different bacterial structures owing to their multicomponent composition. This gives them advantages compared with antibiotics because such complex mechanisms of action make it more difficult for bacteria to develop resistance, compared with single target therapy.

Many components of EOs have been screened for antimicrobial activity. EOs are mainly composed of terpenoids, in particular monoterpenes and sesquiterpenes. Sometimes there are also diterpenes and a number of low molecular weight aliphatic hydrocarbons, acids, alcohols, aldehydes, and acyclic esters, or lactones, coumarins, and homologues of phenylpropanoids present in the oil, depending on how it was prepared [30]. The highest antimicrobial properties have been established in terpenes, such as carvacrol, geraniol, menthol, and thymol. [17]. Carvacrol, citrals, *p*-cymene, and thymol contribute to increased membrane permeability and the swelling of cellular membranes. Carvacrol and thymol are thought to disturb the outer membrane of Gram-negative bacteria, leading to the release of lipopolysaccharides [31]. Furthermore, *p*-cymene was shown to enable carvacrol influx due to its permeabilizing activity, thus resulting in synergistic activity when both components are present in an EO [32]. Another component with a phenolic structure, eugenol, may react with proteins and thus prevent the activity of enzymes in bacterial cells [31].

One of the EOs with the highest antibacterial activity is that of cinnamon and its major component, cinnamaldehyde, has demonstrated strong antimicrobial effects in several studies [33–35]. Cinnamaldehyde also interacts with the bacterial cell membrane and causes its disruption. Along with eugenol, it was shown to inhibit energy metabolism in several Gram-positive bacteria, such as *Listeria monocytogenes* and *Lactobacillus sakei*. Moreover, cell membrane damage caused by these components may lead to loss of the proton motive force and leakage of small ions from the cell. Another reported effect of cinnamaldehyde and eugenol activities is inhibition of both glucose import and glycolysis [36].

In general, the mechanism of action of EOs in strains of bacteria resistant to antibiotics is considered to be the same as in antibiotic-sensitive strains. This is supported by its multicomponent composition and simultaneous effects on different targets in the bacterial cell. At the same time, EOs have some additional activities that are especially useful against MDR strains. EOs have been shown to exhibit antiplasmid activity, which has a great potential for restricting the transfer of resistance between MDR and sensitive bacteria. Schelz et al. [37] demonstrated a rather high antiplasmid activity in peppermint EO and its main component, menthol, against the metabolic plasmid of *E. coli* F′lac K12 LE140. Peppermint oil at a concentration of 0.54 mg/mL caused the elimination of 37.5% of plasmids and 0.325 mg/mL menthol caused 96% plasmid elimination [37].

Some EOs have also been shown to possess promising activity against biofilm-associated microorganisms. Such bacteria are estimated to be 50–500 times more resistant than their planktonic counterparts [30]. Quorum-sensing (QS), a major means of bacterial cooperation during biofilm formation, was shown to be impaired by certain EOs and their active components. Anti-QS activity has been described in clove oil [38]; geranium, lavender, rose, and rosemary EOs; and to a lesser extent in citrus and eucalyptus EOs [39] and those of some endemic Colombian plants, particularly *Lippia alba*; as well as in some EO components, such as citral, carvone, and α-pinene [40].

In spite of the general high activity of many EOs, resistance is an inevitable process for all classes of antimicrobials, and EOs are not an exception. Altered response to one of the most active EO components, thymol, has been observed in *E. coli* [41]. The results of this study showed that random transposon-inserted mutants in strains with relatively higher sensitivity were found in *rfaQ* or *qseC* genes, involved in lipopolysaccharide biosynthesis and quorum-sensing, respectively. Mutant strains of *E. coli* that showed higher levels of resistance had mutations in genes whose products are involved in the degradation of short-lived regulatory and abnormal proteins (encoded by the *lon* gene), menaquinone biosynthesis (*menA*), and the efflux pump of cadaverine and lysine (*cadB*), as well as in an intergene region (between the two small genes *yiiE* and *yiiF*) encoding a small hypothetical protein and a gene encoding a putative membrane protein of unknown function (*yagF*). These findings indicate multitarget modes of activity and multitolerance mechanisms in EOs.

Resistance to antibiotics, and also to EOs, can be more successfully overcome by their combined application. Combined antibiotic and EO treatments may increase the activity of both classes of antimicrobials and, therefore, require more attention as a promising therapeutic approach for coping with MDR bacteria.

COMBINING ESSENTIAL OILS WITH ANTIBIOTICS AND OTHER ANTIMICROBIAL AGENTS

Combining Essential Oils and Antibiotics

Combinations of antimicrobial agents provide many benefits, including increased activity and reduced toxic effects of the combined components [42]. Many studies have been devoted to combining antibiotics both within and between groups [43,44], but studies on combinations of antibiotics and other classes of antimicrobial agents, particularly with EOs, are rather scarce. Table 10-1 summarizes the studies on EO-antibiotic combinations performed using the microdilution or checkerboard methods.

The first report of the combined activity of *Origanum vulgare* EO with antibiotics against a MDR strain of extended-spectrum β-lactamase-producing *E. coli* was published by Si et al. [45]. Most combinations were synergistic [with fluoroquinolones, doxycycline, lincomycin, and mequindox florfenicol, where fractional inhibitory concentration (FIC) indexes ranged from 0.375 to 0.5], additive (for amoxicillin, lincomycin, and polymyxin; FIC indexes of 0.625−0.75), or indifferent (for kanamycin, with an FIC index 1.5); there were no antagonistic effects.

Lorenzi et al. [46] revealed the ability of *Helichrysum italicum* EO to significantly reduce multidrug resistance in *A. baumannii*, *Enterobacter aerogenes*, *E. coli*, and *P. aeruginosa*. Moreover, geraniol, which is a major component of *H. italicum* EO, significantly enhanced the activity of several antibiotic groups, including β-lactams (ampicillin and penicillin), quinolones (norfloxacin), and chloramphenicol. The targets of geraniol and phenylalanine-arginine β-naphthylamide (PaβN) were thought to be different because the activity of geraniol was stronger than that of PAβN in an *acrAB* mutant strain of *E. aerogenes*.

Efflux pump inhibitor activity was also previously found in some other plant-derived compounds [56]. Inhibitor activity against the NorA efflux pump of *S. aureus* was described for 5′-methoxyhydnocarpin isolated from *Berberis fremontii* [57]. Further, an extract of *Afzelia africana* demonstrated the properties of a broad spectrum efflux pump inhibitor [58]. Likewise, the efflux inhibitor ferruginol, which inhibits the activity of three resistance pumps in *S. aureus*, quinolone (NorA), tetracycline (TetK), and erythromycin (MsrA), was isolated from the cones of *Chamaecyparis lawsoniana* [58].

TABLE 10-1 Studies on Combinations of Essential Oils and Antibiotics

Essential oil or source of component	Plant family	Antibiotics	Bacteria tested	Effect	Reference
Origanum vulgare	Lamiaceae	Amoxicillin, ceftiofur, ceftriaxone, doxycycline, florfenicol, kanamycin, levofloxacin, lincomycin, maquindox, polymyxin, sarafloxacin	ESBL-producing *Escherichia coli*	Synergy: doxycycline, florfenicol, levofloxacin, maquindox, sarafloxacin; additive effect: amoxicillin, ceftriaxone, ceftiofur, lincomycin, polymyxin; indifference: kanamycin	[45]
Helichrysum italicum, geraniol	Asteraceae	Ampicillin, norfloxacin, penicillin	*Enterobacter aerogenes*	Enhancement of activity of all antibiotics	[46]
Mentha piperita	Lamiaceae	Ciprofloxacin	*Klebsiella pneumoniae*, *Staphylococcus aureus*	Against *S. aureus*: synergy at five ratios and antagonism at another; against *K. pneumoniae*: synergy at four ratios, antagonism at another	[47]
Rosmarinus officinalis	Lamiaceae	Ciprofloxacin	*K. pneumoniae*, *S. aureus*	Against *S. aureus*: antagonism at all ratios; against *K. pneumoniae*: synergy at seven ratios, antagonism at another	[47]
Thymus vulgaris	Lamiaceae	Ciprofloxacin	*K. pneumoniae*, *S. aureus*	Against *S. aureus*: synergy at two ratios, antagonism at another; against *K. pneumoniae*: synergy at four ratios, antagonism at another	[47]
Melaleuca alternifolia	Myrtaceae	Ciprofloxacin	*K. pneumoniae*, *S. aureus*	Against *K. pneumoniae*: synergy at three ratios, antagonism at another; against *S. aureus*: antagonism at all ratios;	[47]

(Continued)

TABLE 10-1 Studies on Combinations of Essential Oils and Antibiotics (cont'd)

Essential oil or source of component	Plant family	Antibiotics	Bacteria tested	Effect	Reference
Croton zehntneri	Euphorbiaceae	Gentamicin	Pseudomonas aeruginosa, S. aureus	Enhancement of antibiotic activity after contact with gaseous components of essential oil	[48]
Croton zehntneri	Euphorbiaceae	Norfloxacin	S. aureus	Enhancement of antibiotic activity	[49]
Melaleuca alternifolia	Myrtaceae	Tobramycin	E. coli, S. aureus	Synergy against both strains tested	[50]
Aniba rosaeodora	Lauraceae	Gentamicin	Acinetobacter baumannii, Bacillus cereus, B. subtilis, Enterococcus faecalis, E. coli, K. pneumonia, P. aeruginosa, Salmonella enterica serovar Typhimurium, Serratia marcescens, S. aureus, Yersinia enterocolitica	Synergy	[51]
Melaleuca alternifolia	Myrtaceae	Gentamicin	A. baumannii, B. cereus, B. subtilis, E. faecalis, E. coli, K. pneumonia, P. aeruginosa, Salmonella ser. Typhimurium, S. marcescens, S. aureus, Y. enterocolitica	Mostly indifferent effect	[51]
Origanum vulgare	Lamiaceae	Gentamicin	A. baumannii, B. cereus, B. subtilis, E. faecalis, E. coli, K. pneumonia, P. aeruginosa, Salmonella ser. Typhimurium, S. marcescens, S. aureus, Y. enterocolitica	Synergy against B. cereus, B. subtilis, and S. aureus.; Indifference against other bacteria	[51]
Pelargonium graveolens	Geraniaceae	Gentamicin	A. baumannii, B. cereus, B. subtilis, E. faecalis, E. coli, K. pneumonia, P. aeruginosa, Salmonella ser. Typhimurium, S. marcescens, S. aureus, Y. enterocolitica	Synergy	[51]

Zataria multiflora	Lamiaceae	Vancomycin	Clinical isolates of MRSA, MSSA	Synergy with vancomycin.	[52]
Thymus maroccanus, Thymus broussonetii	Lamiaceae	Chloramphenicol	E. aerogenes, E. coli, K. pneumoniae, P. aeruginosa, Salmonella ser. Typhimurium	Enhancement of chloramphenicol activity against resistant isolates.	[53]
Thymus maroccanus, Thymus broussonetii, carvacrol	Lamiaceae	Cefixime, ciprofloxacin, gentamicin, pristinamycin	Gram-positive (Bacillus cereus, B. subtilis, Micrococcus luteus, S. aureus) and Gram-negative (Enterobacter cloacae, E. coli, Salmonella spp., K. pneumoniae, P. aeruginosa, Vibrio cholerae) antibiotic-resistant nosocomial strains	Most combinations were synergistic; several were indifferent	[54]
Citrus limon	Rutaceae	Amikacin, gentamicin, imipenem, meropenem	Acinetobacter spp.	Synergy with amikacin, imipenem and meropenem; indifference with gentamicin	[55]
Cinnamomum zeylanicum	Lauraceae	Amikacin, gentamicin	Acinetobacter spp.	Synergy with amikacin; indifference with gentamicin	[55]

ESBL, extended-spectrum β-lactamase; MRSA, methicillin-resistant *Staphylococcus aureus*; MSSA, methicillin-susceptible *Staphylococcus aureus*.

The activity of four EOs (*M. alternifolia, Mentha piperita, Rosmarinus officinalis,* and *Thymus vulgaris*) in combination with ciprofloxacin at nine different ratios (9:1, 8:2, 7:3, 6:4, 5:5, 4:6, 3:7, 2:8, and 1:9) was studied by Van Vuuren et al. [47] against *S. aureus* and *K. pneumoniae* using the checkerboard method. The combination of *R. officinalis* or *M. alternifolia* with ciprofloxacin had an antagonistic effect against *S. aureus*, while combinations between the other oils and ciprofloxacin against both tested strains were synergistic for some ratios and antagonistic for others. This study demonstrated concentration-dependent interactions between EOs and antibiotics: the same EO and antibiotic combination showed effects ranging from synergistic to indifferent to antagonistic, depending on the ratios. Because combinations between ciprofloxacin and EOs were antagonistic at some ratios, antibiotic-EO combinations should be used with caution and only after obtaining experimental proof of at least the absence of an antagonistic effect. The authors concluded that combinations of *M. alternifolia, M. piperita, R. officinalis,* and *T. vulgaris* with ciprofloxacin should be avoided because of the antagonism observed at some ratios. These results indicate the need for systematic experimental studies on the interaction between EOs and antibiotics in order to reveal combinations that are appropriate for clinical application.

Rodrigues et al. [48] demonstrated enhancement of gentamicin activity against *S. aureus* and *P. aeruginosa* by a volatile component of *Croton zehntneri* EO using the minimal inhibitory dose method and gaseous contact with the bacteria *in vitro*; the activity of gentamicin increased by 42.8% against *P. aeruginosa*. Using the same method, Coutinho et al. [49] assessed the activity of *C. zehntneri* EO in combination with norfloxacin against fluoroquinolone-resistant *S. aureus* overexpressing the *norA* gene encoding the NorA efflux protein: after contact with the volatile component of the oil, activity of norfloxacin increased by 39.5%. The

findings of both studies are especially important for inhalation administration of *C. zehntneri* EO to treat respiratory infections caused by *P. aeruginosa* and *S. aureus*.

D'Arrigo et al. [50] determined the minimal inhibitory concentrations (MICs), time-kill curves of bacterial killing, and postantibiotic effect (PAE) of combinations between tobramycin and the *M. alternifolia* EO against two bacterial strains—*S. aureus* and *E. coli*. When determining the MICs, the effects of this combination ranged from synergistic to indifferent: FIC indexes were 0.37 and 0.62 against *E. coli* and *S. aureus*, respectively. In time-kill curves, the effect was synergistic against both strains, with *E. coli* exhibiting greater susceptibility. The PAE was extended against both tested strains following treatment with the combination compared with single components. The authors attributed the observed synergy to possible conformational changes significantly affecting reactions with cell membranes and enhancing antibiotic uptake by the bacterial cells.

Combinations of gentamicin and four EOs (*Aniba rosaeodora, M. alternifolia, O. vulgare,* and *Pelargonium graveolens*) against 15 reference bacterial strains were studied by Rostato et al. [51] using the checkerboard method. The effect of these tested combinations ranged from synergistic to indifferent, with the best results being obtained for *A. rosaeodora* and *P. graveolens* EOs. Especially pronounced synergy was found for the combination of these EOs with gentamicin: the FIC for both combinations was 0.11, indicating very strong synergy. These observed synergistic effects may be explained by the high content of terpene alcohols in *P. graveolens* (63%) and *A. rosaeodora* (73%) EOs, which may enhance the inhibition of protein synthesis by gentamicin.

The activity of *Zataria multiflora* EO, both alone and in combination with vancomycin, against clinical isolates of MRSA and methicillin-susceptible *S. aureus* (MSSA) was studied by Mahboubi and Bidqoli [52]. This EO demonstrated high activity against both MSSA and MRSA, with an MIC range of 0.25–1 μL/mL and minimal bactericidal concentrations (MBCs) of 0.5–2 μL/mL. Furthermore, *Z. multiflora* EO enhanced the activity of vancomycin. This antibiotic has been commonly used for the treatment of MRSA infections since the early 1980s; however, the appearance of vancomycin-intermediate *S. aureus* (VISA) and vancomycin-resistant *S. aureus* (VRSA) has led to an urgent need for alternatives to vancomycin [59]. One way of extending the usage of vancomycin against MRSA would be to combine EOs and vancomycin.

Fadli et al. [53] searched for efflux pump inhibitors to enhance antibiotic activities against resistant strains among EOs from Moroccan plants. Several EOs exhibited activity against isolates of *E. coli*, *E. aerogenes*, *K. pneumoniae*, *P. aeruginosa*, and *Salmonella enterica* serotype Typhimurium; this activity increased in the presence of the efflux pump inhibitor, PAβN. EOs of *Thymus broussonetii* and *T. maroccanus* could enhance the susceptibility of resistant isolates to chloramphenicol. The authors concluded that EOs may contain efflux pump inhibitor activity. In another study, Fadli et al. [54] examined the effects of combinations of these EOs, and also their major component carvacrol, with different antibiotics (cefixime, ciprofloxacin, gentamicin, and pristinamycin) against antibiotic-resistant nosocomial strains. EO-antibiotic combinations were tested by the checkerboard method, while the carvacrol-ciprofloxacin combinations were studied by determining the MIC of ciprofloxacin in the presence of 1/4 MIC of carvacrol. A total of 80 combinations were tested; among them, 71% showed total synergism, 20% had a partial synergistic interaction, and only 9% showed no effect. The combination of carvacrol and ciprofloxacin also showed a promising synergistic effect. Furthermore, the activity against Gram-positive

bacteria was more pronounced than against Gram-negatives.

Guerra et al. [55] studied two EOs, *Citrus limon* and *Cinnamomum zeylanicum*, in combination with amikacin, gentamicin, imipenem, and meropenem. The authors used the microdilution method for determining antibiotic MICs in medium containing 1/8 MIC of the EO. Synergy was found for combinations of the *C. limon* EO with amikacin, imipenem, and meropenem, and for combinations of the *C. zeylanicum* EO with amikacin. These findings are important because *Acinetobacter* spp. generally show high levels of antibiotic resistance; however, there is a general lack of studies into the activity of EOs in combination with antibiotics against this microorganism.

In reviewing the published studies, it is worth emphasizing that in spite of the strong antimicrobial properties described for many EOs [21,23,24,53] and many EO-antibiotic combinations [51,52,54], the possibility of antagonistic interactions should not be ignored [47,60]. Therefore, precautions should be used in the choice of EO-antibiotics combinations: EO-antibiotic combinations should only be used for the treatment of MDR infections if they have demonstrated at least minimal synergistic interactions *in vitro*.

Directed Delivery of Essential Oils by Incorporation into Polymeric Nanoparticles

Direct transportation of EOs to the site of inflammation may significantly increase their activities and reduce adverse effects. This can be achieved by incorporating EOs into polymeric nanoparticles. Apart from increasing local EO concentrations through directed transport, the incorporation of EOs into nanoparticles can enhance their activity several fold due to the increased area available for microbial-EO contact. Incorporation of antimicrobial agents into polymeric nanoparticles has been well studied for antibiotics [61]; in contrast, studies devoted to the incorporation of EOs into nanoparticles are limited.

One study into the *in vitro* activity of an EO-polymeric nanoparticle complex was performed by Chen et al. [62], who prepared chitosan nanoparticles grafted with either of two EO components, eugenol (a phenylpropene component of EOs of basil, cinnamon, clove, and other plants) and carvacrol (a monoterpenoid phenol component of EOs of oregano, thyme, and other plants). The antibacterial properties of chitosan nanoparticles grafted with EO components were evaluated against *E. coli* and *S. aureus*, and antioxidant activity was assayed with diphenylpicrylhydrazyl (DPPH). The results of the study demonstrated that grafted eugenol and carvacrol conferred antioxidant activity to the chitosan nanoparticles. Furthermore, the antibacterial activity of these grafted nanoparticles was better than or equal to the activity of the unmodified chitosan nanoparticles; however, the cytotoxicity of both types of nanoparticles in the 3T3 mouse fibroblast model was significantly lower than that of the pure EOs [62].

Hu et al. [63] produced chitosan nanoparticles grafted with thymol of five different sizes (21, 35, 134, 167, and 189 nm) and studied their activities against the Gram-positive bacteria *S. aureus* and *Bacillus subtilis*. Thymol-loaded water-soluble chitosan nanoparticles were found to be more active than thymol itself. Moreover, a decrease in particle size lead to enhancement of the antimicrobial activity.

Iannitelli et al. [64] studied alterations to the properties of preformed bacterial staphylococcal biofilms by poly(lactide-co-glycolide), or PLGA, nanoparticles with encapsulated carvacrol. The hydrodynamic diameter of these nanoparticles was approximately 210 nm, which allowed the diffusion of nanoparticles through mucus layers present on the surfaces of anatomical sites. Rheologic tests were used to study the alteration of bacterial biofilms and revealed considerable reduction in both the elasticity and mechanical stability of the biofilms, which in turn may facilitate the penetration of antimicrobial agents deep into bacterial biofilms.

The volatile characteristics of EOs are useful in the treatment of respiratory infections but create problems when treating other types of diseases due to decreased EO activity after evaporation. This problem can also be solved by incorporating EOs into nanoparticles. Lai et al. [65] demonstrated a reduction in the evaporation rate of *Artemisia arborescens* EO in *in vitro* release experiments following incorporation into solid lipid nanoparticles. Zhao et al. [66] showed increased stability of the active components in EO extracted from the dry rhizome of Zedoary (*Curcuma zedoaria*) when introduced into a self-nano-emulsifying drug delivery system containing ethyl oleate, Tween 80, and transcutol P. The size of the nanoparticles was approximately 70 nm and the stability of the EO components was retained during storage for 12 months at 25 °C. Yang et al. [67] studied the insecticidal activity of polyethylene glycol-coated nanoparticles loaded with garlic EO against adult *Tribolium castaneum*. Increased activity, compared with garlic EO not incorporated in nanoparticles, was observed over 5 months, indicating high stability associated with a slow, persistent release of active components from the nanoparticles.

Encapsulating EOs into polymeric nanoparticles and microparticles has also shown potential as a method for increasing their stability. Weisheimer et al. [68] compared the effectiveness of several different materials and techniques in forming lemongrass EO- microparticles. The best results were obtained using a combination of β-cyclodextrin encapsulation and the precipitation technique.

Several studies have also compared the release rate of different EO components from nanoparticles. For example, Sansukcharearnpon et al. [69] showed that limonene had the fastest release, with essentially no retention, from nanoparticles composed of a polymer blend of ethylcellulose, hydroxypropyl methylcellulose, and polyvinyl alcohol, while eucalyptol and menthol showed the slowest release.

These studies show that polymeric nanoparticles are a promising method for increasing the antibacterial activity of antibiotics, and also of EOs and their components. Furthermore, the ability of polymeric nanoparticles to penetrate into the deep layers of bacterial biofilms has great potential for combating biofilm-associated infections, which nowadays represent a serious clinical concern.

Combining Essential Oils and Bacteriophages

Other alternatives to the use of antibiotics in antibacterial treatments are bacteriophages. Their great potential in coping with bacteria, including MDR forms, has been demonstrated in a number of different studies [70–72]. Furthermore, combinations of bacteriophages with antibiotics [73] and with metallic nanoparticles [74] have shown beneficial effects. However, combinations between bacteriophages and EOs have not been sufficiently evaluated.

One study on bacteriophage-EOs interactions was performed by Viazis et al. [75], who studied the survival of enterohemorrhagic *E. coli* 0157:H7 after exposure to the BEC8 bacteriophage cocktail, either alone or in combination with the EO component *trans*-cinnamaldehyde, in a food model of whole baby Romaine lettuce and baby spinach leaves. Activity was assessed depending on the concentration of bacteria and the incubation temperature. At high bacterial concentrations [10^5 colony-forming units (CFU)/mL and 10^6 CFU/mL] there were no survivors after 10 min at all temperature regimens when BEC8 and *trans*-cinnamaldehyde were applied together. In contrast, when the bacteriophages and *trans*-cinnamaldehyde

were applied separately, the increase of bacterial concentration from 10^4 CFU/mL to 10^5–10^6 CFU/mL and/or a decrease in incubation temperature lead to a decrease in the activity of both agents.

The results of this study demonstrated the potentially high beneficial effect of combined bacteriophage-EO therapy [75]. Therefore, bacteriophage-EO combinations merit further investigation using different EOs and bacterial models in order to identify the most beneficial combinations and understand the mechanisms of such interactions.

The antibacterial activity of bacteriophages is attributed to two main mechanisms: 'lysis from within' and 'lysis from without' [76]. Bacteriophages can destroy bacterial cells by the simultaneous adherence of a sufficiently high number of viral particles to the bacterial cell, alterations to the membrane potential, and activation of cell wall-degrading enzymes with subsequent lysis (i.e., 'lysis from without'). Another way is bacterial cell lysis after penetration by bacteriophages, multiplication, and the release of high amounts of new phage particles created within the bacterium (i.e., 'lysis from within'). EOs may be able to contribute to both processes. For example, they may enhance damage to the bacterial cell membranes by bacteriophages or they facilitate phage entry into bacteria owing to independent alterations to the cell membranes (Fig. 10-2).

FUTURE PERSPECTIVES

The activities of EO-antibiotic combinations have mainly been investigated against reference bacterial strains or clinical isolates, regardless of their antibiotic resistance. Further studies are necessary to specifically identify synergistic combinations between EOs and antibiotics against MDR bacteria, such as MRSA, VRE, and antibiotic-resistant isolates of *E. coli*, *K. pneumoniae*, and *P. aeruginosa*, and

FIGURE 10-1 Proposed beneficial effects of essential oil-antibiotic interactions on efflux pumps. (A) The antibiotic is made ineffective by its removal from the bacterial cell by efflux pumps. (B) Certain essential oils (EOs) may inhibit efflux pumps which restores the antibacterial activity of antibiotics within the bacterial cell. *Based on the data of Lorenzi et al. [46] and Fadli et al. [57].*

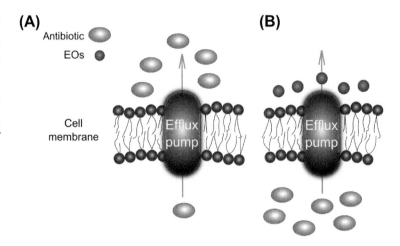

FIGURE 10-2 Proposed beneficial effects of essential oil-bacteriophage interactions on bacterial lysis. (A) The simultaneous action of essential oils (EOs) and bacteriophages (BFs) on the cell membrane leads to its destruction and bacterial lysis ('lysis from without'). (B) Damage to the bacterial cell membrane by EOs facilitates bacteriophage penetration into the bacterial cell, with subsequent replication and bacterial lysis ('lysis from within'). *Based on the data of Andreoletti et al. [76].*

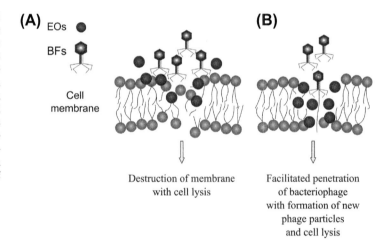

especially against recently emerged MDR pathogens, such as VISA and VRSA. A better understanding of the mechanism of beneficial effects between EOs and antibiotics against MDR bacteria is desirable. A challenge is the production of clinically applicable pharmaceutical forms of such combined drugs with sufficient potency for use in both local and systemic infections.

Published data have demonstrated that bacteriophage-EO combinations can enhance the antibacterial properties of both components; however, there is a significant lack of studies on this topic. A comprehensive evaluation of bacteriophage-EO interactions against MDR bacteria, including an assessment of their concentration- and time-dependent interactions, should be planned for the near future.

CONCLUSIONS

The high prevalence of MDR bacteria in the etiology of different infections has led to a reassessment of the methods used to address antibiotic resistance and a renewed interest in plant-derived antimicrobial products that can be used as alternatives to antibiotics. Many plant EOs have demonstrated high antibacterial activity against both antibiotic-susceptible and resistant isolates. The use of EOs to inhibit bacterial multiplication is not a new topic, but the ineffectiveness of traditional antibiotics has brought new interest to this area. Many of the advantages of EOs, both in their complex mechanism of action and in their multiple healing properties, indicate them to be promising agents for treating MDR bacteria, either alone or as an adjuvant support for antibiotic treatment.

Combining EOs with antibiotics may lead to an increase in the antibacterial activities of both EOs and antibiotics, and may also reduce the toxic effects of both agents against mammalian cells. Many EOs have demonstrated an *in vitro* ability to act synergistically with different antibiotics. EOs may inhibit antibiotic efflux pumps, and thus restore the activity of antibiotics that currently have reduced clinical applications owing to the development of resistance. Likewise, EOs may prolong the clinical use of currently effective drugs. However, it is also worth mentioning that in spite of the high activity of both EOs and antibiotics alone, some EO-antibiotic combinations have shown antagonistic effects *in vitro*; therefore, combined treatment with antibiotics and EOs should be chosen only if the combination has demonstrated a beneficial effect in experimental studies.

EOs may potentiate the effectiveness of antibiotics against MDR bacteria and exhibit synergism with bacteriophages; furthermore, studies have shown that polymeric nanoparticles can be used for the delivery of EOs to the site of infection and, therefore, enhance their activity.

Combining EOs with nonantibiotic agents thus has great potential for fighting MDR microorganisms and requires substantial further investigation.

References

[1] Livermore DM. Has the era of untreatable infections arrived? J Antimicrob Chemother 2009;64(Suppl. 1). i29—36.
[2] Kalan L, Wright GD. Antibiotic adjuvants: multicomponent anti-infective strategies. Expert Rev Mol Med 2011;13:e5.
[3] Hansra NK, Shinkai K. Cutaneous community-acquired and hospital-acquired methicillin-resistant. Staphylococcus aureus Dermatol Ther 2011; 24:263—72.
[4] Barnes BE, Sampson DA. A literature review on community-acquired methicillin-resistant Staphylococcus aureus in the United States: clinical information for primary care nurse practitioners. J Am Acad Nurse Pract 2011;23:23—32.
[5] Antimicrobial resistance surveillance in Europe. Annual report of the European Antimicrobial Resistance Surveillance Network (EARS-Net). Available from: http://www.ecdc.europa.eu/en/publications/Publications/Forms/ECDC_DispForm.aspx?ID=774. [accessed 30.12.11].
[6] Johnson AP. Methicillin-resistant Staphylococcus aureus: the European landscape. J Antimicrob Chemother 2011;66(Suppl. 4). iv43—8.
[7] Mainous 3rd AG, Diaz VA, Matheson EM, Gregorie SH, Hueston WJ. Trends in hospitalizations with antibiotic—resistant infections: U.S., 1997—2006. Public Health Rep 2011;126:354—60.
[8] Köck R, Becker K, Cookson B, van Gemert-Pijnen JE, Harbarth S, Kluytmans J, et al. Methicillin-resistant Staphylococcus aureus (MRSA): burden of disease and control challenges in Europe. Euro Surveill 2010; 15:19688.
[9] Kumar A, Ellis P, Arabi Y, Roberts D, Light B, Parrillo JE, et al, Cooperative antimicrobial therapy of septic shock database research group. Cooperative antimicrobial therapy of septic shock database research group. Initiation of inappropriate antimicrobial therapy results in a fivefold reduction of survival in human septic shock. Chest 2009; 136:1237—48.
[10] Santimaleeworagun W, Wongpoowarak P, Chayakul P, Pattharachayakul S, Tansakul P, Garey KW. Clinical outcomes of patients infected with carbapenem-resistant Acinetobacter baumannii treated with single

or combination antibiotic therapy. J Med Assoc Thai 2011;94:863—70.

[11] Enoch DA, Phillimore N, Mlangeni DA, Salihu HM, Sismey A, Aliyu SH, et al. Outcome for Gram-negative bacteraemia when following restrictive empirical antibiotic guidelines. QJM 2011;104:411—9.

[12] Tenover FC. Mechanisms of antimicrobial resistance in bacteria. Am J Med 2006;119(Suppl. 1—6): S3—10; discussion S62—70.

[13] Janssen AM, Scheffer JJ, Baerheim Svendsen A. Antimicrobial activity of essential oils: a 1976—1986 literature review. Aspects of the test methods. Planta Med 1987;53:395—8.

[14] Kalemba D, Kunicka A. Antibacterial and antifungal properties of essential oils. Curr Med Chem 2003;10:813—29.

[15] Burt S. Essential oils: their antibacterial properties and potential applications in foods—a review. Int J Food Microbiol 2004;94:223—53.

[16] Bakkali F, Averbeck S, Averbeck D, Idaomar M. Biological effects of essential oils—a review. Food Chem Toxicol 2008;46:446—75.

[17] Solórzano-Santos F, Miranda-Novales MG. Essential oils from aromatic herbs as antimicrobial agents. Curr Opin Biotechnol 2012;23:136—41.

[18] Tyagi AK, Malik A. Liquid and vapour-phase antifungal activities of selected essential oils against Candida albicans: microscopic observations and chemical characterization of Cymbopogon citrates. BMC Complement Altern Med 2010;10:65.

[19] Monzote L, García M, Montalvo AM, Scull R, Miranda M, Abreu J. *In vitro* activity of an essential oil against Leishmania donovani. Phytother Res 2007; 21:1055—8.

[20] Garozzo A, Timpanaro R, Stivala A, Bisignano G, Castro A. Activity of Melaleuca alternifolia (tea tree) oil on Influenza virus A/PR/8: study on the mechanism of action. Antiviral Res 2011;89:83—8.

[21] Miguel MG. Antioxidant and anti-inflammatory activities of essential oils: a short review. Molecules 2010;15:9252—87.

[22] Sadlon AE, Lamson DW. Immune-modifying and antimicrobial effects of Eucalyptus oil and simple inhalation devices. Altern Med Rev 2010;15: 33—47.

[23] Serrano C, Matos O, Teixeira B, Ramos C, Neng N, Nogueira J, et al. Antioxidant and antimicrobial activity of Satureja montana L. extracts. J Sci Food Agric 2011;91:1554—60.

[24] Woollard AC, Tatham KC, Barker S. The influence of essential oils on the process of wound healing: a review of the current evidence. J Wound Care 2007;16:255—7.

[25] Ríos JL, Recio MC. Medicinal plants and antimicrobial activity. J Ethnopharmacology 2005;100:80—4.

[26] Orafidiya LO, Agbani EO, Abereoje OA, Awe T, Abudu A, Fakoya FA. An investigation into the wound-healing properties of essential oil of Ocimum gratissimum linn. J Wound Care 2003;12:331—4.

[27] Tumen I, Süntar I, Keleş H, Küpeli Akkol EA. Therapeutic approach for wound healing by using essential oils of Cupressus and Juniperus species growing in Turkey. Evid. Based Complement. Alternat Med 2012;2012:728281.

[28] Süntar I, Akkol EK, Keleş H, Oktem A, Başer KH, Yeşilada E. A novel wound healing ointment: a formulation of Hypericum perforatum oil and sage and oregano essential oils based on traditional Turkish knowledge. J Ethnopharmacol 2011;134: 89—96.

[29] Enshaieh S, Jooya A, Siadat AH, Iraji F. The efficacy of 5% topical tea tree oil gel in mild to moderate acne vulgaris: a randomized, double-blind placebo-controlled study. Indian J Dermatol Venereol Leprol 2007;73:22—5.

[30] Schelz Z, Hohmann J, Molnar J. Recent advances in research of antimicrobial effects of essential oils and plant derived compounds on bacteria. In: Chattopadhyay D, editor. Ethnomedicine: A source of complementary therapeutics; Research Signpost: 2010. p. 179—201. Kerala, India.

[31] Bassolé IH, Lamien-Meda A, Bayala B, Tirogo S, Franz C, Novak J, et al. Composition and antimicrobial activities of Lippia multiflora Moldenke, Mentha x piperita L. and Ocimum basilicum L. essential oils and their major monoterpene alcohols alone and in combination. Molecules 2010;15:7825—39.

[32] Tamgue O, Louis B, Nguefack J, Dongmo JBL, Dakole CD. Synergism and antagonism of essential oil fractions of Cymbopogon citratus, Ocimum gratissimum and Thymus vulgaris against Penicillium expansum. Int J Plant Pathol 2010;2:51—62.

[33] Pei RS, Zhou F, Ji BP, Xu J. Evaluation of combined antibacterial effects of eugenol, cinnamaldehyde, thymol, and carvacrol against E. coli with an improved method. J Food Sci 2009;74:M379—83.

[34] Ravishankar S, Zhu L, Reyna-Granados J, Law B, Joens L, Friedman M. Carvacrol and cinnamaldehyde inactivate antibiotic-resistant Salmonella enterica in buffer and on celery and oysters. J Food Prot 2010;73:234—40.

[35] Brackman G, Celen S, Hillaert U, Van Calenbergh S, Cos P, Maes L, et al. Structure-activity relationship of cinnamaldehyde analogs as inhibitors of AI-2 based quorum sensing and their effect on virulence of Vibrio spp. PLoS One 2011;6:e16084.

[36] Gill AO, Holley RA. Mechanisms of bactericidal action of cinnamaldehyde against Listeria monocytogenes and of eugenol against L. monocytogenes and Lactobacillus sakei. Appl Environ Microbiol 2004;70:5750—5.

[37] Schelz Z, Molnar J, Hohmann J. Antimicrobial and antiplasmid activities of essential oils. Fitoterapia 2006;77:279—85.

[38] Khan MS, Zahin M, Hasan S, Husain FM, Ahmad I. Inhibition of quorum sensing regulated bacterial functions by plant essential oils with special reference to clove oil. Lett Appl Microbiol 2009; 49:354—60.

[39] Szabó MA, Varga GZ, Hohmann J, Schelz Z, Szegedi E, Amaral L, et al. Inhibition of quorum-sensing signals by essential oils. Phytother Res 2010;24:782—6.

[40] Jaramillo-Colorado B, Olivero-Verbel J, Stashenko EE, Wagner-Döbler I, Kunze B. Anti-quorum sensing activity of essential oils from Colombian plants. Nat Prod Res 2012;26:1075—86.

[41] Shapira R, Mimran E. Isolation and characterization of Escherichia coli mutants exhibiting altered response to thymol. Microb Drug Resist 2007;13: 157—65.

[42] Verma P. Methods for determining bactericidal activity and antimicrobial interactions: synergy testing, time-kill curves, and population analysis. In: Schwalbe R, Steele-Moore L, Goodwin AC, editors. Antimicrobial Susceptibility Testing Protocols. New York: CRC Press; 2007. p. 276—7.

[43] Sun C, Falagas ME, Wang R, Karageorgopoulos DE, Yu X, Liu Y, et al. In vitro activity of minocycline combined with fosfomycin against clinical isolates of methicillin-resistant Staphylococcus aureus. J Antibiot (Tokyo) 2011;64:559—62.

[44] Sheng WH, Wang JT, Li SY, Lin YC, Cheng A, Chen YC, et al. Comparative in vitro antimicrobial susceptibilities and synergistic activities of antimicrobial combinations against carbapenem-resistant Acinetobacter species: Acinetobacter baumannii versus Acinetobacter genospecies 3 and 13TU. Diagn Microbiol Infect Dis 2011;70:380—6.

[45] Si H, Hu J, Liu Z, Zeng ZL. Antibacterial effect of oregano essential oil alone and in combination with antibiotics against extended-spectrum beta-lactamase-producing Escherichia coli. FEMS Immunol Med Microbiol 2008;53:190—4.

[46] Lorenzi V, Muselli A, Bernardini AF, Berti L, Pagès JM, Amaral L, et al. Geraniol restores antibiotic activities against multidrug-resistant isolates from Gram-negative species. Antimicrob Agents Chemother 2009;53:2209—11.

[47] Van Vuuren SF, Suliman S, Viljoen AM. The antimicrobial activity of four commercial essential oils in combination with conventional antimicrobials. Lett Appl Microbiol 2009;48:440—6.

[48] Rodrigues FF, Costa JG, Coutinho HD. Synergy effects of the antibiotics gentamicin and the essential oil of Croton zehntneri. Phytomedicine 2009; 16:1052—5.

[49] Coutinho H, Matias E, Santos K, Tintino SR, Souza C, Guedes G, et al. Enhancement of the Norfloxacin Antibiotic Activity by Gaseous Contact with the Essential Oil of Croton zehntneri. J Young Pharm 2010;2:362—4.

[50] D'Arrigo M, Ginestra G, Mandalari G, Furneri PM, Bisignano G. Synergism and postantibiotic effect of tobramycin and Melaleuca alternifolia (tea tree) oil against Staphylococcus aureus and Escherichia coli. Phytomedicine 2010;17:317—22.

[51] Rosato A, Piarulli M, Corbo F, Muraglia M, Carone A, Vitali ME, et al. In vitro synergistic antibacterial action of certain combinations of gentamicin and essential oils. Curr Med Chem 2010;17:3289—95.

[52] Mahboubi M, Bidgoli FG. Antistaphylococcal activity of Zataria multiflora essential oil and its synergy with vancomycin. Phytomedicine 2010;17:548—50.

[53] Fadli M, Chevalier J, Saad A, Mezrioui NE, Hassani L, Pages JM. Essential oils from Moroccan plants as potential chemosensitisers restoring antibiotic activity in resistant Gram-negative bacteria. Int J Antimicrob Agents 2011;38:325—30.

[54] Fadli M, Saad A, Sayadi S, Chevalier J, Mezrioui NE, Pagès JM, et al. Antibacterial activity of Thymus maroccanus and Thymus broussonetii essential oils against nosocomial infection—bacteria and their synergistic potential with antibiotics. Phytomedicine 2012;19(5):464—71.

[55] Guerra FQ, Mendes JM, Sousa JP, Morais-Braga MF, Santos BH, Melo Coutinho HD, et al. Increasing antibiotic activity against a multidrug-resistant Acinetobacter spp. by essential oils of Citrus limon and Cinnamomum zeylanicum. Nat Prod Res 2011;26(23): 2235—8.

[56] Tegos G, Stermitz FR, Lomovskaya O, Lewis K. Multidrug pump inhibitors uncover remarkable activity of plant antimicrobials. Antimicrob Agents Chemother 2002;46:3133—41.

[57] Stermitz FR, Lorenz P, Tawara JN, Zenewicz LA, Lewis K. Synergy in a medicinal plant: antimicrobial action of berberine potentiated by 5′-methoxyhydnocarpin, a multidrug pump inhibitor. Proc Natl Acad Sci USA 2000;97:1433—7.

[58] Smith EC, Williamson EM, Wareham N, Kaatz GW, Gibbons S. Antibacterials and modulators of bacterial

resistance from the immature cones of Chamaecyparis lawsoniana. Phytochemistry 2007;68:210−7.

[59] Senekal M. Current resistance issues in antimicrobial therapy. CME 2010;28:54−7.

[60] Van Vuuren S, Viljoen A. Plant-based antimicrobial studies—methods and approaches to study the interaction between natural products. Planta Med 2011;77:1168−82.

[61] Ungaro F, d'Angelo I, Coletta C, d'Emmanuele di Villa Bianca R, Sorrentino R, Perfetto B, et al. Dry powders based on PLGA nanoparticles for pulmonary delivery of antibiotics: Modulation of encapsulation efficiency, release rate and lung deposition pattern by hydrophilic polymers. J Control Release 2012;157:149−59.

[62] Chen F, Shi Z, Neoh KG, Kang ET. Antioxidant and antibacterial activities of eugenol and carvacrol-grafted chitosan nanoparticles. Biotechnol Bioeng 2009;104:30−9.

[63] Hu Y, Du Y, Wang X, Feng T. Self-aggregation of water-soluble chitosan and solubilization of thymol as an antimicrobial agent. J Biomed Mater Res A 2009;90:874−81.

[64] Iannitelli A, Grande R, Di Stefano A, Di Giulio M, Sozio P, Bessa LJ, et al. Potential antibacterial activity of carvacrol-loaded poly(DL-lactide-co-glycolide) (PLGA) nanoparticles against microbial biofilm. Int J Mol Sci 2011;12:5039−51.

[65] Lai F, Wissing SA, Müller RH, Fadda AM. Artemisia arborescens L essential oil-loaded solid lipid nanoparticles for potential agricultural application: preparation and characterization. AAPS Pharm Sci Tech 2006;7(1):E2.

[66] Zhao Y, Wang C, Chow AH, Ren K, Gong T, Zhang Z, et al. Self-nanoemulsifying drug delivery system (SNEDDS) for oral delivery of Zedoary essential oil: formulation and bioavailability studies. Int J Pharm 2010;383:170−7.

[67] Yang FL, Li XG, Zhu F, Lei CL. Structural characterization of nanoparticles loaded with garlic essential oil and their insecticidal activity against Tribolium

castaneum (Herbst) (Coleoptera: Tenebrionidae). J Agric Food Chem 2009;57:10156−62.

[68] Weisheimer V, Miron D, Silva CB, Guterres SS, Schapoval EE. Microparticles containing lemongrass volatile oil: preparation, characterization and thermal stability. Pharmazie 2010;65:885−90.

[69] Sansukcharearnpon A, Wanichwecharungruang S, Leepipatpaiboon N, Kerdcharoen T, Arayachukeat S. High loading fragrance encapsulation based on a polymer-blend: preparation and release behavior. Int J Pharm 2010;391:267−73.

[70] Wróblewska M. Novel therapies of multidrug-resistant Pseudomonas aeruginosa and Acinetobacter spp. infections: the state of the art. Arch Immunol Ther Exp (Warsz) 2006;54:113−20.

[71] Kutateladze M, Adamia R. Bacteriophages as potential new therapeutics to replace or supplement antibiotics. Trends Biotechnol 2010;28:591−5.

[72] Karamoddini MK, Fazli-Bazzaz BS, Emamipour F, Ghannad MS, Jahanshahi AR, Saed N, et al. Antibacterial efficacy of lytic bacteriophages against antibiotic-resistant Klebsiella species. Sci W J 2011;11: 1332−40.

[73] Bedi MS, Verma V, Chhibber S. Amoxicillin and specific bacteriophage can be used together for eradication of biofilm of Klebsiella pneumoniae B5055. World J Microbiol Biotech 2009;25:1145−51.

[74] You J, Zhang Y, Hu Z. Bacteria and bacteriophage inactivation by silver and zinc oxide nanoparticles. Colloids Surf B Biointerfaces 2011;85:161−7.

[75] Viazis S, Akhtar M, Feirtag J, Diez-Gonzalez F. Reduction of Escherichia coli 0157:H7 viability on leafy green vegetables by treatment with a bacteriophage mixture and trans-cinnamaldehyde. Food Microbiol 2011;28:149−57.

[76] Andreoletti O, Budka H, Buncic S, Colin P, Collins JD, De Koeijer A. Scientific opinion of the panel on biological hazards on a request from European Commission on the use and mode of action of bacteriophages in food production. EFSA J 2009; 1076:1−26.

CHAPTER

11

Medicinal Plants as Alternative Sources of Therapeutics against Multidrug-Resistant Pathogenic Microorganisms Based on Their Antimicrobial Potential and Synergistic Properties

Kalpna D. Rakholiya, Mital J. Kaneria, Sumitra V. Chanda

Phytochemical, Pharmacological, and Microbiological Laboratory, Department of Biosciences, Saurashtra University, Rajkot-360 005, Gujarat, India

GENERAL INTRODUCTION

Infectious diseases are the leading cause of death worldwide; this has become a global concern. The wide use of antibiotics in the treatment of bacterial infections has led to the emergence and spread of resistant strains; even very low concentrations of antibiotics released into the environment can enrich the population of resistant strains [1]. There is an urgent and imperative need to develop novel therapeutics, new practices, and antimicrobial strategies for the treatment of infectious diseases caused by multidrug-resistant microorganisms. This has intensified the search for novel therapeutic leads against fungal, parasitic, bacterial, and viral infections. The discovery of new antibacterial compounds as suitable substitutes for conventional antibiotics might be a possible solution to this problem [2].

For many years, a variety of different chemical and synthetic compounds have been used as antimicrobial agents in food to reduce the incidence of food poisoning and spoiling, and to control the growth of pathogenic microorganisms. However, the widespread indiscriminate use of chemical preservatives has led to many ecologic and medical problems,

Fighting Multidrug Resistance with Herbal Extracts,
Essential Oils and Their Components
http://dx.doi.org/10.1016/B978-0-12-398539-2.00011-2

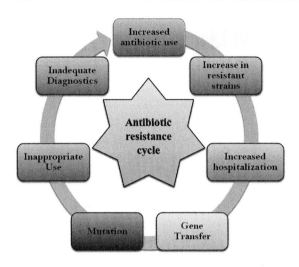

FIGURE 11-1   The antibiotic resistance cycle.

including hypersensitivity, allergic reactions, and immune suppression [3], which make it necessary to search for strategies that are accessible, simple to apply, and nontoxic [4]. There are two main modes of drug discovery: the first is through the use of chemical synthesis for pharmaceutical purposes and the second is the use natural products as a basis for drug discovery [5].

The development of bacterial resistance to many existing antibiotics has serious consequences, as shown in Figure 11-1. Antibiotics are designed to kill microorganisms, which then adapt to antibiotics, making them less effective and resulting in antibiotic resistance through a number of mechanisms (Fig. 11-1).

ANTIMICROBIAL AGENTS FROM PLANT SOURCES

Plant oils and extracts have been used for a wide variety of purposes for many years; recently, they have generated widespread interest as a source of natural antimicrobials [6]. Essential oils (EOs) and plant extracts are of particular interest because they are relatively safe, increase the shelf life of foods, are widely accepted by consumers, and have the potential to be exploited for multiple uses [7–10].

Plant Extracts

Traditionally, humans have used crude extracts of different parts of plants as curative agents. Plant extracts have also been used in the treatment of infectious diseases caused by antibiotic-resistant microbes. In fact, herbal medicines have received much attention as sources of lead compounds, since they are considered to have stood the test of time, be relatively safe for human use, and be environment friendly [11–13]. They are also economical, easily available, and affordable. Lastly, natural products are a treasure chest for new drug discovery because of their chemical diversity. They can be used as a source of pure compounds or as standardized plant extracts. All parts or any single part of the plant, such as bark, leaves, peel, seed, and stem, may possess antimicrobial properties [14–16]. Plants produce secondary metabolites that can inhibit bacteria, fungi, viruses, and pests. The reported antimicrobial activity of some plants is listed in Table 11-1.

There is a general consensus that secondary metabolites present in plant extracts can inhibit Gram-positive bacteria more than Gram-negative bacteria [4,37,63–65], i.e., Gram-positive bacteria are more susceptible to plant extracts. This difference is simply a consequence of the difference in cell wall structure between these major classes of bacteria. The cell wall of a Gram-negative bacterium is surrounded by an additional lipopolysaccharide membrane, which provides a hydrophilic surface and functions as a permeability barrier for many plant extracts [66]. However, this is not always true; some plant extracts inhibit Gram-negative bacteria more than Gram-positive bacteria [67–69].

TABLE 11-1 List of Medicinal Plants, Solvents Used for Extraction, and Assays Employed in Antimicrobial Studies

Plants/isolated compounds	Extract/antibiotic	Microorganism(s)	Methods	Reference(s)
ESSENTIAL OILS				
Lycopus lucidus Turcz. var.	EO	CA, EC, ECl, MRSA, PA, SA, SL, SSi	DD, MBC, MIC	[17]
Geranium robertianum, G. sanguineum L.	EO	AF, BS, CA, CS, EC, KP, MF, SA, SC, SEn, SLU	GC, GC-MS, MBC MIC	[18]
Citrus × *bergamia*	EO	MFe, MH, MP	GC, MIC	[18]
Achillea millefolium	EO	LI, LM	DD, GC-MS, MBC, MIC	[19]
Aegle marmelos L. Corrêa	EO	AA, AF, AFL, AND, ANi, ATE, CH, CLU, FO, HOR, TVI	GC-MS, LD$_{50}$, MIC, poison food technique	[20]
HUMAN PATHOGENS				
Hibiscus rosa-sinensis	AQ, crude protein, ET, hot AQ, ME	BS, EC, PA, SA, *Salmonella* spp., *Streptococcus* spp.	AW, DD	[21]
Eugenia jambolana	AC, AQ, ET	EC, KP, PA, SA	AW, MIC, MIC50, TK, TLCB	[22]
ORAL PATHOGENS				
Sorghum bicolor (L.) Moench	ET (70% in water)	SM, SSo	MBC, MIC	[23]
Rheum undulatum	Fraction, ME	AN, SM, SO	HPLC, MBC, MIC	[24]
Quercus infectoria G.Olivier	AC, ME	FN, PG, SM, SSal,	AW, MBC, MIC	[25]
Calotropis gigantiea (L.) R. Br. ex Schult.	AQ, CH, EA, HE, ME	AV, LA, LC, SM, SMi	AW, GC-MS, FTIR, MBC, MIC, NMR	[26]
ANIMAL PATHOGENS				
Propolis	ET	SA	AFM, MIC, TK	[27]
PLANT EXTRACTS				
Euphorbia hirta L., *Withania somnifera* L.	Alkaloid	AT, BS, EA, KP, RP	DD, MBC, MIC	[28]
Phyllanthus muellerianus (Kuntze) Exell	AQ, ME	CA, CS, EC, SA, SM, SP	MBC, MIC	[29]

(Continued)

TABLE 11-1 List of Medicinal Plants, Solvents Used for Extraction, and Assays Employed in Antimicrobial Studies (cont'd)

Plants/isolated compounds	Extract/antibiotic	Microorganism(s)	Methods	Reference(s)
Hippophae rhamnoides L.	AQ, EA, HE	BC, CA, EC, ED, PA, SA	MIC	[30]
Cedrus deodara	AQ	BC, BS, EC, PV, SA	DD, HPLC, MBC, MIC, NMR	[31]
Clausena excavata Burm.f.	Compound, DCM	AF, ANi, CC, CEr, CGl, CT, FNe, FO, LT, MC, RS, RSt, SSC, TCu	MIC, TLCB	[32]
Vaccinium angustifolium	Compound, ME	EC, LM, SA, ST, YE	DD, MBC, MIC	[33]
Vitis vinifera L.	AC (50% in water), AQ, ET (50% in water), ME	CA, EA, SA	MIC	[34]
Aralia nudicaulis	AQ	BT, EC, LI, PA, SA, SEn	MIC	[35]
	ME	MT	MIC	[36]
Psoralea corylifolia L.	AQ, DMF, ME	AFi, BM, EA, PMo, SEp	AW	[37]
Syzygium cumini (L.) Skeels	AC, AQ, EA, PE, TO	BM, BS, CA, CF, CG, CL, CN, CR, EA, KP, PM, SA, SEp, ST	AW	[38]
SYNERGISTIC				
Allyl isothiocyanate, carvacrol, cinnamaldehyde, eugenol, thymol	A, B, E, N, P, T	EC, SA, ST	CK, MIC	[39]
Syzygium aromaticum	A, EO, G	FN, LM, PG, PI, SAn, SCr, SG, SM, SR, SS	CK, MBC, MIC, TK	[40]
Artemisia afra Jacq. ex Willd., *Eucalyptus globulus*, *Osmitopsis asteriscoides*	EO	CN, EFa, KP, MOC	FIC, isobologram, MIC	[41]
Coriandrum sativum L.	C, CP, CZ, EO, G, PP, T	AB	CK, MIC	[42]
Ocimum sanctum	EO, Fe, K	CA, CG, CK, CT	DD, IC	[43]
Eucalyptus globulus, aromadendrene, 1,8-cineole, globulol	EO	BS, CA, CG, EC, EFa, KP, PA, SA, SAg, SEp, SP, SSP	CK, isobologram, MBC, MIC	[44]
Ocimum sanctum Linn.	C, ET, TR	SEn	DD	[45]

Plant source				
Thymus broussonetii, T. maroccanus	CE, CP, EO, G, PR	BC, BS, EC, ECl, KP, MLU, PA, SA, Salmonella sp., VC	CK, DD, MBC, MIC	[46]
Tea	Silver ions	CA, PA, SA	CK, MIC, MLC, TK	[47]
Torilis anthriscus (L.) C.C. Gmel.	AQ, C, EA, ET (80% in water), S	ACh, AF, AT, BM, BS, ECa, ECl, FO, KP, PF, PGL, PP, PVe, SA, TH, TRo	CK, DD, MIC	[48]
Ocimum gratissimum L.	A, CP, K, NS, S, SN	CA, EC, PA, PM, SA	AW, DD	[49]
Cymbopogon citrates, Cymbopogon giganteus	EO	EA, EF, LM, PA, SEn, ST	CK, DD, MIC	[50]
Cytisus capitatus Scop., *Cytisus nigricans* L.	AC, CP, EA, ET, G	BS, EC, EF, KP, PA, PM, SA	CK, MIC	[51]
Rhus coriaria, Rosa × damascena, Sarcopoterium spinosum	CP, FNF, P, SFM, T	PA	MIC	[52]
Vangueria spinosa (Roxb. ex Link) Roxb.	D, ET, OF	EC, KP, PA, SA	CK, MBC, MIC, TK	[53]
Berberine, ellagic acid, gallic acid, myricetin, protocatechuic acid, rutin	CEF, CP, PIP POL B, SMX, T, TMP	PA	CK, MIC, TK	[54]
α-amyrin, betulinic acid, betulinaldehyde (pentacyclic triterpenoids)	M, VA	SA	CK, MBC, MIC	[55]
Galla Rhois (methyl gallate)	EO, NA	EC, ECl, KO, SD, SENT, SMIN, ST	CK, MIC	[56]
Afzelia africana Smith.	A, AMO, C, CP, E, ME, P, T	BP, BS, KK, KP, MLU, PV, SA, SEp	AW, MIC, TK	[57]
Sasa veitchii	NaOH (3% in water)	MRSA, VRE	MBC, MIC, SEM	[58]
Garcinia kola	AC, AMO, CH, E, P, T	SA, SFA, EC, EFa, KP	CK, MIC, TK	[59]
Croton zehntneri Pax & K.Hoffm.	EO, G	PA, SA	MID	[60]
Momordica charantia L.	AMI, CL, ET, G, KAN, NEO, TOB	SA	MIC	[61]

(Continued)

TABLE 11-1 List of Medicinal Plants, Solvents Used for Extraction, and Assays Employed in Antimicrobial Studies (cont'd)

Plants/isolated compounds	Extract/antibiotic	Microorganism(s)	Methods	Reference(s)
Allium cepa, Allium sativum, Citrus aurantifolia, Coriandrum sativum, Piper nigrum, Zingiber officinale	AM, AQ, C, CC, CP, CTX, E, FOX, K, NA, NEO, NOR, P, S, T	EC	DD	[62]

Solvents: AC, acetone; AQ, aqueous; CH, chloroform; DCM, dichloromethane; DMF, dimethylformamide; EA, ethyl acetate; EO, essential oil; ET, ethanol; HE, hexane; ME, methanol; PE, petroleum ether; TO, toluene.

Assays: AW, agar well diffusion assay; CK, checker board; DD, disc diffusion assay; FIC, fractional inhibitory concentration; FTIR, Fourier transformer infra-red spectroscopy; GC, gas chromatography; HPLC, high performance liquid chromatography; LD50, lethal dosage 50%; MBC, minimum bactericidal concentration; MIC, minimum inhibitory concentration; MIC 50, minimum inhibitory concentration required to inhibit the growth of organisms by 50%; MID, minimal inhibitory dose; MLC, minimum lethal concentration; MS, mass spectroscopy; NMR, nuclear magnetic resonance; TK, time-kill assay; TLCB, thin layer chromatography-bio-autography; SEM, scanning electron microscopy.

Antibiotics: A, ampicillin; AMI, amikacin; AMO, amoxicillin; B, bacitracin; C, chloramphenicol; CC, clindamycin; CE, cefixime; CEF, ceftazidine; CL, chlorprom-azine; CP, ciprofloxacin; CTX, cefotaxime; CZ, cefoperazone; D, doxycycline; E, erythromycin; FOX, cefoxitin; G, gentamicin; K, kanamycin; M, methicillin; N, novobiocin; NA, nalidixic acid; NEO, neomycin; NOR, norfloxacin; NS, nystatin; OF, ofloxacin; P, penicillin G; PIP, piperacillin; POL B, polymyxin B; PP, piperacillin; PR, pristinamycin; S, streptomycin; SFM, sulfadimethoxine; SMX, sulfamethoxazole; SN, septrin; T, tetracycline; TMP, trimethoprim; TOB, tobramycin; TR, trimethoprim; VA, vancomycin.

Microorganisms: AA, Alternaria alternata; AB, Acinetobacter baumannii; ACh, Azotobacter chroococcum; AF, Aspergillus fumigatus; AFi, Alcaligenes faecalis; AFL, Aspergillus flavus; AN, Actinomyces naeslundii; AND, Aspergillus nidulans; ANi, Aspergillus niger; AT, Agrobacterium tumefaciens; ATE, Aspergillus terreus; AV, Actinomyces viscosus; BC, Bacillus cereus; BM, Bacillus mycoides; BP, Bacillus pumilus; BS, Bacillus subtilis; BT, Brochothrix thermosphacta; CA, Candida albicans; CC, Cryptococcus curvatus; CEr, Curvularia eragrostidis; CF, Citrobacter freundii; CG, Candida glabrata; CGI, Colletotrichum gloeosporioides; CH, Cladosporium herbarum; CK, Candida krusei; CL, Cryptococcus luteolus; CLU, Curvularia lunata; CN, Cryptococcus neoformans; CR, Corynebacterium rubrum; CS, Clostridium sporogenes; CT, Candida tropicalis; EA, Enterobacter aerogenes; EC, Escherichia coli; ECa, Erwinia carotovora; ECl, Enterobacter cloacae; ED, Enterococcus durans; EF, Epidermophyton floccosum; EFa, Enterococcus faecalis; FN, Fusarium nucleatum; FNe, Filobasidiella neoformans; FO, Fusarium oxysporum; HOR, Helmin-thosporium oryzae, KK, Kocuria kristinae; KO, Klebsiella oxytoca; KP, Klebsiella pneumoniae; LA, Lactobacillus acidophilus; LC, Lactobacillus casei; LI, Listeria innocua; LM, Listeria monocytogenes; LT, Lasiodiplodia theobromae; MC, Mucor circinelloides; MF, Micrococcus flavus; MFe, Myco-plasma fermentans; MH, Mycoplasma hominis; MLU, Micrococcus luteus; MP, Mycoplasma pneumoniae; MRSA, methicillin-resistant Staphylococcus aureus; MT, Mycobacterium tuberculosis; PA, Pseudomonas aeruginosa; PF, Pseudomonas fluorescens; PG, Porphyromonas gingivalis; PGI, Pseudomonas glycinea; PI, Prevotella intermedia; PM, Proteus mirabilis; PMo, Proteus morganii; PP, Pseudomonas phaseolicola; PV, Proteus vulgaris; PVe, Penicillium verrucosum; RP, Raoultella planticola; RS, Rhizoctonia solani; RSt, Rhizopus stolonifer; SA, Staphylococcus aureus; SAg, Streptococcus agalactiae; SAn, Streptococcus anginosus; SC, Saccha-romyces cerevisiae; SCr, Streptococcus criceti; SD, Salmonella derby; SEn, Salmonella enterica; SENT, Salmonella enteritidis; SEp, Staphylococcus epidermidis; SFA, Streptococcus faecalis; SG, Streptococcus gordonii; SL, Serratia liquefaciens; SLU, Sarcina lutea; SM, Streptococcus mutans; SMi, Streptococcus mitis; SMIN, Salmonella minnesota; SO, Streptococcus oralis; SP, Streptococcus pyogenes; SR, Streptococcus ratti; SS, Streptococcus sanguinis; SSal, Streptococcus salivarius; SSC, SSi, Staphylococcus simulans, Sclerotinia sclerotiorum; SSP, Staphylococcus saprophyticus; SSo, Streptococcus sobrinus; ST, Salmonella enterica serovar Typhimurium; TCu, Trichosporon cutaneum; TH, Trichoderma harzianum; TRo, Trichothecium roseum; TVI, Trichoderma viride; VC, Vibrio cholerae; VRE, vancomycin-resistant enterococci; YE, Yersinia enterocolitica.

Essential Oils and Volatile Oils

EOs are complex mixtures of volatile compounds that result from the secondary metabolic pathways of plants [70]. The antimicrobial properties of volatile aromatic oils and medium-chain fatty acids derived from edible plants are considered valuable therapeutic alternatives to treat various diseases caused by microorganisms. EOs are composed of secondary metabolites that are commonly concentrated in the bark, fruit, or leaves of aromatic plants. Major sources of EOs include plants of the carrot (Apiaceae), citrus (Rutaceae), mint (Lamiaceae), and myrtle (Myrtaceae) and families. EOs are also used as fragrances and flavoring agents in foods and beverages [71]. Among the great variety of EOs, citrus fruit EOs and their major components have gained acceptance in the food industry, since they are generally recognized as safe by the Food and Drug Administration and many foods tolerate their presence.

Palmeira-deOliveira et al. [72] reported that *Thymbra capitata* EO, which is rich in carvacrol (75%), showed a potent anti-candida effect [minimum inhibitory concentration (MIC) of 0.32 μL/mL). The EO was able to disrupt the biomass and inhibit the metabolic activity of preformed biofilms of distinct *Candida* spp. Nosocomial infections are hospital-acquired infections and hence easily transmitted, resulting in high morbidity and mortality rates. Fadli et al. investigated synergism between *Thymus maroccanus* and *T. broussonetii* EOs and conventional antibiotics against nosocomial bacteria [46]. Their results indicate that these oils have a high inhibitory activity against the bacteria tested.

Acinetobacter spp. have recently emerged as a serious cause of healthcare-associated infections [73]. *A. baumannii* remains an important and difficult-to-treat pathogen whose resistance patterns result in a significant challenge for clinicians. It is a nosocomial pathogen, which is resistant to several antibiotics. It is most commonly found in hospital environments, which increases the risk of infection. [74]. Duarte et al. reported the synergistic activity of coriander oil and conventional antibiotics against *A. baumannii* [42]. In addition, Settanni et al. [75] reported inhibition of foodborne pathogen bacteria by EOs extracted from citrus fruits.

FUTURE DIRECTIONS: SYNERGISTIC THERAPY

Many organisms such as methicillin-resistant *Staphylococcus aureus*, Mycobacterium tuberculosis, *Neisseria gonorrhoeae*, vancomycin-resistant enterococci (VRE) are antibiotic resistant. Until now, natural plant extracts and EOs have been used in the treatment of such resistant organisms owing to their highly significant and important antibiotic properties [43]. However, treatment with many synthetic antimicrobial drugs is complicated by their high toxicity, low tolerability, ineffectiveness against new or reemerging microbes, and the development of drug-resistant strains in patients undergoing treatment. Therefore, there is a need for more intensive research efforts to develop new antimicrobial drugs.

One possible approach to improve the range and scope of current antimicrobial therapy is the use of combinations of antimicrobials. The use of combination therapy in clinical practice is very common and is employed for the therapeutic advantages that it may provide over single agents. Screening studies using medicinal plants with antimicrobial activity aiming to identify synergistic interactions with antimicrobial drugs provide an important source of bioactive compounds that could be exploited in combination therapies. Such compounds or active fractions may not necessarily have strong antibacterial activities themselves but may synergize with classical antibiotics through known or novel modes of action [76,77].

As new antimicrobial compounds are discovered, there is a need to assess their potential in combination therapies with older antibiotics that have been rendered ineffective by the development of resistant strains, even when it is evident that such compounds are not directly inhibitory. The use of agents that do not kill pathogenic bacteria but instead modify them to produce a phenotype that is susceptible to antibiotics could be an alternative approach for the treatment of infectious diseases. The curative effect of plant extracts in combination therapy has been referred to as resistance modifying/modulating activity [78]. In synergistic interaction between two agents, one agent enhances the activity of the other and together they may act more effectively than a single agent. This could be a new approach to solve the problem of bacterial resistance and reduced susceptibility.

High-intensity marine aquaculture of fish and invertebrates often results in outbreaks of bacterial diseases that severely limit productivity. Mousavi et al. successfully showed the efficacy of EO combinations against marine bacteria [79]. The various levels at which antimicrobial interactions take place need to be explored and identified. Drug synergism between known antimicrobial agents; bioactive plant extracts and EOs; or plant extracts and EOs is a novel concept, as recently reported (Table 11-1). Some studies concluded that combinations of antibiotics, EOs, extracts, and phytochemicals have greater antimicrobial activity against multidrug-resistant microorganisms than do the individual components. The combination of less active components with more highly active components has resulted in synergism and lower MIC values.

Potential Benefits

Combination therapy or synergistic therapy may result in improved efficacy over the use of single drugs, an increased spectrum of antimicrobial activity, prevention of treatment failure when antimicrobial resistance is suspected, prevention of resistance development, a decrease in dose-related toxicity caused by the need to use less of a toxic antimicrobial agent, low costs, and enhanced antimicrobial killing or growth inhibition compared with monotherapy [80]. Drugs used in combination may have different mechanisms of action, as well as affecting different sites of the body, but the overall effect of the treatment combination may be one of the following.

(1) Synergism. The literal meaning is *working together*. "Synergism occurs when two or more compounds interact in ways that mutually enhance, amplify or potentiate each other's effect more significantly than the simple sum of these ingredients"[81].
(2) Antagonism. A combination of compounds is antagonistic if their joint effect is weaker than the sum of effects of the individual agents or weaker than the effect of either individual agent.
(3) Additive. An additive interaction is "the effect where the combined action is equivalent to the sum of the activities of each drug when used alone" [81].
(4) Indifferent. An indifferent interaction between treatments occurs "if their joint effect is equal to the effect of either of the individual agents" [82].

Experimental Approaches

A number of different methodologies have been proposed to explain antimicrobial interactions. The different methodologies used in plant-based antimicrobial studies are described next.

Diffusion Assay (Agar Well/Disc Diffusion Assay)

Each independent test sample, as well as the combination, is placed in an individual well or on a disc. The inhibition zone of the combination is compared with those of the individual test samples. Although simple, these assays are subject to many variables that may influence the results and should, at most, be used as a qualitative guide only [83].

Minimum Inhibitory Concentration Assays

The microdilution method is used and combinations are comparatively assessed by combining the individual inhibitors at selected concentrations. This arrangement of combinations formed by multiple dilutions is referred to as the *checkerboard method*.

Measurement of Synergy

This involves the use of an algebraic equation to determine synergy using the fractional inhibitory concentration (FIC), a widely accepted means of measuring interactions by calculating the FIC index (ΣFICI) using the following equations:

$$\sum \text{FICI} = \text{FIC}_A + \text{FIC}_B$$

where

$$\text{FIC}_A = \text{MIC}_A \text{ combination}/\text{MIC}_A \text{ alone}$$

and

$$\text{FIC}_B = \text{MIC}_B \text{ combination}/\text{MIC}_B \text{ alone}$$

The ΣFICI values are interpreted as follows: $\leq 0.5 =$ synergistic; $0.5-0.75 =$ partial synergy; $0.76-1.0 =$ additive; $> 1.0-4.0 =$ indifferent (noninteractive); $> 4.0 =$ antagonistic [50,84].

The Isobole Method

The use of isobolograms, which take into account treatment combinations at various concentrations, provides a more realistic means of measurement. The isobole is a curve constructed by plotting coordinates consisting of values that represent the fractional effect for each of the two components.

Time-Kill Assay

Time-kill studies provide descriptive information on the relationship between the bactericidal activity and concentration of the test substance. The time-kill method has been recommended as one of the best methodologies for studying synergy. Antagonism in time-kill methods may be defined as at least a100-fold increase in colony counts, whereas synergism is indicated by a 100-fold decrease in colony counts. An advantage of this method is the possibility of identifying a direct correlation between exposure time to a plant test material and the extent of pathogen death [85].

Applications

Food Industry

Food authorities are paying increasing attention to the regulation and uses of so-called antimicrobial food additives. The US Food and Drug Administration, for instance, requires the chemical and biologic identification of the antimicrobial agents and the group of target microbes before approval of the antimicrobial food additive [86]. Foodborne disease is one of the major concerns of both food producers and consumers and food spoilage remains a major problem in different parts of the world. Ensuring food safety and a long shelf life relies on minimizing the initial levels of microbial contamination and preventing the growth of microorganisms. In an effort to meet this demand, the food industry has developed a great interest in the use of natural antimicrobial compounds [87]. Foodborne illnesses resulting from the consumption of food contaminated with pathogenic bacteria have been of vital concern to public health authorities. Among the reported outbreaks in the United States during the 1993–1997 period for which the etiology was determined, bacterial pathogens caused the largest percentage of outbreaks (75%) and the largest percentage of cases (86%) [88]. Lacombe et al. reported that lowbush blueberry (*Vaccinium angustifolium*) could be used against foodborne pathogens [89]. Some EO-based preservatives are already commercially available. *DMC Base Natural*, comprising 50% EOs from rosemary, sage, and citrus and 50% glycerol and carvone, a monoterpene present in the EO of *Carum carvi*, is widely used as a safe food additive [90]. *Salmonella* spp. have become the major cause of foodborne diseases, which has raised safety concerns in public health authorities. In several geographic regions, a large proportion of foodborne diseases result from hazardous salmonellosis caused by infectious *Salmonella* spp. [91]. Salmonella food poisoning can affect anyone, especially those with weakened immune systems [92]. This pathogen can therefore have a large socioeconomic impact due to illness, medical costs, loss of productivity, disability, death, litigation, and recalls due to contaminated products [93].

Cosmetic Industry

Many cosmetic products contain parabens, a group of preservatives widely used to prevent contamination by microorganisms. A growing number of chemical-free cosmetics and personal care items are now prepared with safe and nontoxic compounds. The application of plants and plant extracts in cosmetics is widespread; they are used for purposes such as antioxidants, coloring cosmetics, immunostimulants, moisturizing, radical scavenging, sunscreens, tanning, washing, whitening, and as preservatives and thickeners. [94].

Pharmaceutical Industry

Infectious diseases remain a leading cause of the global disease burden, with especially high morbidity and mortality in developing countries. The large pharmaceutical companies spend millions of dollars searching for new drug that can combat resistant microorganisms. Antibiotics that work today may not work tomorrow. Therefore, there is an urgent need

to identify new drugs to treat multiple drug-resistant microorganisms [13]. Plants and their parts are put to many uses in industries ranging from cosmetics to oral care, and from agriculture to medicine.

ORAL DISEASES

Dental caries and periodontitis are the most common chronic oral diseases worldwide, and affect most of the younger population [25]. The formation of dental caries is caused by the colonization and accumulation of oral microorganisms; adherence is the first step in the colonization process. Many antibiotics are available in the market to treat oral diseases but, again, oral bacteria have been reported to show increased antibiotic resistance because of excessive antibiotic use. This can result in a disturbance to the normal oral and intestinal flora, causing side effects such as vomiting, diarrhea, and tooth staining. Indians have less dental caries compared with the Western population [26]. The microorganisms responsible for dental diseases are shown in Table 11-1. Ishnava et al. reported that the formation of dental caries is caused by the colonization and accumulation of oral microorganism like streptococci, especially *Streptococcus mutans* [25]. Natural products are providing a ray of hope for preventing oral diseases like dental caries.

FUTURE PERSPECTIVES

The development of new antimicrobial agents is a research area of the utmost importance. Products derived from plant extracts may control microbial growth in diverse situations and in specific diseases. The use of treatment combinations is likely to reduce the minimum effective dose of the drugs, thus minimizing potential toxic side effects and treatment costs. Various methods of assessing the potential of different treatment combinations covered in this review may be helpful for the cosmetic, food, and pharmaceutical industries. However, further studies are necessary to elucidate the mechanisms of action of synergistic combinations. The mechanism and role of plant extracts and EOs on microbial cells has not been completely investigated; hence, this should be an important aspect of future studies. Moreover, an evaluation of the mechanism of action of plants and their products will increase their practical applications and promote further research.

Acknowledgments

The authors thank Prof. S.P. Singh, Head of the Department of Biosciences, Saurashtra University, Rajkot, Gujarat, India for providing excellent research facilities. The authors also thank Mr. V. Chanda for English language editing of this manuscript. Ms. Kalpna Rakholiya and Mr. Mital Kaneria are thankful to the University Grants Commission, New Delhi, India for providing financial support.

References

[1]   Andersson DI, Hughes D. Persistence of antibiotic resistance in bacterial populations. FEMS Microbiol Rev 2011;35:901−11.

[2]   Stabili L, Acquaviva MI, Biandolino F, Cavallo RA, De Pascali SA, et al. The lipidic extract of the seaweed Gracilariopsis longissima (Rhodophyta, Gracilariales): a potential resource for biotechnological purposes? New Biotechnol 2012;29:443−50.

[3]   Cakir A, Kordali S, Kilic H, Kaya E. Antifungal properties of essential oil and crude extracts of Hypericum linarioides Bosse. Biochem Syst Ecol 2005;33:245−56.

[4]   Ait-Ouazzou A, Loran S, Arakrak A, Laglaoui A, Rota C, Herrera A, et al. Evaluation of the chemical composition and antimicrobial activity of Mentha pulegium, Juniperus phoenicea, and Cyperus longus essential oils from Morocco. Food Res Int 2012;45:313−9.

[5]   Chanda S, Dudhatra S, Kaneria M. Antioxidative and antibacterial effects of seeds and fruit rind of nutraceutical plants belonging to the Fabaceae family. Food Funct 2010;1:308−15.

[6]   Kamazeri T, Samah OA, Taher M, Susanti D, Qaralleh H. Antimicrobial activity and essential oils of Curcuma aeruginosa, Curcuma mangga, and Zingiber cassumunar from Malaysia. Asian Pac J Trop Med 2012;5:202−9.

[7]   Chanda S, Nair R. Antimicrobial activity of Polialthia longifolia (Sonn.) Thw. Var. Pendula leaf extracts against 91 clinically important pathogenic microbial strains. Chinese Med 2010;1:31−8.

[7a] Chanda S, Vyas BRM, Vaghasiya Y, Patel H, a. Global resistance trends and the potential impact of Methicillin Resistant Staphylococcus aureus (MRSA) and its solutions. 2nd Series. In: Mendez-Vilas A, editor. Current Research, Technology and Education Topics in Applied Microbiology and Microbial Biotechnology. Spain: Formatex; 2010. p. 529−36.

[8]   Kumar P, Bhatt RP, Singh L, Sati OP, Khan A, Ahmad A. Antimicrobial activities of essential oil and methanol extract of Coriaria nepalensis. Nat Prod Res 2011;25:1074−81.

[9]   Ivanovic J, Misic D, Zizovic I, Ristic M. In vitro control of multiplication of some food-associated bacteria by thyme, rosemary and sage isolates. Food Control 2012;25:110−6.

[10] Hayrapetyan H, Hazeleger WC, Beumer RR. Inhibition of Listeria monocytogenes by pomegranate (Punica granatum) peel extract in meat pate at different temperatures. Food Control 2012;23:66−72.

[11] Fazly Bazzaz BS, Khajehkaramadin M, Shokooheizadeh HR. In vitro antibacterial activity of Rheum ribes extract obtained from various plant parts against clinical isolates of Gram-negative pathogens. Iranian J Pharmaceut Res 2005;2:87−91.

[12] Chanda S, Kaneria M. Indian nutraceutical plant leaves as a potential source of natural antimicrobial agents. In: Mendez-Vilas A, editor. Science against Microbial Pathogens: Communicating Current Research and Technological Advances, 2. Spain: Formatex Research Center; 2011. p. 1251−9.

[13] Chanda S, Rakholiya K. Indian Combination therapy: Synergism between natural plant extracts and antibiotics against infectious diseases. In: Mendez-Vilas A, editor. Science against Microbial Pathogens: Communicating Current Research and Technological Advances, 1. Spain: Formatex Research Center; 2011. p. 520−9.

[14] Kaneria M, Baravalia Y, Vaghasiya Y, Chanda S. Determination of antibacterial and antioxidant potential of some medicinal plants from Saurashtra region, India. Indian J Pharm Sci 2009;71:406−12.

[15] Parekh J, Jadeja D, Chanda S. Efficacy of aqueous and methanol extracts of some medicinal plants for potential antibacterial activity. Turk J Biol 2005;29:203−10.

[16] Chanda S, Kaneria M, Baravalia Y. Antioxidant and antimicrobial properties of various polar solvent

extracts of stem and leaves of four Cassia species. Afr J Biotechnol 2012;11:2490—503.

[17] Yu JQ, Lei JC, Zhang XQ, Yu HD, Tian DZ, Liao ZX. Anticancer, antioxidant and antimicrobial activities of the essential oil of Lycopus lucidus Turcz. var. hirtus Regel. Food Chem 2011;126:1593—8.

[18] Furneri PM, Mondello L, Mandalari G, Paolino D, Dugo P, Garozzo A, et al. In vitro antimycoplasmal activity of Citrus bergamia essential oil and its major components. Eur J Med Chem 2012;52:66—9.

[19] Jadhav S, Shah R, Bhave M, Palombo EA. Inhibitory activity of yarrow essential oil on Listeria planktonic cells and biofilms. Food Control 2012; 29:125—30.

[20] Singh D, Kumar A, Dubey NK, Gupta R. Essential oil of Aegle marmelos as a safe plant-based antimicrobial against postharvest microbial infestations and aflatoxin contamination of food commodities. J Food Sci 2009;74:302—7.

[21] Ruban P, Gajalakshmi K. In vitro antibacterial activity of Hibiscus rosa-sinensis flower extract against human pathogens. Asian Pac J Trop Biomed 2012; 2:399—403.

[22] Bag A, Bhattacharyya SK, Pal NK, Chattopadhyaya RR. In vitro antibacterial potential of Eugenia jambolana seed extracts against multidrug-resistant human bacterial pathogens. Microbiol Res 2012;167:352—7.

[23] Xu L, Liu R, Li D, Tu S, Chen J. An in vitro study on the dental caries preventing effect of oligomeric procyanidins in sorghum episperm. Food Chem 2011;126:911—6.

[24] Kim J, Kim H, Pandit S, Chang K, Jeon J. Inhibitory effect of a bioactivity-guided fraction from Rheum undulatum on the acid production of Streptococcus mutans biofilms at sub-MIC levels. Fitoterapia 2011;82:352—6.

[25] Basri DF, Tan LS, Shafiei Z, Zin NM. In vitro antibacterial activity of galls of Quercus infectoria Olivier against oral pathogens. J Evid Based Complementary Altern Med 2012;2012:632796.

[26] Henrique FS, Ana Andrea TB, Sukarno OF, Hilario CM. Bactericidal activity of ethanolic extracts of propolis against Staphylococcus aureus isolated from mastitic cows. World J Microbiol Biot 2012;28:485—91.

[27] Santana HF, Andrea A, Barbosa T, Ferreira SO, Mantovani HC. Bactericidal activity of ethanolic extracts of propolis against Staphylococcus aureus isolated from mastitic cows. World J Microbiol Biot 2012;28:485—91.

[28] Singh G, Kumar P. Antibacterial potential of alkaloids of Withania somnifera L. & Euphorbia hirta L. Int J Pharm Pharma Sci 2012;4:78—81.

[29] Brusotti G, Cesari I, Frassa G, Grisoli P, Dacarro C, Caccialanza G. Antimicrobial properties of stem bark extracts from Phyllanthus muellerianus (Kuntze) Excell. J Ethnopharmacol 2011;135:797—800.

[30] Michel T, Destandau E, Floch GL, Lucchesi ME, Elfakir C. Antimicrobial, antioxidant and phytochemical investigations of sea buckthorn (Hippophae rhamnoides L.) leaf, stem, root and seed. Food Chem 2012;131:754—60.

[31] Zeng W, He Q, Sun Q, Zhong K, Gao H. Antibacterial activity of water-soluble extract from pine needles of Cedrus deodara. Int J Food Microbiol 2012;153:78—84.

[32] Kumar R, Saha A, Saha D. A new antifungal coumarin from Clausena excavata. Fitoterapia 2012;83:230—3.

[33] Lacombe A, Wu CHV, White J, Tadepalli S, Andre EE. The antimicrobial properties of the lowbush blueberry (Vaccinium angustifolium) fractional components against foodborne pathogens and the conservation of probiotic Lactobacillus rhamnosus. Food Microbiol 2012;30:124—31.

[34] Cheng VJ, Bekhit AED, McConnell M, Mros S, Zhao J. Effect of extraction solvent, waste fraction and grape variety on the antimicrobial and antioxidant activities of extracts from wine residue from cool climate. Food Chem 2012;134:474—82.

[35] Adamez JD, Samino EG, Sanchez EV, Gonzalez-Gomez D. In vitro estimation of the antibacterial activity and antioxidant capacity of aqueous extracts from grape-seeds (Vitis vinifera L.). Food Control 2012;24:136—41.

[36] Li H, O'Neill T, Webster D, Johnson JA, Gray CA. Anti-mycobacterial diynes from the Canadian medicinal plant Aralia nudicaulis. J Ethnopharmacol 2012;140:141—4.

[37] Nair R, Chanda S. In-vitro antimicrobial activity of Psidium guajava L. leaf extracts against clinically important pathogenic microbial strains. Braz J Microbiol 2007;38:452—8.

[38] Kaneria M, Chanda S. Evaluation of antioxidant and antimicrobial capacity of Syzygium cumini L. leaves extracted sequentially in different solvents. J Food Biochem 2011;. http://dx.doi.org/10.1111/j.1745—4514. 2011.00614. x.

[39] Palaniappan K, Holley RA. Use of natural antimicrobials to increase antibiotic susceptibility of drug resistant bacteria. Int J Food Microbiol 2010; 140:164—8.

[40] Moon S, Kim H, Cha J. Synergistic effect between clove oil and its major compounds and antibiotics against oral bacteria. Arch Oral Biol 2011;56:907—16.

[41] Suliman S, Van Vuuren SF, Viljoen AM. Validating the in vitro antimicrobial activity of Artemisia afra in

polyherbal combinations to treat respiratory infections. South Afr J Bot 2010;76:655—61.

[42] Duarte A, Ferreira S, Silva F, Domingues FC. Synergistic activity of coriander oil and conventional antibiotics against Acinetobacter baumannii. Phytomedicine 2012;19:236—8.

[43] Amber K, Aijaz A, Immaculata X, Luqman KA, Nikhat M. Anticandidal effect of Ocimum sanctum essential oil and its synergy with fluconazole and ketoconazole. Phytomedicine 2010;17:921—5.

[44] Mulyaningsih S, Sporer F, Reichling SZ, Wink M. Synergistic properties of the terpenoids aromadendrene and 1,8-cineole from the essential oil of Eucalyptus globulus against antibiotic-susceptible and antibiotic-resistant pathogens. Phytomedicine 2010; 17:1061—6.

[45] Mandal S, Mandal MD, Pal NK. Enhancing chloramphenicol and trimethoprim in vitro activity by Ocimum sanctum Linn. (Lamiaceae) leaf extract against Salmonella enterica serovar Typhi. Asian Pac J Trop Med 2012;5:220—4.

[46] Fadli M, Saad A, Sayadi S, Chevalier J, Mezrioui N, Pages J, et al. Antibacterial activity of Thymus maroccanus and Thymus broussonetii essential oils against nosocomial infection—bacteria and their synergistic potential with antibiotics. Phytomedicine 2012;19:464—71.

[47] Low WL, Martin C, Hill DJ, Kenward MA. Antimicrobial efficacy of silver ions in combination with tea tree oil against Pseudomonas aeruginosa, Staphylococcus aureus and Candida albicans. Int J Antimicrob Ag 2011;37:162—5.

[48] Stefanovic O, Stanojevic D, Comic L. Inhibitory effect of Torilis anthriscus on growth of microorganisms. Cent Eur J Biol 2009;4:493—8.

[49] Nweze EI, Eze EE. Justification for the use of Ocimum gratissimum L in herbal medicine and its interaction with disc antibiotics. BMC Complem. Altern Med 2009;9:37.

[50] Bassole IHN, Lamien-Meda A, Bayala B, Obame LC, Ilboudo AJ, Franz C, et al. Chemical composition and antimicrobial activity of Cymbopogon citratus and Cymbopogon giganteus essential oils alone and in combination. Phytomedicine 2011;18:1070—4.

[51] Stefanovic O, Comic L. Inhibitory effect of Cytisus nigricans L. and Cytisus capitatus Scop. on growth of bacteria. Afr J Microbiol Res 2011;5:4725—30.

[52] Adwan G, Abu-Shanab B, Adwan K. Antibacterial activities of some plant extracts alone and in combination with different antimicrobials against multidrug-resistant Pseudomonas aeruginosa strains. Asian Pac J Trop Med 2010;3:266—9.

[53] Chatterjee SK, Bhattacharjee I, Chandra G. In vitro synergistic effect of doxycycline & ofloxacin in combination with ethanolic leaf extract of Vangueria spinosa against four pathogenic bacteria. Indian J Med Res 2009;130:475—8.

[54] Jayaraman P, Sakharkar MK, Lim CS, Tang TH, Sakharkar KR. Activity and interactions of antibiotic and phytochemical combinations against Pseudomonas aeruginosa in vitro. Int J Biol Sci 2010;6:556—68.

[55] Chung PY, Navaratnam P, Chung LY. Synergistic antimicrobial activity between pentacyclic triterpenoids and antibiotics against Staphylococcus aureus strains. Ann Clin Microbiol Antimicrob 2011;10:20—5.

[56] Choi J, Kang O, Lee Y, Oh Y, Chae H, Jang H, et al. Antibacterial activity of methyl gallate isolated from Galla Rhois or carvacrol combined with nalidixic acid against nalidixic acid resistant bacteria. Molecules 2009;14:1773—80.

[57] Aiyegoro O, Adewusi A, Oyedemi S, Akinpelu D, Okoh A. Interactions of antibiotics and methanolic crude extracts of Afzelia africana (Smith.) against drug resistance bacterial isolates. Int J Mol Sci 2011;12:4477—87.

[58] Shirotake S, Nakamura J, Kaneko A, Anabuki E, Shimizu N. Screening bactericidal action of cytoplasm extract from kumazasa bamboo (Sasa veitchii) leaf against antibiotics- resistant pathogens such as MRSA and VRE strains. J Bioequiv Bioavailab 2009;1:80—5.

[59] Sibanda T, Okoh AI. In vitro evaluation of the interactions between acetone extracts of Garcinia kola seeds and some antibiotics. Afr J Biotechnol 2008;7:1672—8.

[60] Rodrigues FFG, Costa JGM, Coutinho HDM. Synergy effects of the antibiotics gentamicin and the essential oil of Croton zehntneri. Phytomedicine 2009;16:1052—5.

[61] Coutinho HDM, Costa JGM, Falcao-Silva VS, Siqueira-Junior JP, Lima EO. Effect of Momordica charantia L. in the resistance to aminoglycosides in methicilin-resistant Staphylococcus aureus. Comp Immunol Microb 2010;33:467—71.

[62] Rahman S, Parvez AK, Islam R, Khan MH. Antibacterial activity of natural spices on multiple drug resistant Escherichia coli isolated from drinking water. Bangladesh Ann Clin Microbiol Antimicrob 2011;10:7—10.

[63] Parekh J, Chanda S. In-vitro antimicrobial activities of extracts of Launaea procumbens Roxb. (Labiatae), Vitis vinifera L. (Vitaceae) and Cyperus rotundus L. (Cyperaceae). Afr J Biomed Res 2006;9:89—93.

[64] Chanda S, Rakholiya K, Nair R. Antimicrobial activity of Terminalia catappa L. leaf extracts against some

clinically important pathogenic microbial strains. Chinese Med 2011;2:171—7.

[65] Chanda S, Kaneria M, Vaghasiya Y. Evaluation of antimicrobial potential of some Indian medicinal plants against some pathogenic microbes. Indian J Nat Prod Res 2011;2:225—8.

[66] Veras HNS, Rodrigues FFG, Colares AV, Menezes IRA, Coutinho HDM, Botelho MA, et al. Synergistic antibiotic activity of volatile compounds from the essential oil of Lippia sidoides and thymol. Fitoterapia 2012;83:508—12.

[67] Chanda S, Baravalia Y, Kaneria M, Rakholiya K. Fruit and vegetable peels—strong natural source of anti-microbics. 2nd Series. In: Mendez-Vilas A, editor. Current Research, Technology and Education Topics in Applied Microbiology and Microbial Biotechnology. Spain: Formatex; 2010. p. 444—50.

[68] Akter A, Neela FA, Khan MSI, Islam MS, Alam MF. Screening of ethanol, petroleum ether and chloroform extracts of medicinal plants, Lawsonia inermis L. and Mimosa pudica L. for antibacterial activity. Indian J Pharm Sci 2010;72:388—92.

[69] Parekh J, Chanda S. In vitro antibacterial activity of the crude methanol extract of Woodfordia fruticosa Kurz. flower (Lythraceae). Braz J Microbiol 2007;38:204—7.

[70] Bakkali F, Averbeck S, Averbeck D, Idaomar M. Biological effects of essential oils—a review. Food Chem Toxicol 2008;46:446—75.

[71] Jafari S, Esfahani S, Fazeli MR, Jamalifar H, Samadi M, Samadi N, et al. Antimicrobial activity of lime essential oil against food-borne pathogens isolated from cream-filled cakes and pastries. Int J Biol Chem 2011;5:258—65.

[72] Palmeira-de-Oliveira A, Gaspar C, Palmeira-de-Oliveira R, Silva-Dias A, Salgueiro A, Cavaleiro C, et al. The anti-Candida activity of Thymbra capitata essential oil: Effect upon pre-formed biofilm. J Ethnopharmacol 2012;140:379—83.

[73] Sheng W, Wang J, Lib S, Lin Y, Cheng A, Chen Y, et al. Comparative in vitro antimicrobial susceptibilities and synergistic activities of antimicrobial combinations against carbapenem-resistant Acinetobacter species: Acinetobacter baumannii versus Acinetobacter genospecies 3 and 13TU. Diagn Micr Infec Dis 2011;70:380—6.

[74] Fishbain J, Peleg AY. Treatment of Acinetobacter infections. Clin Infect Dis 2010;51:79—84.

[75] Settanni L, Palazzolo E, Guarrasi V, Aleo A, Mammina C, Moschetti G, et al. Inhibition of food-borne pathogen bacteria by essential oils extracted from citrus fruits cultivated in Sicily. Food Control 2012;26:326—30.

[76] Pyun MS, Shin S. Antifungal effects of the volatile oils from Allium plants against Trichophyton species and synergism of the oils with ketaconazole. Phytomedicine 2006;13:394—400.

[77] Ushimaru PI, Barbosa LN, Fernandes AAH, Di Stasi LC, Fernandes Júnior A. In vitro antibacterial activity of medicinal plant extracts against Escherichia coli strains from human clinical specimens and interactions with antimicrobial drugs. Nat Prod Res 2011;26(16):1553—7.

[78] Kurek A, Nadkowska P, Pliszka S, Wolska KI. Modulation of antibiotic resistance in bacterial pathogens by oleanolic acid and ursolic acid. Phytomedicine 2012;19:515—9.

[79] Mousavi SM, Wilson G, Raftos D, Mirzargar SS, Omidbaigi R. Antibacterial activities of a new combination of essential oils against marine bacteria. Aquacult Int 2011;19:205—14.

[80] Harris MR, Coote PJ. Combination of caspofungin or anidulafungin with antimicrobial peptides results in potent synergistic killing of Candida albicans and Candida glabrata in vitro. Int J Antimicrob Ag 2010;35:347—56.

[81] Ncube B, Finnie JF, Staden JV. In vitro antimicrobial synergism within plant extract combinations from three South African medicinal bulbs. J Ethnopharmacol 2012;139:81—9.

[82] Satish KP, Moellering RC, Eliopoulos GM. Antimicrobial combinations. In: Lorian V, editor. Antibiotics in Laboratory Medicine. Philadelphia: Williams & Wilkins, Lippincott; 2005.

[83] Hewitt W, Vincent S. The agar diffusion assay. In: Theory and application of microbiological assay. New York: Academic Press; 1989. p. 38—79.

[84] Sabate DC, Gonzalez MJ, Porrini MP, Eguaras MJ, Audisio MC, Marioli JM. Synergistic effect of surfactin from Bacillus subtilis C4 and Achyrocline satureioides extracts on the viability of Paenibacillus larvae. World J Microb Biot 2012;28:1415—22.

[85] Saiman L. Clinical utility of synergy testing for multidrug-resistant Pseudomonas aeruginosa isolated from patients with cystic fibrosis: 'the motion for. Paediatr Resp Rev 2007;8:249—55.

[86] Shene C, Reyes AK, Villarroel M, Sineiro J, Pinelo M, Rubilar M. Plant location and extraction procedure strongly alter the antimicrobial activity of murta extracts. Eur Food Res Technol 2009;228:467—75.

[87] Hayashi M, Naknukool S, Hayakawa S, Ogawa M, Ni'matulah AA. Enhancement of antimicrobial activity of a lactoperoxidase system by carrot extract and β-carotene. Food Chem 2012;130:541—6.

[88] Caillet S, Cote J, Sylvain J, Lacroix M. Antimicrobial effects of fractions from cranberry products on the growth of seven pathogenic bacteria. Food Control 2012;23:419−28.

[89] Lacombe A, Wu CHV, White J, Tadepalli S, Andre EE. The antimicrobial properties of the lowbush blueberry (Vaccinium angustifolium) fractional components against foodborne pathogens and the conservation of probiotic Lactobacillus rhamnosus. Food Microbiol 2012;30:124−31.

[90] Prakash B, Singh P, Mishra PK, Dubey NK. Safety assessment of Zanthoxylum alatum Roxb. essential oil, its antifungal, antiaflatoxin, antioxidant activity and efficacy as antimicrobial in preservation of Piper nigrum L. fruits. International Journal of Food Microbiol 2012;153:183−91.

[91] Rabsch W, Tschape H, Andreas J, Baumler AJ. Review, non-typhoidal salmonellosis: Emerging problems. Microbes Infect 2001;3:237−47.

[92] Bajpai VK, Baek K, Kang SC. Control of Salmonella in foods by using essential oils: A review. Food Res Int 2012;45:722−34.

[93] Mani-Lopez E, Garcia HS, Lopez-Malo A. Organic acids as antimicrobials to control Salmonella in meat and poultry products. Food Res Int 2012;45:713−21.

[94] Schurch C, Blum P, Zulli F. Potential of plant cells in culture for cosmetic application. Phytochem Rev 2008;7:599−605.

[95] Radulovic N, Dekic M, Stojanovic-Radic Z. Chemical composition and antimicrobial activity of the volatile oils of Geranium sanguineum L. and G. robertianum L. (Geraniaceae). Med Chem Res 2012;21:601−15.

12

Perspectives and Key Factors Affecting the Use of Herbal Extracts against Multidrug-Resistant Gram-Negative Bacteria

Ali Parsaeimehr[1], Elmira Sargsyan[1], Amir Reza Jassbi[2]

[1]G. S. Davtyan Institute of Hydroponics Problems, National Academy of Sciences Republic of Armenia, Noragyukh 108, Yerevan 375082, Armenia and [2]Medicinal and Natural Products Chemistry Research Center, Shiraz University of Medical Science, Shiraz 71345−3388, Iran

INTRODUCTION

There have been many threats to the existence of mankind and, currently, the reduction in antibiotic effectiveness has raised a new challenge in our world. Since the discovery of antibiotics, there has been a deeply held belief in the eradication of infectious diseases by the application of antibiotics and the intensive development of new classes of antibiotics, but unfortunately, the overuse of antibiotics has led to multidrug resistance in several classes and strains of microbes [1−3]. Multidrug resistance in microorganisms has been defined as "the circumstance that the microorganism has obtained ability of resistance against more than one antimicrobial agent" [4]. In fact, alterations in the bacterial genome through vertical evolution (referred to as the Darwinian principles of natural selection) or horizontal evolution (gene transfer via plasmids or transposons) are the main causes of antibiotic resistance in bacteria [5,6]. These types of alterations ability have been survival factors for microorganisms for millions of years; however, the severity of diseases and infections caused by pathogens can have catastrophic effects on human societies. Therefore, the management, prevention, and control of multidrug-resistant (MDR) microorganisms should be international priorities. Reports presented by the European Antimicrobial Resistance Surveillance Network (EARS-NET) indicate that 400,000 patients suffer annually from infections caused by resistant microorganisms. Evidence alarmingly points to multidrug resistance of *Enterococcus*,

Fighting Multidrug Resistance with Herbal Extracts,
Essential Oils and Their Components
http://dx.doi.org/10.1016/B978-0-12-398539-2.00012-4

Mycobacterium, *Pseudomonas*, *Staphylococcus*, and *Streptococcus* spp. to the most available antimicrobial agents. Unfortunately, *Klebsiella pneumoniae*, a Gram-negative bacterium responsible for common infectious problems in hospitals, has shown significant resistance to carbapenems, a class of β-lactam antibiotics and one of the strongest available antibiotics. In addition, species of *Enterobacter*, a genus of Gram-negative bacteria, has shown resistance to cephalosporins [7–12].

For centuries, extracts and volatile oils of aromatic and medicinal plants have been used in traditional medicine to treat diseases and infections; indeed, the antimicrobial efficacy of some of these medicinal plants has been beyond our expectations. This ability is rooted in the synthesis of a variety of secondary metabolites via biosynthetic pathways [13,14]. Even though secondary metabolites are not involved in primary metabolic processes, including growth and reproduction, they play vital roles under harsh biotic and abiotic conditions. Currently, different classes of secondary metabolites, such as alkaloids, flavonoids, tannins, and terpenoids have been discovered, isolated, and used as phytomedicines against bacteria (Table 12-1). For example, berberine, an isoquinoline alkaloid, is synthesized in plants such as *Berberis aquifolium*, *Hydrastis canadensis*, *Phellodendron amurense*, and *Tinospora cordifolia* and is used as an anticancer, antidepressant, antidiabetic, anti HIV, anti-inflammatory, and antimicrobial agent, as well as a low-density lipoprotein reducing drug [15,16]. Plant extracts and volatile oils contain different classes of secondary metabolites with a broad range of biologic activities. Ephedra, a medicinal plant belonging to the Ephedraceae family, has been used for many years in traditional Chinese medicine to treat allergies, bronchial asthma, chills, colds, coughs, edema, fever, flu, headaches, and nasal congestion. Ephedra also exhibits antioxidant activities and reliable antimicrobial activity against *Escherichia coli*,

K. pneumoniae, *Pseudomonas aeruginosa*, and *Salmonella enterica* serovar Typhimurium [17].

Ocimum gratissimum L. (Labiatae) is commonly used to cure diseases such as coughs, diarrhea, and ophthalmic and skin diseases in tropical regions. Nakamura and colleagues [29] reported the inhibitory effect of *Ocimum gratissimum* essential oils on *E. coli*, *Klebsiella* spp., *Proteus mirabilis*, *P. aeruginosa*, *Salmonella enteritidis*, and *Shigella flexneri* [30,31]. By screening 54 plant extracts (methanol and aqueous) against MDR *Salmonella* ser. Typhimurium, they revealed that methanol extracts of *Aegle marmelos*, *Holarrhena antidysenterica*, *Myristica fragrans*, *Punica granatum*, *Salmalia malabarica*, and *Terminalia arjuna* have strong antibacterial activities against *Salmonella* ser. Typhimurium, while moderate antimicrobial activities were recorded in *Acacia catechu*, *Acacia nilotica*, *Apium graveolens*, *Butea monosperma*, *Carum copticum*, *Cichorium intybus*, *Embelia ribes*, *Ocimum sanctum*, *Peucedanum graveolens*, *Picrorhiza kurroa*, and *Solanum nigrum* [32]. Although views about the safety and effectiveness of medicinal plants vary widely among scientists, 25% of present day drugs come from plants. During their inhibitory actions on the growth of human pathogenic bacteria, some phytochemicals clearly have fewer side effects than their chemical alternatives [33–35]. Apigenin (4′,5,7-trihydroxyflavone) is a flavone that occurs naturally in plants such as *Achillea millefolium*, *Anthemis nobilis*, *Apium graveolens*, *Chamaemelum nobile*, and *Matricaria recutita*; it has been shown to inactivate bacterial enzymes and has anticarcinogenic, anti-infective, and antiviral activities [36].

MECHANISMS OF ANTIBACTERIAL RESISTANCE IN BACTERIA

Bacteria are great survivors that have successfully maintained their existence over millions of years through severe environmental conditions. Through the ages, some species

TABLE 12-1 Phytochemicals Derived from Medicinal Plants with Inhibitory Influences on Gram-Negative Bacteria

Class of secondary metabolites	Phytochemicals	Plant source	Inhibitory Against	Reference(s)
Phenolic	Luteolin	*Anredera cordifolia, Elephantorrhiza burkei, Senna petersiana, Struchium sparganophorum, Thymus vulgaris*	*Enterobacter aerogenes, Enterobacter cloacae, Escherichia coli, Klebsiella pneumoniae, Pseudomonas aeruginosa*	[18,19]
	Kaempferol	*Balanites aegyptiaca, Dodonaea viscosa, Impatiens balsamina, Oxytropis falcata*	*E. coli, P. aeruginosa*	[20,21]
	Galangin	*Alpinia officinarum, Helichrysum auronitens, Rubia cordifolia*	*E. cloacae*	[22]
	Fisetin	*Fragaria* spp., *Rhus verniciflua*	*E. coli, P. aeruginosa*	[14]
	Myricetin	*Bleekeria vitiensis, Cephalotaxus harringtonia, Dysoxylum binectariferum*	*K. pneumoniae, Salmonella enterica*	[23,24]
Alkaloids	Quinine	*Woodfordia fruticosa*	*K. pneumoniae, P. aeruginosa, Pseudomonas pseudoalcaligenes*	[11]
	Berberine	*Hydrastis canadensis*	*E. cloacae, E. coli, P. aeruginosa*	[25]
	Piperine	*Piper nigrum*	*E. coli, P. aeruginosa*	[26]
Terpenes	Asiaticoside	*Centella asiatica*	*E. coli, Helicobacter pylori, P. aeruginosa*	[27]
	Trichorabdal	*Rabdosia trichocarpa*	*H. pylori*	[28]

have gained significance as useful tools for the development of science, such as genetic engineering and molecular biology. *Thermus aquaticus*, a thermophilic bacterium that lives in hot springs at temperatures ranging from 45 °C to 80 °C, is the source of Taq DNA polymerase—a heat-resistant enzyme with multiple applications in molecular biology. In contrast, *Helicobacter pylori* is a common problem in our societies. *H. pylori* is a Gram-negative bacterium that presents in the stomach of patients suffering from chronic gastritis, gastric ulcers, and stomach cancer. This bacterium can survive in the harsh acidic condition of the stomach at pH 1−2 by means of synthesizing urease and thus hydrolyzing urea to ammonia and carbon dioxide [37].

Antibiotics can influence the growth of bacterial by two pathways: killing them (bactericidal) or diminishing their ability to grow or reproduce (bacteriostatic). Bacteria use several mechanism of resistance in response to antibiotics.

(1) They may have the genetic (plasmid borne) or acquired (genetic transformation leading to an increased mutation rate as a survival response) ability to synthesize enzymes that can denature, destroy, or deactivate antibacterial agents. For instance, β-lactamases are enzymes synthesized by almost all Gram-negative bacteria and are responsible for resistance to β-lactam antibiotics (i.e., carbapenems, cephalosporins, cephamycin, and penicillins) [38].

(2) They may pump out the antibiotic agent before it reaches lethal levels or its cellular site of action via efflux pumps (proteinaceous transporters). This activity has been reported in Enterobacteriaceae, *E. coli*, and *P. aeruginosa* against aminoglycosides and fluoroquinolones [39–41].

(3) They may obtain the ability to use alternative metabolic pathways from resistance genes in order to bypass the activity of antimicrobials and thus become insensitive. This mechanism is reported against some class of glycopeptide antibiotics, such as bleomycin, decaplanin, ramoplanin, teicoplanin, telavancin, and vancomycin in *Enterococcus gallinarum*, *Lactobacillus casei*, and *Pediococcus pentosaceus* [1].

(4) Bacteria may change the conformation of the cellular target site or cell wall structure, thus making it impossible for antibiotics to bind to these targets.

In addition, the outer membrane of Gram-negative bacteria can provide an intrinsic barrier to block antimicrobials penetrating the cell and reaching their destinations [40].

REVIEW OF THE CHALLENGES OF ANTIMICROBIAL RESISTANCE IN GRAM-NEGATIVE BACTERIA

Antibiotics often show considerable inhibition of Gram-positive bacteria, but insufficient inhibition of Gram-negative bacteria. Gram-negative bacteria are structurally different from Gram-positive bacteria and are responsible for a number of serious hospital-acquired infectious diseases. In Gram-negative bacteria, (1) the thin, but complex, peptidoglycan outer membrane (5–20% of the cell wall) containing a second phospholipid bilayer (the outer membrane) with a hydrophilic surface and (2) efflux pumps are the key factors in drug resistance.

Salmonella enterica Serovar Typhimurium

This human pathogenic bacterium is responsible for systemic febrile illness and typhoid fever (growth range of 7–45 °C, pH 3.8–9.5). *Salmonella* ser. Typhimurium has shown significant resistance to ampicillin, chloramphenicol, nalidixic acid, and trimethoprim-sulfamethoxazole [44,45]. Kariuki and colleagues reported that commonly used antibiotics were completely effective against all 16 strain of *Salmonella* ser. Typhimurium collected during 1988–1993 [minimum inhibitory concentrations (MICs) ranging from 0.012–0.016 mg/L for ciprofloxacin to 1–3 mg/L for chloramphenicol]; in contrast, strains collected during 1997–1999 showed considerable resistance to the same antibiotics (all MICs were > 32 mg/L) [46]. Also, an increasing trend of resistance to commonly used antityphoid drugs was reported in *Salmonella* ser. Typhimurium isolates and researchers concluded that the overuse of ciprofloxacin for the treatment of typhoid fever had influenced the development of ciprofloxacin-resistant strains of *Salmonella* ser. Typhimurium in and around Kolkata [47].

Shigella dysenteriae

This bacterium is responsible for the Shiga toxin and its severe disease symptoms are collectively called shigellosis. *S. dysenteriae* has shown significant resistance to β-lactam antibiotics. Dutta and colleagues claimed that 97%

of *S. dysenteriae* strains showed resistance to amoxicillin, ampicillin, ciprofloxacin, nalidixic acid, norfloxacin, tetracycline, and trimethoprim-sulfamethoxazole [48]. Another study of 17 *Shigella* strains revealed that after an interruption of 14 years, this pathogen reemerged with an altered antibiotic resistance pattern. Unfortunately, besides being resistant to ampicillin, chloramphenicol, nalidixic acid, tetracycline, and trimethoprim-sulfamethoxazole, they were also resistant to lomefloxacin, norfloxacin, ofloxacin, and pefloxacin [49].

Escherichia coli

The serotypes of this bacterium are harmful, being responsible for food poisoning, and produce lethal toxins that lead to serious infections and diarrhea. Reports have shown significant resistance of *E. coli* to amoxicillin, levofloxacin, piperacillin, and tetracycline; second-generation cephalosporin antibiotics (cefuroxime); and third-generation cephalosporin antibiotics (ceftazidime and ceftriaxone). Sahm and colleagues [50] reported that 2763 out of 38,835 (7.1%) urinary isolates of *E. coli* were resistant to three or more antibiotic agents; the highest resistance was recorded against ampicillin (97.8%) and the lowest against nitrofurantoin (7.7%). Raju and Ballal also claimed that 75% of isolated *E. coli* strains showed multidrug resistance to ampicillin, nalidixic acid, and trimethoprim-sulfamethoxazole on a disk diffusion-based test [51].

Klebsiella Species

These are rod-shaped bacteria with a polysaccharide capsule that are responsible for neonatal septicemia, urinary tract infections, and wound infections. *Klebsiella* spp. have shown significant levels of antibiotics resistance to ampicillin, cefoxitin, ceftazidime, ceftriaxone, erythromycin, gentamicin, sparfloxacin, and trimethoprim. There is evidence to indicate that AmpC β-lactamases are responsible for the resistance of *Klebsiella* spp. to β-lactam drugs [52,53].

Pseudomonas Species

These bacteria have highly virulent pathogenic activities associated with a widespread distribution. Porins (β-barrel proteins) are present in their cell walls. *Pseudomonas* spp. are naturally resistant to β-lactam antibiotics and penicillin, and their metabolic diversity enables them to overcome potentially lethal conditions. Poole and colleagues reported that a single operon (mex-AmexBoprK) provides resistance to chloramphenicol, quinolones, and tetracycline in *Pseudomonas* spp. [54].

Xanthomonas maltophilia (Stenotrophomonas maltophilia)

This is an opportunistic, colonizing, and, increasingly, nosocomial pathogen that is difficult to treat. *Xanthomonas maltophilia* is responsible for some serious infections in hospitalized and immunodeficient patients. This bacterium is genetically resistant to antibiotics such as antipseudomonal penicillins, aztreonam, cephalosporins, imipenem, and quinolones [55,56].

A study by Liaw and colleagues at National Taiwan University Hospital found significant correlations between integrons, efflux pumps, melanin-like pigment, and biofilm formation in resistant *X. maltophilia* [57].

Acinetobacter baumannii

This is one of the most important pathogenic bacterium, although some *Acinetobacter* ssp. are nonpathogenic in healthy individuals. *A. baumannii* was responsible for a series of infections in immunodeficient patients. The evidence shows that horizontal gene transfer is a key factor in its resistance to amikacin, carbapenem, chloramphenicol, penicillin, and tigecycline

[58]. Reid and colleagues reported that the MIC of tigecycline for *A. baumannii* isolates was increasing alarmingly through its therapeutic use [59]. Therefore, it is important that patients being treated with tigecycline for *Acinetobacter* infections should be monitored for any signs of antibiotic resistance.

Campylobacter Species

These are spiral-shaped Gram-negative bacteria. Infections caused by *Campylobacter* spp. are collectively called *campylobacteriosis* and symptoms include cramps, fever, and pain. These bacteria produce a toxin (cytolethal distending toxin) that affects the patient's immune system. The literature indicates significant resistance to ciprofloxacin, erythromycin, fluoroquinolone, and β-lactams in campylobacter. *Campylobacter* spp. use several resistance strategies, including chromosomal mutation, alterations to the structure of the membrane or of porins, and the use of efflux pumps [11,21,41,60]

HERBAL EXTRACTS ARE A PRICELESS SOURCE OF ANTIMICROBIALS

Since the discovery of multidrug resistance in bacteria, the efficacy of some common antimicrobials is unfortunately falling, but herbal medicines can offer some alternatives. The literature shows the antimicrobial activities of herbal extracts can range from 100 to 1000 µg/mL (using the MIC test). Phytochemicals present in herbal extracts can inhibit or disrupt the activities of MDR bacteria by influencing the activities of efflux pumps, β lactamase enzymes, resistance plasmids, and bacterial gene transposition. Photochemicals such as carnosic acid, epigallocatechin gallate, 5'-methoxyhydnocarpin D, pheophorbide, and totarol have potential as efflux pump blockers and others such as baicalin, rugosin B, and tellimagrandin I act as β-lactamase

inhibitors. Targeting the R-plasmid with phytochemicals is a new strategy for combating MDR bacteria. In this regard, 5-hydroxy-2-methyl-1,4-naphthaquinone (plumbagin), derived from the root of *Plumbago zeylanica*, has shown an inhibitory influence on R-plasmids from resistant *E. coli*. In addition, 1'-acetoxychavicol acetate, derived from *Alpinia galanga*, is an antiplasmid with an inhibitory influence on *E. faecalis*, *E. coli*, and *Salmonella* ser. Typhimurium [10,61]. There is evidence of a synergistic relationship between extracts of medicinal plants and certain antibiotics. Usage of methanol extracts of some edible medicinal plants (*Anagyris foetida*, *Euphorbia macroclada*, *Gundelia tournefortii*, *Hibiscus sabdariffa*, *Lepidium sativum*, *Origanum syriacum*, *Pimpinella anisum*, and *Trigonella foenum-graecum* L.) combined with antibiotics (amoxicillin, cephalexin, chloramphenicol, doxycycline, nalidixic acid, and neomycin) against resistant *E. coli* led to increased inhibition by cephalexin, chloramphenicol, doxycycline, nalidixic acid, and neomycin.

Moreover, *Eruca sativa* (Cruciferae), *Gundelia tournefortii* (Compositae), and *Origanum syriacum* (Labiateae) enhanced the activity of clarithromycin against resistant *E. coli* strains [62]. Gislene and colleagues reported that combinations of *Caryophyllus aromaticus* (syn. *Syzygium aromaticum*), *Punica granatum*, *Syzygium jambolanum* (syn. *Syzygium cumini*), or *Thymus vulgaris* with antibiotics had a considerable inhibitory influence on *P. aeruginosa* (resistant to 19 different antibiotics) [63].

Phenylalanine arginine β-naphthylamide (PAβN) is an effective inhibitor of bacterial efflux pump systems. There is evidence of its specific effect on the MexAB-OprM and AcrAB-TolC efflux pumps, present in *P. aeruginosa* and *Enterobacter aerogenes*, respectively. Fankam and colleagues reported that *Dichrostachys glomerata* extracts had synergy with chloramphenicol, norfloxacin, and tetracycline and significantly increased the activity of PAβN as an efflux pump inhibitor (69.2% with an MIC of $\leq 1024\,\mu g/mL$) [63]. Further, extracts from *Aframomum citratum*, *Beilschmiedia cinnamomea*, *Echinops giganteus*, *Fagara zanthoxyloides* (syn. *Zanthoxylum zanthoxyloides*), *Olax subscorpioidea*, and *Piper capense* had inhibitory effects (MIC of $\leq 32-1024\,\mu g/mL$) on Gram-negative bacteria (*E. coli* MC4100 and W3110, *K. pneumoniae* KP63 and K24, and *P. stuartii* NAE16).

EXTRACTION AND IDENTIFICATION OF HERBAL ANTIMICROBIAL AGENTS

The intelligent design of a test system is a key factor for obtaining reliable results, and many different features should be considered, from solvent selection to compound purification. The quantitative and qualitative yields of extracts, the amount of active compounds, and the inhibitory and toxicity of isolated compounds are important factors to consider [65]. The use of different solvents results in the isolation of different compounds; once an extract has shown antimicrobial activity, the next step is the identification of the active phytochemical [4].

The hydrophilic or lipophilic characteristics of phytochemicals are considered to be important for the inhibition of resistant bacteria. Terpenoids are lipid-soluble agents and can act as membrane-catalyzed enzymes regulators and uncouplers. Terpenoids can interfere with proton translocation and interrupt ADP phosphorylation in primary energy metabolism. Eloff and colleagues analyzed the influence of 10 different solvents on *Combretum woodii* (syn. *Combretum kraussii*): they reported that aqueous extracts did not inhibit the microorganisms tested, whereas extracts made with other solvents did show inhibitory activity [66]. Moreover, extracts made using intermediate polar solvents exhibited higher antimicrobial activity compared to those made using more polar or nonpolar solvents. The influence of ethanol extracts of *Prosopis spicigera* (syn. *Prosopis*

cineraria), *Trachyspermum ammi*, and *Zingiber officinale* on MDR microbes revealed that these plants all had an inhibitory potential, although the highest MIC was recorded when petroleum ether had been used as a solvent [8]. Antibacterial analysis of 35 aqueous herbal extracts against resistant *Klebsiella* spp. by Sharmeen and colleagues showed that crude extracts of *S. aromaticum* (leaf) and *Citrus limon* L. (fruit; 90% for both) and *Spondias pinnata* (leaf; 85%) had the highest antibacterial activities [52].

Different parts of plants may accumulate different compounds, and specific phytochemicals can exist in roots (i.e., glycyrrhizic acid), stems (i.e., ephedrine and pseudoephedrine), leaves (i.e., cuminaldehyde, eugenol, and marmesin), fruits (i.e., auraptene, luvangetin, psoralen, and tannin), or flowers (i.e., cliotides).

PERSPECTIVES

Our understanding of the mechanisms of multidrug resistance is expected to have immense implications on the research efforts involved in screening phytochemicals as a valuable source of novel therapeutic agents. This goal will be impossible without cooperation among scientists in the areas of biotechnology, microbiology, molecular biology, pharmacology, phytochemistry, and phytotherapy. New technologies and methods should be used to investigate and identify new phytochemicals with antibacterial activities. These must then be assessed clinically and potential synergistic activities with commonly used antibiotics must also be considered. Finally, standardized therapies must be used to treat patients infected by MDR bacteria.

References

[1] Dutka-Malen S, Blaimont B, Wauters G, Courvalin P. Emergence of high-level resistance to glycopeptides in Enterococcus gallinarum and Enterococcus casseliflavus. Antimicrob Agents Chemother 1994;38: 1675–7.

[2] Guz NR, Stermitz FR, Johnson JB, Beeson TD, Willen S, Hsiang J, et al. Flavonolignan and Flavone Inhibitors of a Staphylococcus aureus Multidrug Resistance Pump: Structure-Activity Relationships. J Med Chem. 2001;44:261–8.

[3] Karchmer AW. From theory to practice: resistance in Staphylococcus aureus and new treatments. Clin Microb Infect 2006;12:15–21.

[4] Nascimento GGF, Locatelli J, Freitas PC, Silva GL. Antibacterial activity of plant extracts and phytochemicals on antibiotic-resistant bacteria. Braz J Microbiol. 2000;31:247–56.

[5] Chien A, Edgar DB, Trela JM. Deoxyribonucleic Acid Polymerase from the Extreme Thermophile Thermus aquaticus. J Bacteriol 1976;127:1550–7.

[6] McInerney JO, Pisani D, Bapteste E, O'Connell MJ. The public goods hypothesis for the evolution of life on. Earth Biol Direct 2011;6:41.

[7] Kaye KS, Cosgrove S, Harris A, Eliopoulos GM, Carmeli Y. Risk Factors for Emergence of Resistance to Broad-Spectrum Cephalosporins among Enterobacter spp. Antimicrob Agents Chemother 2001;45:2628–30.

[8] Khan R, Zakir M, Afaq SH, Latif A, Khan AU. Activity of solvent extracts of Prosopis spicigera, Zingiber officinale and Trachyspermum ammi against multidrug resistant bacterial and fungal strains. J Infect Dev Ctries 2010;4:292–300.

[9] Khera MK, Cliffe IA, Prakash O. Synthesis and in vitro activity of novel 1, 2, 4-triazolo [4, 3-a] pyrimidine oxazolidinone antibacterial agents. Bio And Med Chemi Lett. 2011;21:5266–9.

[10] Lakhmi VV, Padma S, Polasa H. Elimination of multidrug-resistant plasmid in bacteria by plumbagin, a compound derived from a plant. Cur Microb 1987;16:3159–61.

[11] Payot S, Bolla JM, Corcoran D, Fanning S, Mégraud F, Zhang Q. Mechanisms of fluoroquinolone and macrolide resistance in Campylobacter spp. Microbes Infect 2006;8:1967–71.

[12] Teffo LS, Aderogba MA, Eloff JN. Antibacterial and antioxidant activities of four kaempferol methyl ethers isolated from Dodonaea viscosa Jacq. var. angustifolia leaf extracts. South African. J Bot 2009;50:25–9.

[13] Jimenez-Arellanes A, Meckes M, Ramirez R, Torres J, Luna-Herrera J. Activity against multidrug-resistant Mycobacterium tuberculosis in Mexican plants used to treat respiratory diseases. Phytother Res. 2003;17:903–8.

[14] Kannathasan K, Senthilkumar A, Venkatesalu V. In vitro antibacterial potential of some Vitex species against human pathogenic bacteria. Asian Pac J Trop Med 2011;4:645–8.

[15] Lin S, Tsai SC, Lee CC, Wang BW, Liou JW, Shyu KG. Berberine inhibits HIF-1α expression via enhanced proteolysis. Molecular Pharmacology 2004;66:612–9.

[16] O'Hara M, Kiefer D, Farrell K, Kemper K. A review of 12 commonly used medicinal herbs. Arch Fam Med 1998;7:523–36.

[17] Parsaeimehr A, Sargsyan E, Javidnia K. A Comparative Study of the Antibacterial, Antifungal and Antioxidant Activity and Total Content of Phenolic Compounds of Cell Cultures and Wild Plants of Three Endemic Species of Ephedra. Molecules 2010;15: 1668–78.

[18] Basilea A, Giordanoa S, LoÂ pez-SaÂ ezb JA, Cobianch RC. Antibacterial activity of pure flavonoids isolated from mosses. Phytochemistry 1999;52:1479–82.

[19] Tomasz A. Multiple-Antibiotic-Resistant Pathogenic Bacteria—A Report on the Rockefeller University Workshop. N Engl J Med 1994;330:1247–51.

[20] Lim YH, Kim IH, Seo JJ. In vitro Activity of Kaempferol Isolated from the Impatiens balsamina alone and in Combination with Erythromycin or Clindamycin against Propionibacterium acnes. J Microb 2007;45: 473–7.

[21] Taylor DE, Courvalin P. Mechanisms of antibiotic resistance in Campylobacter species. Antimicrob. Agents Chemother 1988;32:1107–12.

[22] Cushnie TP, Lamb, A. J. Assessment of the antibacterial activity of galangin against 4-quinolone resistant strains of Staphylococcus aureus. Phytomedicine 2006;13:187–91.

[23] HX X, Lee SF. Activity of plant flavonoids against antibiotic-resistant bacteria. Phytother Res 2001;15:39–43.

[24] Alakomi HL, Puupponen-Pimiä R, Aura AM, Helander IM, Nohynek L, Oksman-Caldentey KM, et al. Weakening of Salmonella with Selected Microbial Metabolites of Berry-Derived Phenolic Compounds and Organic Acids. J Agric Food Chem 2007;55:3905–12.

[25] Ball AR, Casadei G, Samosorn S, Bremner JB, Ausubel FM, et al. Conjugating Berberine to a Multidrug Resistance Pump Inhibitor Creates an Effective Antimicrobial. Acs Chemical Biology 2006;1:594–600.

[26] Silva DR, Endo EH, Filho BPD, Nakamura CV, Svidzinski TIE, et al. Chemical Composition and Antimicrobial Properties of Piper ovatum Vahl. Molecules 2009;14:1171–82.

[27] Norzaharaini MG, Wan Norshazwani WS, Hasmah A, Nor Izani NJ, Rapeah S. A Preliminary Study on Antimicrobial Activities of Asiaticoside and Asiatic Acid against Selected Gram Positive and Gram Negative Bacteria. Health Environ J 2011;2:23–6.

[28] Sun HD, Huang SX, Han QB. Diterpenoids from Isodon species and their biological activities. Nat Prod Rep 2006;23:673–98.

[29] Nakamura CV, Nakamura TU, Bando E, Melo AFN, Cortez DAG, et al. Antibacterial Activity of Ocimum gratissimum L. Essential Oil. Mem. Inst. Oswaldo. Cruz, Rio de Janeiro 1999;94:675−8.

[30] Onajobi FD. Smooth muscle contracting lipid-soluble principles in chromatographic fractions of Ocimum gratissimum. J Ethnopharmacol 1986;18:3−11.

[31] Sainsbury M, Sofowora EA. Essential oil from the leaves and inflorescence of Ocimum gratissimum. Phytochemistry 1971;10:3309−10.

[32] Rani P, Khullar N. Antimicrobial evaluation of some medicinal plants for their anti-enteric potential against multi-drug resistant Salmonella typhi. Phytother Res 2004;18:670−3.

[33] Ahmad I, Beg AZ. Antimicrobial and phytochemical studies on 45 Indian medicinal plants against multi-drug resistant human pathogens. J Ethnopharmacol 2001;74:113−23.

[34] Breinholt V, Lauridsen ST, Dragsted LO. Differential effects of dietary flavonoids on drug metabolizing and antioxidant enzymes in female rat. Xenobiotica 1999;29:1227−40.

[35] Nielsen SE, Young JF, Daneshvar B, Lauridsen ST, Knuthsen P, Sandström B, et al. Effect of parsley (Petroselinum crispum) intake on urinary apigenin excretion, blood antioxidant enzymes, and biomarkers for oxidative stress in human subjects. Br J Nutr 1999;81:447−55.

[36] Tahir NI, Shaari K, Abas F, Parveez GKA, Ishak Z, Ramli US. Characterization of Apigenin and Luteolin Derivatives from Oil Palm (Elaeis guineensis Jacq.) Leaf Using LC−ESI-MS/MS. J Agric Food Chem 2012;60:11201−10.

[37] Montecucco C, Rappuoli R. Living dangerously: How Helicobacter pylori survives in the human stomach. Nature Rev Mol Cell Biol 2001;2:457−66.

[38] Davies J. Inactivation of antibiotics and the dissemination of resistance genes. Science 1994;264:375−82.

[39] Fralick JA. Evidence that TolC is required for functioning of the Mar/AcrAB efflux pump of Escherichia coli. J Bacteriol 1996;178:5803−5.

[40] Kaye KS, Engemann JJ, Fraimow HS, Abrutyn E. Pathogens resistant to antimicrobial agents: epidemiology, molecular mechanisms, and clinical management. Infect Dis Clin N Am 2004;18:467−511.

[41] Ozer B, Duran N, Onlen Y, Savas L. Efflux pump genes and antimicrobial resistance of Pseudomonas aeruginosa strains isolated from lower respiratory tract infections acquired in an intensive care unit. J Antibiot 2012;65:9−13.

[42] Walsh CT, Fisher SL, Park IS, Prahalad M, Wu Z. Bacterial resistance to vancomycin: five genes and one missing hydrogen bond tell the story. Chem Biol 1996;3:21−8.

[43] Chopra I, Roberts M. Tetracycline Antibiotics: Mode of Action, Applications, Molecular Biology, and Epidemiology of Bacterial Resistance. Microbiol Mol Biol 2001;65:232−60.

[44] Rowe B, Ward LR, Threlfall EJ. Multidrug-Resistant Salmonella typhi: A Worldwide Epidemic. Clin Infect Dis 1997;24:106−9.

[45] Takkar VP, Kumar R, Takkar R, Khurana S. Resurgence of Chloramphenicol Sensitive Salmonella typhi. Indian pediatrics 1995;32:586−7.

[46] Kariuki S, Gilks C, Revathi G, Hart CA. Genotypic Analysis of Multidrug-Resistant Salmonella enterica Serovar Typhi, Kenya. Emerg Infect Dis 2000;6:649−51.

[47] Saha MR, Dutta P, Niyogi SK, Dutta S, Mitra U, Ramamurthy T, et al. Decreasing trend in the occurrence of Salmonella enterica serotype Typhi amongst hospitalised children in Kolkata, India during 1990−2000. Indian J Med Res 2002;115:46−8.

[48] Dutta S, Ghosh A, Ghosh K, Dutta D, Bhattacharya SK, Nair GB, et al. Newly Emerged Multiple-Antibiotic-Resistant Shigella dysenteriae Type 1 Strains in and around Kolkata, India, Are Clonal. J Clin Microbiol. 2003;41:5833−4.

[49] Pazhani GP, Sarkar B, Ramamurthy T, Bhattacharya SK, Takeda Y, Niyogi SK. Clonal Multidrug-Resistant Shigella dysenteriae Type 1 Strains Associated with Epidemic and Sporadic Dysenteries in Eastern India. Antimicrob. Agents Chemother 2004;48:681−4.

[50] Sahm DF, Thornsberry C, Mayfield DC, Jones ME, Karlowsky JA. Multidrug-resistant urinary tract isolates of Escherichia coli: prevalence and patient demographics in the United States in 2000. Antimicrob Agents Chemother 2001;45:1402−6.

[51] Raju B, Ballal M. Multidrug resistant enteroaggregative Escherichia coli diarrhoea in rural southern Indian population. Scand J Infect Dis 2009;41:105−8.

[52] Sharmeen R, Hossain N, Rahman M, Foysal J, Miah F. In-vitro antibacterial activity of herbal aqueous extract against multi-drug resistant Klebsiella sp. isolated from human clinical samples. Int curr pharm J 2012;1:133−7.

[53] Subha A, Renuka Devi V, Ananthan S. AmpC b-lactamase producing multidrug resistant strains of Klebsiella spp. and Escherichia coli isolated from children under five in Chennai. Indian J Med Res 2003;117:13−8.

[54] Poole K, Krebes K, McNally C, Neshat S. Multiple antibiotic resistance in Pseudomonas aeruginosa: evidence

for involvement of an efflux operon. J Bacteriol 1993;175:7363—72.

[55] Marshall WF, Keating MR, Anhalt JP, Steckelberg JM. Xanthomonas maltophilia: an emerging nosocomial pathogen. Mayo Clin Proc 1989;64:1097—104.

[56] Penzak SR, Abate BJ. Stenotrophomonas (Xanthomonas) maltophilia: a multidrug-resistant nosocomial pathogen. Pharmacotherapy 1997;17:293—301.

[57] Liaw SJ, Lee YL, Hsueh PR. Multidrug resistance in clinical isolates of Stenotrophomonas maltophilia: roles of integrons, efflux pumps, phosphoglucomutase (SpgM), and melanin and biofilm formation. Int J Antimicrob Agents 2010;35:126—30.

[58] Perez F, Hujer AM, Hulten EA, Fishbain J, Hujer KM, Aron D, et al. Antibiotic resistance determinants in Acinetobacter spp. and clinical outcomes in patients from a major military treatment facility. Am J Infect Control 2010;38:63—5.

[59] Reid GE, Grim SA, Aldeza CA, Janda WM, Clark NM. Rapid development of Acinetobacter baumannii resistance to tigecycline. Pharmacotherapy 2007;27:1198—201.

[60] Luo N, Sahin O, Lin J, Michel LO, Zhang Q. *In vivo* selection of Campylobacter isolates with high levels of fluoroquinolone resistance associated with gyrA mutations and the function of the CmeABC efflux pump. Antimicrob Agents Chemother 2003;47:390—4.

[61] Latha C, Shriram VD, Jahagirdar SS, Dhakephalkar PK, Rojatkar SR. Antiplasmid activity of 1'-acetoxychavicol acetate from Alpinia galanga against multi-drug resistant bacteria. J Ethnopharmacol 2009;123:522—5.

[62] Darwish RM, Aburjai TA. Effect of ethnomedicinal plants used in folklore medicine in Jordan as antibiotic resistant inhibitors on Escherichia coli. BMC. Complement Altern Med 2010;10:9.

[63] Gislene GF, Juliana L, Paulo CF, Giuliana LS. Antibacterial activity of plant extracts and phytochemicals on Antibiotic Resistant Bacteria. Braz J Microbiol 2000;31:247—56.

[64] Fankam AG, Kuete V, Voukeng IK, Kuiate JR, Pages JM. Antibacterial activities of selected Cameroonian spices and their synergistic effects with antibiotics against multidrug-resistant phenotypes. BMC. Complement Altern Med 2011;11:104.

[65] Meena AK, Uttam Singh AK, Yadav B, Singh M, Rao M. Pharmacological and Phytochemical Evidences for the Extracts from Plants of the Genus Vitex—A Review. Int J Pharm Clinical Res 2010;2:1—9.

[66] Eloff JN, Famaken JO, Katerere DRP. Combretum woodii (Combretaceae) leaf extracts have high activity against Gram-negative and Gram-positive bacteria. Afr J Biotechnol 2005;4:1161—6.

13

Use of Plant-Derived Extracts and Essential Oils against Multidrug-Resistant Bacteria Affecting Animal Health and Production

Lyndy McGaw

Phytomedicine Programme, Department of Paraclinical Sciences, Faculty of Veterinary Science, University of Pretoria, Onderstepoort 0110, South Africa

INTRODUCTION

Antibiotics have been extensively used in animals as both therapeutic veterinary drugs and growth promoters, and this widespread use of antimicrobial agents may be a significant factor contributing to the emergence and spread of antibiotic resistance [1]. Following a review of antimicrobial resistance in the food chain, it was concluded that the introduction of antibiotics led to selective pressure, which has allowed the development of resistance in animal pathogens [2]. These bacteria can then be transmitted animals to humans via unpasteurized milk, meat, and some vegetables, thus transferring antimicrobial resistance to bacteria colonizing humans [1]. As early as 1969, a report concluded that the administration of antibiotics to livestock, especially at subtherapeutic levels

as growth promoters, poses a hazard to human and animal health [3]. Subtherapeutic use of antimicrobials in animals can lead to resistance selection and concomitant treatment failure; therefore, it was considered necessary to implement a veterinary antibiotic use policy with systematic monitoring so as to minimize public health risks [1]. Subsequently, recommendations were made to monitor the emergence of drug-resistant microbes and monitoring programs were established in various European countries [1], followed by the use of antibiotics as growth promoters in production animals being banned in European Union (EU) countries [4].

Animals are susceptible to a wide range of infectious organisms, including bacteria, fungi, protozoa, and viruses, which cause diseases with varying impacts. Disease control strategies in production animals, as well as in companion

Fighting Multidrug Resistance with Herbal Extracts,
Essential Oils and Their Components
http://dx.doi.org/10.1016/B978-0-12-398539-2.00013-6

animals, must of necessity be effective, and preferably proactive, to deal with various diseases appropriately and promptly. Animal diseases affect livestock productivity by means of reduced feed intake; altered digestion and metabolism; a drop in reproduction; low weight gain and milk production; and increased morbidity and mortality [5]. Livestock management to reduce the risk of disease is essential to control animal diseases. Simultaneously, alternative strategies to antibiotic growth promoters (AGPs) that will reduce incidences of infection and enhance growth are being researched. A major component of this effort is devoted to studies investigating the use of plant extracts and essential oils (EOs) with efficacy in reducing microbial load and enhancing the immunity of animals. The emphasis is placed in this chapter on resistance developed by bacterial pathogens of livestock, in comparison to those of companion animals, owing to the importance of such disease-causing organisms and the use of production animals as a source of food and income for the world's burgeoning population.

ANIMAL HEALTH AND MULTIDRUG RESISTANCE IN PRODUCTION ANIMALS

Infectious Diseases in Animals

A variety of parasitic organisms, such as bacteria, chlamydiae, fungi, helminths, protozoa, rickettsiae, and viruses, cause disease in animals. Resistance to orthodox drugs used to treat these diseases has been noted in many of these organisms, but the focus here is on bacterial diseases. Various diseases are caused by Gram-positive bacterial species, including those belonging to the genera *Bacillus*, *Clostridium*, *Enterococcus*, *Staphylococcus*, and *Streptococcus*. Some species occur naturally in the microbial flora associated with the skin or mucous membranes, but these bacteria may cause disease when disruption to

the normal microflora occurs (e.g., caused by a weakened immune system or a wound). Common in such instances are cases of abscesses, mastitis, and pyoderma. Important animal pathogenic bacteria in the Gram-negative group include *Arcobacter*, *Bartonella*, *Brucella*, *Campylobacter*, *Helicobacter*, *Lawsonia*, *Neisseria*, *Pseudomonas*, *Salmonella*, and *Vibrio* species.

Salmonella ssp. cause salmonellosis in livestock, resulting in enteritis, septicemia, and abortion. *Campylobacter jejuni* causes enteritis and sometimes mastitis and abortion in cattle, and can cause abortion in sheep and goats [6]. Brucellosis, caused by *Brucella abortus*, is an important pathogen that results in abortion in cows [7]. *B. abortus*, *C. jejuni*, and *Salmonella* ssp. are all important zoonoses, implicated in infections in humans coming into contact with affected animals. Cattle in feedlots are susceptible to respiratory diseases, caused by *Histophilus somni* (formerly *Haemophilus somnus*), *Mannheimia* (formerly *Pasteurella*) *hemolytica*, *Mycoplasma* spp., and *Pasteurella multocida*, as well as mastitis, usually caused by *Staphylococcus aureus* [8].

Antibiotic Resistance in Animal Bacterial Pathogens

The rapidity with which bacteria develop resistance to currently available antibiotics counteracts efforts to control diseases of economic importance. As concisely summarized by Tenover [9], the principal mechanisms of action of antimicrobial compounds include interference with cell wall synthesis (e.g., lactams and glycopeptide agents), protein synthesis inhibition (macrolides and tetracyclines), interference with nucleic acid synthesis (fluoroquinolones and rifampin), inhibition of metabolic pathways (e.g., trimethoprim-sulfamethoxazole), and disruption of bacterial membrane structure (polymyxins and daptomycin).

Antibiotic resistance may occur by means of three general mechanisms, namely prevention of interaction of the drug with the target, efflux of the antibiotic from the cell, and direct destruction or modification of the compound [10]. Bacteria may be intrinsically resistant to one or more classes of antibacterials, or they can acquire resistance by mutation or by acquiring resistance genes from other organisms [9]. Of more concern is the transfer of genetic material from resistant to susceptible organisms via conjugation, transformation, or transduction [11]. This phenomenon may occur between strains of the same species or between different bacterial species or genera [9]. Such acquired resistance genes may then enable the bacteria to synthesize enzymes to destroy the antibacterial agent or to express efflux systems that prevent the drug from reaching its target, as mentioned previously. New genetic material may also allow the bacteria to modify the drug's target site or to produce alternative metabolic pathways that bypass the drug's mechanism of action [9].

Antibiotic resistance is often associated with selective pressure caused by the use of antibacterial agents, which promotes the emergence of resistant strains. The processes of mutation and selection, combined with the ability to exchange genetic information, allow many bacterial species to quickly adapt to antibacterial agents following exposure. A single mutation in a crucial bacterial gene may be sufficient to confer resistance, but it is more likely that this merely allows the organism to reduce its susceptibility enough to survive until it is able to acquire additional mutations or extra genetic material to result in adequate resistance [11].

The use of antibiotics is generally high in intensively farmed poultry (broilers for meat, as well as layers for eggs) and pigs. These animals are often kept indoors at a high population density, which is conducive to the spread of bacterial infections, particularly of the respiratory and intestinal tracts [8]. Cattle kept in feedlots and dairy cows are also relatively high

consumers of antibiotics [8]. Infectious diseases predictably have a negative effect on profitability, but the high cost of administering antibiotics to all the animals on a farm or feedlot (either as metaphylactics to both sick and healthy animals or as prophylactics, where antibiotics are administered to prevent disease) also affects profits [8]. Resistance in many bacterial pathogens affecting animals has been noted, including *C. jejuni*, *E. coli*, *Salmonella* spp., and *S. aureus* isolates from poultry, and *E. coli* and *S. aureus* in cattle [8]. *Enterococcus* spp., among others, have also shown resistance to antimicrobial agents [8].

ANTIBIOTIC GROWTH PROMOTERS

History of the Use of Antibiotic Growth Promoters

Antibiotics and agricultural animal production have been close companions for decades. Some medications were used therapeutically to treat disease, but most were used prophylactically and to improve growth rate and feed conversion efficiency as AGPs [12]. Antibiotic growth promotion in production animals has been in use since the early 1960s in the United States and other countries [4]. As early as 1946, reports emerged of the beneficial effects of antibiotic supplementation on production efficiency in poultry and pigs [13]. However, soon afterwards, following feeding trials of streptomycin in turkeys, resistance in food animals became evident [14]. Resistance to tetracycline in chickens after feeding of growth-promoting quantities was also noted [15,16]. In a report to the British parliament, concerns about antibiotic resistance development in human pathogens and recommendations to ban subtherapeutic use of antibiotics in animal feed were stated [3]. Public concern subsequently grew regarding the use of AGPs and the possible threat to the efficacy of therapeutic antibiotics used in humans [17].

Banning Antibiotic Growth Promoters

There is evidence to support the theory that antibiotic resistance genes can be transmitted from animal to human microbial pathogens [18]. However, there has been much vigorous debate on this topic [4]. The identification of resistance mechanisms and the means by which they move into the food chain has been reviewed recently [19]. Worldwide, pathogenic bacteria resistant to many antimicrobial agents emerged in the 1980s [20]. With this growing threat to human health, recommendations for taking the precautionary measure of banning antimicrobial use in food animals proliferated.

The World Health Organization (WHO) suggested an epidemiologic link between medical impacts and the use of antimicrobials in food animals [21]. In another WHO report, it was suggested as precautionary measures that countries take proactive steps to establish surveillance mechanisms of antimicrobial use and resistance development and to reduce the need for antimicrobials in animals [22]. The report also recommended that the use of AGPs belonging to classes of antimicrobials used by humans should be halted or phased out by legislation, unless or at least until risk assessments were conducted [22]. Also included were suggestions for routine practices of animal health management to avoid the prophylactic use of antibiotics, and for limiting antimicrobial availability by limiting their prescription to therapeutic use. These precautionary recommendations were based on the possible existence of a reservoir of antibiotic-resistant bacteria (mainly enterococci) in food animals that could be transferred to humans.

The first country to ban the use of antimicrobials for promoting animal growth was Sweden in 1986 [20]. Since 2000, Denmark has restricted the use of antibiotics to prescription only

therapeutic use [4]. The use of avoparcin was banned in all EU member states by the EU Commission in 1997 [4], and this was followed by several bans on other classes of antibiotics in subsequent years. The final phase of the EU-wide ban on AGPs in animal feed became effective on 1 January 2006 (EC Regulation No. 1831/2003).

In other countries, this precedent of legislative curbing of the use of AGPs has not been so judiciously followed. However, it is expected that non-EU producers wishing to export to EU countries will have to reject AGPs [4]. There has been little regulatory activity on the use of AGPs in the USA, but the practice is under scrutiny [23]. The use of AGPs is under review in Australia [24]. In the USA, and potentially other countries too, pressure from consumers is driving poultry producers to rear animals without AGPs [4].

Effect of the Antibiotic Growth Promoter Ban on Animal Production

The impact of the removal of AGPs from animal production can certainly be minimized if alternative disease-preventing measures are taken, such as improving husbandry practices and nutrition [12]. In Denmark, the productivity of broilers appears to have been unaffected by the AGP ban, but feed conversion rates have increased [25]. Mortalities caused by necrotic enteritis did not increase following the AGP ban, but consumption of salinomycin, an ionophore anticoccidial drug with activity against *Clostridium perfringens*, has risen steadily in Denmark since the ban [4]. The use of anticoccidials in the poultry industry is still permitted. Although the effects of the ban on poultry appear not to be as bad as expected, this contrasts with the case for pigs, where withdrawal of AGPs in weaners was associated with a decrease in the average daily weight gain and an increase in mortality [26]. Interestingly, the AGP ban resulted in significantly lower resistance of enterococci isolated from animal feces to various antibiotics [27,28].

ALTERNATIVES TO ANTIBIOTIC GROWTH PROMOTERS

The worldwide reduction in the use of AGPs in animal production has led to exciting challenges and opportunities for veterinarians and scientists, stimulating research into innovative strategies and alternative products [29]. Basic research on possible alternatives is required, in conjunction with independent and carefully controlled studies that assess the effectiveness of the alternative product in enhancing performance and health [29]. In developed countries, where rigorous standards of hygiene are adhered to as part of modern farming methods, the need for AGPs or acceptable alternatives is perhaps not as necessary as in developing countries, where animals may be kept under suboptimal conditions. In these countries, animal production needs to be enhanced by any means that is not harmful so as to sustain the rapidly growing human population [30].

It is still not clear how AGPs exert their beneficial effect [12], but modeling the effects of AGPs using germ-free animals suggests that the benefits of antimicrobials originate from their effects on intestinal microflora [4]. Replacements for AGPs are most likely to involve the use of multiple products in the diet, with each constituent possessing some of the benefits of an AGP; management changes will also play a vital role [4]. It is obvious that the major criterion for a good alternative should be practicality—it must simply improve performance and feed efficiency, and be economically viable [12,4].

Currently proposed alternatives to AGPs include probiotics, prebiotics, enzymes, vaccinations, and plant extracts, including EOs. Probiotics have been defined as living microbial feed supplements that beneficially affect the host animal by improving the balance of the existing microflora [31]. These supplements are currently viewed as being able to enhance production by positively affecting the digestive microflora to promote performance and protect against colonization of the host by enteropathogens and other harmful bacteria [32]. Prebiotics generally refer to nondigestible feed ingredients with selective effects on the intestinal microbiota, for example oligosaccharides [32]. These compounds may improve health by selectively stimulating the growth or activity, or both, of certain bacteria present in the colon [33]. Beneficial effects following the addition of enzymes to monogastric diets are mostly related to the increased amounts of nutrients that can be released from the diet and therefore be absorbed from the gut [34]. Other potential AGP alternatives have been reviewed and include bacteriophages (viral transmissible lytic agents) and bacteriocins (bacteria-killing proteins) [32], as well as organic acids and other micronutrient absorption enhancers [34]. The use of plant extracts and EOs will be addressed in the following section.

THERAPEUTIC EFFICACY OF PLANT EXTRACTS

Plants comprise complex mixtures of chemicals, many of which possess interesting and distinctive pharmacologic activities. Numerous chemical compounds from plants have unique structures that are impossible for human chemists to design or manufacture. The combined existence of an extensive variety of constituents in plants leads to the possibility that, in addition to those compounds with significant biologic activity, there may be other chemicals that enhance the activity of the bioactive compounds. These chemicals may have unexpected or unusual effects that are not easily discernible, but may play an important role in promoting health. As a result of the presence of nutritional elements in herbal preparations, and because constituent compounds interact in various ways with one another, the clinical efficacy of plant mixtures may have an increased depth and breadth compared to that seen with conventional drug therapy [35].

The compounds responsible for efficacy may not only be the generally recognized pharmacologically active classes of phytochemicals such as alkaloids, saponins, and terpenoids but may also include amino acids, minerals, and vitamins, for example. This notion is borne out by a situation that is often encountered during bioassay-guided fractionation used to isolate biologically active constituents of plant extracts. In many instances, the activity of the purified compound is found to be less than expected in comparison to the activity of the crude extract, which comprises thousands of different chemicals. It would usually be anticipated that if the concentration of the active compound is very low in a crude extract, its activity should increase accordingly when purified, but this is often not the case, even when several constituents with varying degrees of efficacy in a particular bioassay are isolated and their combined activities taken into account.

For the most part, herbal products are currently used by the feed industry as flavoring and sensory additives [36]. It is known that aromatic compounds and EOs act in the animal digestive tract to improve appetite and modify the host's bacterial flora, and they have other benefits too, such as dose-dependent antioxidant, bactericidal, or bacteriostatic activity [36]. In response to the EU ban on AGPs in animal feed, much interest in the development of herbal products led to several EU-(co)financed projects from 2001, including Rumen-up, Replace, Feed for Pig Health, Healthy Pig Gut, and Safewastes [37]. In 2006, the Dutch Fyto-V project was initiated to promote the development of phytotherapy for preventing or reducing farm animal diseases [37]. Several hundred potential products were reviewed and, of these, 10 were

selected for quality and efficacy tests. At the conclusion of the project, it was determined that some herbal products could be beneficial to animal health, but that more clinical studies were necessary [37]. Currently, herbal products without medical claims can be used in animal feed if they are safe, not mentioned on the list of undesirable substances of the European Directive 2002/32/EC, and do not contain toxic substances above the allowed levels [37].

The mechanism of action of herbal medicine differs from that of conventional pharmacologic drugs [35]. Many traditional healing systems employ combinations of plant species and plant parts in efforts to enhance efficacy, and possibly also to reduce or ameliorate toxicity or adverse side effects. Plant medicines combine a range of chemicals that may have additive, antagonistic, or synergistic effects; the mechanism of action of synergy is of considerable importance and is receiving increasing scientific attention [38]. Reports on the antagonistic interactions of different plant species are relatively few in comparison with those supplying evidence of synergistic effects [38]. Synergy between compounds in plant extracts may have a pharmacokinetic or pharmacodynamic basis [35]. Where one component enhances intestinal absorption or utilization of another constituent, pharmacokinetic synergy is apparent, but where two compounds interact with a single target or system, pharmacodynamic synergy comes into play [35].

In recent years, there has been a marked shift from a deep-seated reliance on single drug therapy to multitarget treatment or prevention of disease. Alternative ways of viewing the maintenance of health (in animals and humans) by promoting the body's own protective and repair mechanisms, as well as via stimulating the immune system, are increasing in popularity, but further supportive scientific evidence is desirable. Natural product-based remedies may not act to completely destroy all infectious organisms in a host's body, but may support the immune system, thus allowing the host to strengthen its immunity to mount a defense against pathogen invasion [30].

Research in our group (Phytomedicine Programme, University of Pretoria; www.up.ac.za/phyto) has led to some promising leads for AGP alternatives developed from plant products [39,40]. One approach has been to combine extracts of plants containing known antibacterial compounds with antioxidant plant-based preparations. Immune system competence is highly important in the absence of growth-promoting antibiotics, and antioxidant substances are known to protect the immune system [41,42] by various mechanisms, including protecting against excessive free radical production that may lead to disease. The extracts are refined or potentiated to increase the antibacterial and antioxidant activity *in vitro*, and mixtures of the extracts in different combinations have been tested on poultry challenged with *C. perfringens*. Some of the combinations tested thus far have given better results in chicken studies than the widely used antibiotic feed additive Zn-bacitracin, and results are being patented. A drawback to the experiments already conducted is that conditions were concluded to be too hygienic to comprehensively assess the efficacy of the herbal preparations against *C. perfringens* challenges in poultry [39,40]. Further research is also desirable in connection with the possible effects of plant extract treatment on the taste of the chicken meat.

In a review on the use of turmeric (*Curcuma longa*) in poultry as a feed supplement, it was stated that the plant species can be used as a natural growth promoter as it has a wide safety margin and several beneficial pharmacologic properties, including immunomodulatory and hepatoprotective effects [43]. Homeopathy is used as a supplement in veterinary medicine, and growing numbers of farmers across Europe are using it, particularly in organic farming [44]. However, detailed, good quality research studies are lacking, and more research is needed concerning veterinary homeopathy [44] to provide evidence of efficacy.

Echinacea ssp. are widely used as medicinal plants around the world, both in human and veterinary medicine, and much research has concentrated on the immunostimulatory properties of plants belonging to this genus. Interest has recently grown concerning the potential benefits of echinacea in food-producing animals as an alternative to AGPs [45]. Several studies in animals documenting the immune-stimulating effects of echinacea extracts have been reviewed [45]. Most of these studies were undertaken in swine, where weanling diarrhea is a major problem, and the results of these investigations suggest that echinacea may be useful for stimulating nonspecific immunity, protecting animals from infections, and enhancing production [45].

When investigating the use of plant extracts as AGP alternatives, it is necessary to develop standardized techniques for the extraction procedure. The extracts must be subjected to stringent quality control testing for efficacy and safety using appropriate analytical methods, as the composition of an extract may vary owing to seasonal or genetic variations. A plant product also needs to be palatable and easily incorporated into the feed (or water supply) of the target animals.

Essential Oils

A large number of EO-containing plants have been used throughout history in ethnoveterinary medicine and as part of animal health management strategies [36]. An advantage of EOs is that they are complex mixtures, so resistance is less likely to develop following their use, as is the case with single synthetic compounds [36]. These substances have been used to improve the flavor and palatability of feed, as well as to increase weight gain and feed conversion efficiency [36]. The biologic activity of EOs has been viewed as a combined or synergistic effect of both active (e.g., antibacterial) and inactive constituents, with the inactive components potentially influencing factors such as the reaction rate and bioavailability of the active chemicals [46].

Many studies conducted in pigs, poultry, and ruminants on the efficacy of EOs and aromatic plant extracts have been comprehensively reviewed [36]. As may be anticipated, the reported results differ widely according to the type and origin of the EO or herb, the quantity added to the feed, and the trial environmental conditions, such as the level of hygiene. There were improved responses to treatments tested under practical conditions of large-scale animal production rather than under controlled experimental conditions with a higher standard of hygiene [36]. The most common EOs used in published studies include carvacrol, cinnamon, clove, lemongrass, oregano, peppermint, and rosemary, while the aromatic herbs evaluated include coriander, garlic, oregano, sage, thyme, and yarrow. It was concluded that, notwithstanding some uncertainties in the reporting of studies performed using pigs and poultry, there is sufficient evidence that EOs and herbs can improve performance in these animals [36]. Far less information is available on trials assessing feed intake and palatability of aromatic herbs and oils conducted using ruminants.

Additives like aromatic herbs, EOs, and spices are known to have digestive or carminative activity, owing to their stimulation of digestive secretions such as saliva and mucus; they also enhance enzyme activity, thus providing an added nutritional benefit [47]. Increased intestinal mucus secretion in chicks fed with a mixture of carvacrol, cinnamaldehyde, and capsaicin may be responsible for the reduced adherence of pathogens, including C. perfringens and E. coli, to the gut epithelium [48]. The same mixture of compounds given to piglets in feed resulted in an increased gastric retention time of ingested feed, which in turn resulted in improved nutrient absorption and enhanced intestinal stability against digestive disorders [49]. Supplementing poultry feed with a commercial blend of EO components (CRINA poultry) led to a reduced concentration of C. perfringens in the intestinal contents [50]. In another study in which specific blends of EO components were shown to control C. perfringens proliferation in broiler intestines, it was suggested that the EOs may have antibacterial activity, as well as effects on stimulation of digestive enzymes, stabilization of intestinal microflora, and inactivation of C. perfringens toxins [51].

The effect of EOs, aromatic plants, and other alternatives on ruminant nutrition was the topic of a special issue of the Animal Feed Science and Technology journal in 2008 [52]. Several papers in this issue dealt with the potential of plants and their constituents as feed additives in ruminant nutrition to improve feed efficiency and to control the spread of pathogens in livestock [53–56]. It has been suggested that EOs could restrict rumen ammonia concentrations, which would lead to more efficient utilization of dietary nitrogen sources [57]. The use of natural plant products, including EOs, saponins, and related compounds, in altering rumen fermentation has also been reviewed by other authors; it was concluded that plant products have a potential in this regard [58].

Adverse Effects of Plant Products on Animal Production

Certain plant constituents are undesirable in a plant extract mixture designed to promote animal health. For example, tannins, particularly condensed tannins, have been shown to inhibit a wide range of enzymes in laboratory-based assays, which reflects their ability to bind and inactivate proteins [59]. This potentially has an antinutritional effect, as proteins are therefore less available for animal digestion. Tannins also tend to bind or chelate other classes of molecules, notably divalent cations and alkaloids, and these bound agents are rendered inactive and insoluble [59]. It has been noted that tannins can cause mild constipation and nausea, perhaps because of protein binding in the stomach and duodenum [59]. Tannins are widely known to be effective against mild diarrhea, and it is possible that some tannins may be useful in counteracting diarrhea without affecting nutrition too negatively. There are many types of tannins and some may be more specific and hence more useful than others; however, there is a lack of knowledge in this area [52].

In ruminants, high concentrations of tannins absorbed into the bloodstream can cause constipation, hepatotoxicity, and organ damage [59]. Proanthocyanidins have the potential to bind nutritive amino acids and digestive enzymes, thus leading to indigestibility and poor absorption [39]. In general, therefore, tannins, or plant extracts containing high levels of tannins, should not be used in combination with other nutrients or medicinal herbs, as they may reduce their absorption and activity, and have other negative side effects. Their generally bitter taste may also reduce the palatability of animal feed. On the other hand, however, tannins may reduce the potential toxicity of some plant-based preparations [59].

Human safety is an important consideration when investigating feed supplements to be given to production animals to be used for human nutrition. Pharmacokinetics is an integral part of studies on potential AGP replacements, as substances fed to the animal may be distributed to any part of the body and could persist for some time after treatment. Generally, it takes about 10 half-lives for the product to be 99% eliminated from the body, so the consumption of animal products containing residues may pose human health risks [60]. There are significant complexities involved in conducting pharmacokinetic studies on extracts as they comprise combinations of many different chemical substances. Although plant extracts and EOs may have been used in clinical practice or ethnoveterinary medicine without evidence of adverse effects in animals, there is a need for safety testing, since events not immediately apparent during clinical use need to be considered, such as subclinical side effects, chronic and delayed toxicity, and effects of overdose [60,61].

CONCLUSIONS

The increased problems experienced with antibiotic resistance in recent years have focused attention on the potential mechanisms responsible for this phenomenon. Contributing to bacterial resistance against many classes of therapeutic antibiotics in current use is the misuse of antibiotics, such as their inappropriate and indiscriminate use and failure to complete a prescribed course of antibiotics in human and animal therapeutic treatment, which encourages the survival of resistant strains. The use of subtherapeutic doses of antibiotics to promote growth and prevent disease in poultry and production animals has also been implicated in the emergence of drug-resistant and multidrug-resistant bacteria, which may be passed along the food chain to infect humans. Following the precautionary principle, EU bans on the use of AGPs as growth promoters have resulted. Public pressure in other countries has also resulted in closer scrutiny of the use of antibiotics in animal feed, particularly with regard to those classes used in human therapy. It is vital to protect the ever-decreasing arsenal of effective antibiotics currently available for therapeutic use. Antibiotics should be used only when other methods have failed [62] and should be employed as treatments rather than as a preventative measure.

The increased focus on the adverse effects of AGPs has led to much research being concentrated on effective, economically viable alternatives, such as prebiotics, probiotics, enzymes, and plant-based extracts. The animal production industry requires a reliable way of enhancing food animal performance while ensuring food safety. Future research will also need to focus on elucidating the exact mechanisms by which antibiotics have their growth-promoting effects.

The problem of resistance development to antimicrobials by economically important pathogens in the field of animal production may be circumvented by the use of complex plant extract mixtures. These mixtures can be potentiated, or beneficiated, to reduce the number of bulky inactive constituents and increase the number of antimicrobial, antioxidant, immune-stimulating, nutritive, and other beneficial compounds with potential synergistic or additive efficacy against common threats to animal health. With a large variety of individual chemical compounds with different effects constituting a feed supplement, the chances of resistance development by infectious bacteria are far lower than against a single antibiotic compound. Bacteria already showing multidrug resistance to known antibiotics may therefore be susceptible to complex plant mixtures composed of a variety of compounds with different activities, possibly involving new mechanisms of action.

The development of new plant-based additives as alternatives or replacements for AGPs requires rigorous efficacy and safety testing. Informative and well-planned laboratory testing can provide indications of *in vitro* efficacy, for example antibacterial activity against pathogenic, including resistant, bacterial strains of economic importance. A long record of traditional use of a plant does not confirm its safety; comprehensive toxicity tests are also required. Clinical tests, preferably those accurately reflecting field conditions of animal housing and care, will provide useful indications of effectiveness and lack of toxicity. There are also other concerns to be addressed in studies relating to the development of AGP alternatives. As well as safety issues and the possible presence of residues in animals fed with natural growth-promoting substances, the interaction of the supplemented plant materials with one another and with therapeutically administered veterinary medications needs to be studied.

A variety of herbal extracts have been tested for their potential to enhance growth and reduce the occurrence of bacterial diseases in production animals, including poultry, pigs, and ruminants. EOs derived from aromatic plants are also promising candidates for AGP alternatives, as they have long been used traditionally as therapies against infections. Many EOs have been shown to have antimicrobial properties and some can enhance flavor and fragrance, although others may negatively affect feed palatability.

Much work remains to be done on the development of alternatives to AGPs derived from plants. The use of ethnoveterinary medicine to treat companion animals is also growing in popularity, and there are many possibilities for the therapeutic use of plant remedies to treat infections, particularly those where multidrug-resistant bacteria are implicated, in all animals. Possible leads to inform targeted searches for new AGP alternatives could be sourced from the extensive lists of medicinal plants used in traditional ethnoveterinary medicine around the world and reviewed in several recent publications [63–73].

This chapter is by no means an exhaustive review of all of the tests and clinical trials that have been performed on plant extracts and EOs with potential value as AGP alternatives, but merely aims to provide an idea of the alternatives available. There are fertile possibilities for future research, and it is heartening to note the increasing scientific interest and activity in this area.

References

[1] Gnanou JC, Sanders P. Antibiotic resistance in bacteria of animal origin: methods in use to monitor resistance in EU countries. Int J Antimicrob Agents 2000;15:311–22.

[2] Anonymous. A review of antimicrobial resistance in the food chain. A technical report for MAFF; 1998.

[3] Swann MM. Report of Joint Committee on the use of antibiotics in animal husbandry and veterinary medicine. London: HMSO; 1969.

[4] Dibner JJ, Richards JD. Antibiotic growth promoters in agriculture: history and mode of action. Poult Sci 2005;84:634—43.

[5] Morris RS. The application of economics in animal health programmes: a practical guide. Scientific and Technical Review of the OIE. Office International des Epizooties 1999:305—14.

[6] Vander Walt ML. Campylobacter jejuni infection. In: Coetzer JAW, Tustin RC, editors. Infectious Diseases of Livestock. Cape Town: Oxford University Press; 2004. p. 1479—83.

[7] Godfroid J, Bosman PP, Herr S, Bishop GC. Bovine brucellosis. In: Coetzer JAW, Tustin RC, editors. Infectious Diseases of Livestock. Cape Town: Oxford University Press; 2004. p. 1510—27.

[8] Henton MM, Eagar HA, Swan GE, van Vuuren M. Part VI. Antibiotic management and resistance in livestock production. S Afr Med J 2011;101:583—6.

[9] Tenover FD. Mechanisms of antimicrobial resistance in bacteria. Am J Med 2006;119:S3—S10.

[10] Mendonça-Filho RR. Bioactive phytocompounds: new approaches in the phytosciences. In: Ahmad I, Aqil F, Owais M, editors. Modern Phytomedicine: turning medicinal plants into drugs. Weinheim: Wiley-VC; 2006. p. 1—24.

[11] McManus MC. Mechanisms of bacterial resistance to antimicrobial agents. Am J Health System-Pharm 1997;54:1420—33.

[12] Huyghebaert G, Ducatelle R, Immerseel F. An update on alternatives to antimicrobial growth promoters for broilers. Vet J 2011;187:182—8.

[13] Jukes TH, Stokstad ELR, Taylor RR, Combs TJ, Edwards HM, Meadows GB. Growth promoting effect of aureomycin on pigs. Arch Biochem 1950;26: 324—30.

[14] Starr MP, Reynolds DM. Streptomycin resistance of coliform bacteria from turkeys fed streptomycin. In: Proceedings of the 51st General Meeting. Chicago, IL: Society of American Bacteriology; 1951. p. 15—34.

[15] Barnes EM. The effect of antibiotic supplements on the faecal streptococci (Lancefield group D) of poultry. Br Vet J 1958;114:333—44.

[16] Elliott SD, Barnes EM. Changes in serological type and antibiotic resistance on Lancefield group D streptococci in chickens receiving dietary chlortetracycline. J Gen Microbiol 1959;20:426—33.

[17] Witte W. Selective pressure by antibiotic use in livestock. Int J Antimicrob Agents 2000;16(Suppl. 1):S19—24.

[18] Greko C. Safety aspects on non-use of antimicrobials as growth promoters. In: Piva A, Bach Knudsen KE, Lindberg JE, editors. Gut Environment of Pigs. Nottingham, UK: Nottingham University Press; 2001. p. 219—30.

[19] Roe MT, Pillai SD. Monitoring and identifying antibiotic resistance mechanisms in bacteria. Poult Sci 2003;82:622—6.

[20] Aarestrup FM. Effects of termination of AGP use on antimicrobial resistance in food animals. In: Working papers for the WHO international review panels evaluation. Geneva, Switzerland: World Health Organization; 2003. p. 6—11. Document WHO/CDS/ CPE/ZFK/2003.1a.

[21] World Health Organization. The medical impact of the use of antimicrobials in food animals. In: Report of a WHO meeting; 1997. p. 1—39. Berlin, Germany.

[22] World Health Organization. WHO Global Principles for the Containment of Antimicrobial Resistance in Animals Intended for Food. In: Document WHO/ CDS/CSR/APH/2000.4. Geneva, Switzerland: WHO; 2000. p. 1—23.

[23] Angulo FJ. Impacts of antimicrobial growth promoter termination in Denmark. In: Proceedings of the 53rd Western Poultry Disease Conference; 2004. p. 16—9. Sacramento, CA.

[24] Durmic Z, McSweeney CS, Kemp GW, Hutton P, Wallace RJ, Vercoe PE. Australian plants with potential to inhibit bacteria and processes involved in ruminal biohydrogenation of fatty acids. Anim Feed Sci Technol 2008;145:271—84.

[25] Emborg HD, Ersboll AK, Heuer OE, Wegener HC. Effects of termination of antimicrobial growth promoter use for broiler health and productivity. In: Working papers for the WHO international review panel's evaluation. Geneva, Switzerland: World Health Organization; 2002. p. 38—42. Document WHO/CDS/CPE/ZFK/2003.1a.

[26] Callesen J. Effects of termination of AGP use on pig welfare and productivity. In: Working papers for the WHO international review panels evaluation; 2003. p. 43—6. WHO/CDS/CPE/ZFK/2003.1a.

[27] Aarestrup FM, Seyfarth AM, Emborg HD, Pedersen K, Hendriksen RS, Bager F. Effect of abolishment of the use of antimicrobial agents for growth promotion on occurrence of antimicrobial resistance in fecal enterococci from food animals in Denmark. Antimicrob. Agents Chemother 2001;45:2054—9.

[28] Boerlin P, Wissing A, Aarestrup FM, Frey J, Nicolet J. Antimicrobial growth promoter ban and resistance to macrolides and vancomycin in enterococci from pigs. J Clin Microbiol 2001;39:4193—5.

[29] Millet S, Maertens L. The European ban on antibiotic growth promoters in animal feed: from challenges to opportunities. Vet J 2011;187:143—4.

[30] Eloff JN, McGaw LJ. Application of plant extracts and products in veterinary infections. In: Ahmad I, Aqil F, editors. New strategies combating bacterial infection. Weinheim: Wiley-VCH; 2009. p. 205–28.

[31] Fuller R. Probiotics in man and animals. J Appl Bacteriol 1989;66:365–78.

[32] Hume ME. Food safety symposium: potential impact of reduced antibiotic use and the roles of prebiotics, probiotics and other alternatives in antibiotic-free broiler production. Historic perspective: prebiotics, probiotics, and other alternatives to antibiotics. Poult Sci 2011;90:2663–9.

[33] Gibson GR, Roberfroid MB. Dietary modulation of the human colonic microbiota: introducing the concept of prebiotics. J Nutr 1995;125:1401–12.

[34] Verstegen MWA, Williams BA. Alternatives to the use of antibiotics as growth promoters for monogastric animals. Anim Biotechnol 2002;13:113–27.

[35] Wynn SG, Fougére BJ. Introduction: why use herbs?. In: Wynn SG, Fougére BJ, editors. Veterinary Herbal Medicine. Missouri, USA: Mosby Inc. (Elsevier); 2007. p. 1–4.

[36] Franz C, Başer KHC, Windisch W. Essential oils and aromatic plants in animal feeding—a European perspective. A review. Flavour and Fragrance J 2010;25:327–40.

[37] Van Asseldonk T. Ethnoveterinary medical practice in the European Union (EU). A case study of the Netherlands. In: Katerere DR, Luseba D, editors. Ethnoveterinary botanical medicine: herbal medicines for animal health. Boca Raton: CRC Press (Taylor and Francis Group); 2010. p. 373–87.

[38] Van Vuuren S, Viljoen A. Plant-based antimicrobial studies—methods and approaches to study the interaction between natural products. Planta Med 2011;77:1168–82.

[39] Chikoto H. Development of a product derived from plant extracts to replace antibiotic feed additives used in poultry production. PhD thesis. Pretoria: University of Pretoria; 2006.

[40] Van Heerden I. Evaluation of a phytogenic product from two western herbal medicines to replace an antimicrobial growth promoter in poultry production. PhD thesis. Pretoria: University of Pretoria; 2009.

[41] Surai PF. Natural antioxidants in avian nutrition and reproduction. Nottingham: Nottingham University Press; 2002.

[42] Grimble RF. Effect of antioxidative vitamins on immune function with clinical applications. Int J Vitam Nutr Res 1997;67:312–20.

[43] Khan RU, Naz S, Javdani M, Nikousefat Z, Selvaggi M, Tufarelli V, et al. The use of turmeric (Curcuma longa) in poultry feed. World's Poult Sci J 2012;68:97–103.

[44] Viksveen P. Antibiotics and the development of resistant microorganisms. Can homeopathy be an alternative? Homeopathy 2003;92:99–107.

[45] Gehring R, Kindscher K. The medicinal use of native North American plants in domestic animals. In: Katerere DR, Luseba D, editors. Ethnoveterinary botanical medicine: herbal medicines for animal health. Boca Raton: CRC Press (Taylor and Francis Group); 2010. p. 213–29.

[46] Baby S, George V. Essential oils and new antimicrobial strategies. In: Ahmad I, Aqil F, editors. New strategies combating bacterial infection. Weinheim: Wiley-VCH; 2009. p. 165–203.

[47] Platel K, Srinivasan M. Digestive stimulant action of spices: A myth or reality? Indian J Med Res 2004;119: 167–79.

[48] Jamroz D, Wertelecki T, Houszka M, Kamel C. Influence of diet type on the inclusion of plant origin active substances on morphological and histochemical characteristics of the stomach and jejunum walls in chicken. J Anim Physiol Anim Nutr (Berl) 2006;90: 255–68.

[49] Manzanilla EG, Perez JF, Martin M, Kamel C, Baucells F, Gasa J. Effect of plant extracts and formic acid on the intestinal equilibrium of early-weaned pigs. J Anim Sci 2004;82:3210–8.

[50] Losa R, Kohler B. Prevention of colonization of Clostridium perfringens in broilers intestine by essential oils. In: 13th European Symposium on Poultry Nutrition. Blankenberge: WPSA; 2001.

[51] Mitsch P, Zitterl-Eglseer K, Kohler B, Gabler C, Losa R, Zimpernik I. The effect of two different blends of essential oil components on the proliferation of Clostridium perfringens in the intestines of broiler chickens. Poult Sci 2004;83:669–75.

[52] Wallace RJ, Colombatto D, Robinson PH. Enzymes, direct-fed microbials and plant extracts in ruminant nutrition. Anim Feed Sci Technol 2008;145:1–4.

[53] Alexander G, Singh B, Sahoo A, Bhat TK. In vitro screening of plant extracts to enhance the efficiency of utilization of energy and nitrogen in ruminant diets. Anim Feed Sci Technol 2008;145:229–44.

[54] Benchaar C, Calsamiglia S, Chaves AV, Fraser GR, Colombatto D, McAllister TA, et al. A review of plant-derived essential oils in ruminant nutrition and production. Anim Feed Sci Technol 2008;145:209–28.

[55] Spanghero M, Zanfia C, Fabbro E, Scicutella N, Camellini C. Effects of a blend of essential oils on some end products of in vitro rumen fermentation. Anim Feed Sci Technol 2008;145:364–74.

[56] Wang Y, Xua Z, Bachb SJ, McAllister TA. Effects of phlorotannins from Ascophyllum nodosum (brown seaweed) on in vitro ruminal digestion of mixed

forage or barley grain. Anim Feed Sci Technol 2008;145:375—95.

[57] Greathead H. Plants and plant extracts for improving animal productivity. Proc Nutr Soc 2003;62:279—90.

[58] Hart KJ, Yanez-Ruiz DR, Duval SM, McEwan NR, Newbold CJ. Plant extracts to manipulate rumen fermentation. Anim Feed Sci Technol 2008;147:8—35.

[59] Yarnell E. Plant chemistry in veterinary medicine: medicinal constituents and their mechanisms of action. In: Wynn SG, Fougére BJ, editors. Veterinary Herbal Medicine. Missouri: Mosby Inc. (Elsevier); 2007. p. 159—82.

[60] Naidoo V, Seier J. Preclinical safety testing of herbal remedies. In: Katerere DR, Luseba D, editors. Ethnoveterinary botanical medicine: herbal medicines for animal health. Boca Raton: CRC Press (Taylor and Francis Group); 2010. p. 69—93.

[61] Bandaranayake WM. Quality control, screening, toxicity, and regulation of herbal drugs. In: Ahmad I, Aqil F, Owais M, editors. Modern phytomedicine: turning medicinal plants into drugs. Weinheim: Wiley-VCH; 2006. p. 25—57.

[62] Wierup M. The control of microbial diseases in animals: alternatives to the use of antibiotics. Int J Antimicrob Agents 2000;14:315—9.

[63] McGaw LJ, Eloff JN. Ethnoveterinary use of southern African plants and scientific evaluation of their medicinal properties. J Ethnopharmacol 2008;119: 559—74.

[64] Balakrishnan Nair MN, Unnikrishnan PM. Revitalizing ethnoveterinary medical traditions: a perspective from India. In: Katerere DR, Luseba D, editors. Ethnoveterinary botanical medicine: herbal medicine for animal health. Boca Raton: CRC Press (Taylor and Francis Group); 2010. p. 95—124.

[65] Iqbal Z, Jabbar A. Inventory of traditional veterinary botanicals from around the world. In: Katerere DR, Luseba D, editors. Ethnoveterinary botanical medicine: herbal medicines for animal health. Boca Raton: CRC Press (Taylor and Francis Group); 2010. p. 125—64.

[66] Sithambaram S, Marimuthu M, Panchadcharam C. The current status and future prospects of medicinal and aromatic plants in veterinary health care in

southeast Asia. In: Katerere DR, Luseba D, editors. Ethnoveterinary botanical medicine: herbal medicines for animal health. Boca Raton: CRC Press (Taylor and Francis Group); 2010. p. 165—94.

[67] Pearson W, Lindinger MI. Evidence-based botanicals in North America. In: Katerere DR, Luseba D, editors. Ethnoveterinary botanical medicine: herbal medicines for animal health. Boca Raton: CRC Press (Taylor and Francis Group); 2010. p. 195—211.

[68] Da Nóbrega Alves RR, Barboza RRD, Souto WMS. Plants used in animal health care in South and Latin America: an overview. In: Katerere DR, Luseba D, editors. Ethnoveterinary botanical medicine: herbal medicines for animal health. Boca Raton: CRC Press (Taylor and Francis Group); 2010. p. 231—55.

[69] Maphosa V, Tshisikhawe P, Thembo K, Masika P. Ethnoveterinary medicine in southern Africa. In: Katerere DR, Luseba D, editors. Ethnoveterinary botanical medicine: herbal medicines for animal health. Boca Raton: CRC Press (Taylor and Francis Group); 2010. p. 257—88.

[70] Githiori J, Gathumbi PK. Ethnoveterinary plants used in East Africa. In: Katerere DR, Luseba D, editors. Ethnoveterinary botanical medicine: herbal medicines for animal health. Boca Raton: CRC Press (Taylor and Francis Group); 2010. p. 289—302.

[71] Boukraa L, Benbarek H, Benhanifa M. Herbal medicines for animal health in the Middle East and North Africa (MENA) region. In: Katerere DR, Luseba D, editors. Ethnoveterinary botanical medicine: herbal medicines for animal health. Boca Raton: CRC Press (Taylor and Francis Group); 2010. p. 303—19.

[72] Okoli IC, Tamboura HH, Hounzangbe-Adote MS. Ethnoveterinary medicine and sustainable livestock management in West Africa. In: Katerere DR, Luseba D, editors. Ethnoveterinary botanical medicine: herbal medicines for animal health. Boca Raton: CRC Press (Taylor and Francis Group); 2010. p. 321—51.

[73] Teng L, Shaw D, Barnes J. Traditional Chinese veterinary medicine. In: Katerere DR, Luseba D, editors. Ethnoveterinary botanical medicine: herbal medicines for animal health. Boca Raton: CRC Press (Taylor and Francis Group); 2010. p. 353—72.

Essential Oils from the Asteraceae Family Active against Multidrug-Resistant Bacteria

María José Abad, Luis Miguel Bedoya, Paulina Bermejo

Department of Pharmacology, Faculty of Pharmacy, University Complutense, Ciudad Universitaria s/n, Madrid 28040, Spain

INTRODUCTION

Infectious diseases are a significant cause of morbidity and mortality worldwide, accounting for approximately 50% of all deaths in tropical countries and as much as 20% of deaths in the Americas [1]. With the advent of globalization, health threats have become much more serious in an increasingly interconnected world, characterized by the higher mobility of people, animals and goods, and economic interdependence. In addition, the impact of global climate change on the development of infectious diseases is well depicted by the classic "disease triangle." In this paradigm, the pathogen, with its virulence factors and inoculum effect, serves as a component cause for disease only in the presence of a susceptible host, via immune modulation and behavioral changes, and a favorable environment. The net emission of greenhouse gases resulting from fossil fuel consumption, deforestation, and altered land use has led to a conducive environment for the emergence of several infectious diseases beyond their previously defined niches. According to the World Health Organization, at least 39 new pathogens have been identified since 1967. Despite the significant progress made in microbiology and the control of microorganisms, sporadic epidemics due to drug-resistant microorganisms and hitherto unknown disease-causing microbes pose an enormous threat to public health. These negative threat trends require a global initiative for the development of new strategies to prevent and treat infectious diseases.

Today, the importance of microbes and medicinal plants as a source of new antimicrobials is well established [2–4]. In 1877, Pasteur and Joubet were among the first to recognize the potential of microbial products as therapeutic agents and demonstrated that common microorganisms could inhibit the growth of Anthrax bacilli. However, the watershed in the

Fighting Multidrug Resistance with Herbal Extracts,
Essential Oils and Their Components
http://dx.doi.org/10.1016/B978-0-12-398539-2.00014-8

field of antimicrobial agents was the advent of penicillin in 1928 by Alexander Fleming, obtained from a strain of the *Penicillium* mold. Since then, fungi and higher plants have been scoured for the production and preparation of novel antibacterial compounds, including aminoglycosides and cephalosporins. However, due to their increasing use and consequent selection pressure, bacteria have started exhibiting resistance to these compounds; hence the urgent need to review and search for new antibacterial compounds derived from plant species. This is a timely exercise, given the continuing and developing problems of bacterial resistance, in particular multidrug resistance (MDR). MDR results in part because of the activation of efflux pumps. The infectious diseases caused by MDR bacteria such as methicillin-resistant *Staphylococcus aureus* (MRSA), penicillin-resistant *Streptococcus pneumoniae*, and vancomycin-resistant enterococci are a very serious problem, as in many cases these bacteria are resistant to a number of antibacterials.

Additionally, food-borne illnesses associated with Gram-positive and Gram-negative bacteria such as *Bacillus cereus*, *Escherichia coli*, *Listeria monocytogenes*, *Salmonella enteritidis*, and *S. aureus* present a major public health concern throughout the world. Microbial contamination and oxidation in foods and drinks not only result in food deterioration and shelf-life reduction, but also lead to disease and economic losses. At present, the food industry is facing enormous pressure due to food deterioration caused by microbial contamination and oxidation. The growth of microorganisms in food products may cause intestinal disorders, vomiting, and diarrhea. In addition to the difficulties that antimicrobial resistance presents to the healthcare sector and pharmaceutical industry, recent reports on antibiotic resistance among food-borne bacteria show that the food and drink industry is increasingly developing interventions to reduce pathogens in foods and protect consumers.

Since the early 1990s, the search for antibacterial potential in medicinal plants has grown tremendously [5]. Plant-based drugs have been used worldwide in traditional medicines for the treatment of a variety of diseases. Approximately 60% of the world population relies on medicinal plants for primary health care. These medicinal plant species serve as a rich source of many biologically active compounds, although the medicinal properties of only a few plant species have been thoroughly investigated. Among them, aromatic plants have been used since ancient times for their preservative and medicinal properties, as well as to impart aroma and flavor to food. For instance, Hippocrates, the so-called *father of medicine*, prescribed aromatic fumigations. The pharmaceutical properties of aromatic plants are partly attributable to essential oils (EOs) [6].

EOs are very complex mixtures of volatile molecules produced by the secondary metabolism of aromatic and medicinal plants, and can be obtained by different methods, including low- or high-pressure distillation of different plant parts, or the use of liquid carbon dioxide or microwaves. Several factors can influence the quality and quantity of the extracted product, in particular the soil composition, plant organ, vegetative cycle phase, and climate. EO composition can be divided into two component groups: the main group is usually of terpenoid origin, and the second comprises aromatic and aliphatic components. In general, monoterpenes and sesquiterpenes, as well as their oxygenated derivatives, are the predominant constituents, but phenylpropanoids and both fatty acids and their esters may also be present. Aromatic plants and their EOs have been used since antiquity for their biologic properties (e.g., antibacterial, antifungal, antiparasitic, antiviral, and insecticidal), as well as for cosmetic and medicinal applications [7]. In particular, the antibacterial activity of plant oils and the compounds isolated from these oils has formed the basis of many applications, including the preservation of raw and processed food, pharmaceuticals, alternative medicines, and natural therapies [8,9].

Numerous members of the Asteraceae family are important crop species of cut flowers and ornamentals, as well as being medicinal and aromatic plants, many of which produce EOs used in folk medicine and in the cosmetic and pharmaceutical industries [10,11]. EOs generally have a broad spectrum of bioactivity owing to the presence of several active ingredients with various modes of action. Extensive studies of Asteraceae species have led to the identification of many EOs and chemical compounds with interesting antibacterial activities. This review summarizes some of the main reports on the chemistry and antibacterial activity of Asteraceae EOs from recent literature (from 2005 to May 2012).

BACTERIA WITH CLINICAL IMPORTANCE

Gram-Positive Bacteria

The human body, in particular the skin and the oral cavity, is constantly exposed to a variety of bacteria, but most individuals maintain a healthy homeostasis. Members of the *Staphylococcus* genus are Gram-positive cocci measuring 0.5–1.5 µm in diameter. They appear as individual cells or as irregular grape-like clusters that are motionless, nonspore forming, and catalase positive. Moreover, most species are facultative anaerobes. *S. aureus* is the most important species and the agent of a variety of infections. In the past, nosocomial infections caused by such microorganisms have been major causes of morbidity and mortality; they are currently among the 10 main causes of death worldwide and the underlying cause of death in 1% of cases [12]. *S. aureus* causes ailments ranging from superficial skin infections to life-threatening diseases such as endocarditis, sepsis and infections of the bloodstream, intestinal tract, respiratory tract, soft tissues, and urinary tract. The ability of *S. aureus* to survive in the eukaryotic intracellular environment may explain several aspects of chronic staphylococcal diseases and long-term colonization. Internalization may provide a means of protection against host defenses and certain classes of antibiotics (methicillin, penicillin, and vancomycin) [13]. Staphylococcal infections are typically associated with tissue death, and evidence suggests that intracellular bacteria are capable of inducing apoptosis. *S. aureus*-mediated apoptosis has been reported in epithelial cells, keratinocytes, endothelial cells, and osteoblasts. In addition, some strains are responsible for food poisoning through the production of an enterotoxin, and pathogenicity is also associated with coagulase positivity.

Another member of the *Staphylococcus* genus, *Staphylococcus epidermidis*, is now recognized as a leading causative agent of nosocomial infections primarily due to its biofilm-forming tendencies. *S. aureus* and *S. epidermidis* form communities (called biofilms) on inserted medical devices, leading to infections that affect many millions of patients worldwide, causing substantial morbidity and mortality. These healthcare-associated infections are a particular problem in both adult and neonatal intensive care units. As biofilms are resistant to antibiotics, device removal is often required to resolve the infection.

Resistance of *S. aureus* to antibiotics appeared within a few years of the start of the antibiotic era, and this problem has reached epic proportions owing to the overuse and improper use of antibiotics. Staphylococci resistance to antibiotics has resulted in the emergence of MDR *Staphylococcus* strains. These facts highlight the need for the development of new and novel antibiotics or improvements in the efficacy of established antibiotics by developing new agents capable of enhancing their antibiotic activity.

Other Gram-positive bacteria include *Streptococcus* spp. *Streptococcus* is a genus of spherical

Gram-positive bacteria belonging to the phylum Firmicutes, which includes food-associated, commensal, and pathogenic species [12]. Acute *Streptococcus pyogenes* infections may take the form of cellulitis, erysipelas, impetigo, pharyngitis, or scarlet fever (rash). Invasive infections can result in myositis, necrotizing fasciitis, and streptococcal toxic shock syndrome. Patients may also develop immune-mediated sequelae such as acute glomerulonephritis and acute rheumatic fever. *S. pneumoniae* causes pneumonia, meningitis, and sometimes occult bacteremia. Community-acquired bacteremia with *S. pneumoniae* is a frequent disease with an annual reported incidence of 9–17 cases per 100,000 persons [14]. The mortality associated with this disease has been reported to vary between 5 and 26%. In addition, *Streptococcus* spp. have been implicated as primary causative agents of dental caries. Caries are caused by a number of microbial species, among which the Gram-positive *Streptococcus mutans* plays a leading role. It has recently been suggested that oral bacteria are also associated with many systemic diseases, such as cardiovascular disease and pneumonia. This highlights the importance of developing an effective mouthwash that is acceptable to the general public and without side effects. Several antibiotics have numerous side effects and resistance has developed in *Streptococcus* ssp. Most of these pathogens have resistant mechanisms that include efflux pumps.

Other Gram-positive bacteria are *B. cereus* and *L. monocytogenes*, which cause food-borne illnesses [12]. Food- and water-borne pathogens continue to be a major cause of mortality in developing countries and cause significant morbidity in developed nations. *L. monocytogenes* is a widespread food-borne pathogen that causes a severe, potentially fatal illness (listeriosis) in animals and humans [15]. Most cases of human listeriosis are associated with the consumption of contaminated food. This microorganism has the ability to invade many types of nonphagocytic cells and spread from cell to cell, crossing important barriers in host organisms. Despite intensified surveillance in food manufacturing, serious cases of listeriosis are still reported. *B. cereus* is a well-known cause of food poisoning; however, it can also cause serious invasive diseases including bacteremia, septicemia, endocarditis, osteomyelitis, pneumonia, brain abscess and meningitis in severely immunocompromised patients such as those with a hematologic malignancy and in patients with indwelling vascular catheters [16]. Prior contamination of medical fluids and devices with *B. cereus* has been reported.

Gram-Negative Bacteria

Diarrhea caused by *E. coli* infection is an emergent problem in both developing and developed countries, and is responsible for high rates of mortality in newborns [17]. Although commensal representatives found in the intestinal flora of humans and animals are nonpathogenic, certain strains are highly pathogenic. All diarrheagenic strains of *E. coli* were initially termed *enteropathogenic*, but as their pathogenic mechanisms became known, they were grouped into five categories: enterotoxigenic, enteropathogenic, enterohemorrhagic, enteroinvasive, and enteroaggregative. These have evolved from nonpathogenic commensal strains by the acquisition of specific virulence genes through mobile genetic elements. They have different pathogeneses, produce distinct clinical syndromes and pathologic lesions, and have different epidemiologic characteristics. Enterotoxigenic *E. coli* are nowadays considered a major cause of *E. coli*-associated diarrhea worldwide, producing one or both types of enterotoxins: heat-stable enterotoxin and heat-labile enterotoxin. The diarrhea is often treated with antibacterial drugs, but this treatment is generally ineffective, due in part to the presence of drug-resistant strains and a failure to identify drug sensitivity. This fact underscores the need for the rapid development of antibacterial drugs that are more effective than those currently in use. In addition, *E. coli* is an important food-borne bacterial pathogen.

Important food- and drink-borne pathogens also include species or strains of *Salmonella*. *Salmonella enterica* is the causative agent of a spectrum of diseases, including enteric fever (typhoid) and self-limiting gastroenteritis, and remains a significant food-borne pathogen throughout both developed and developing countries. *S. enterica* comprises generalist serovars, such as *Salmonella enterica* serovar Typhimurium, and host-adapted strains such as *S. enterica* serovar Typhi. *S. enterica* enters the host through the gastrointestinal tract and translocates by multiple mechanisms to systemic tissues. The ability to actively invade and reside within gut epithelia and macrophages is an important process in the establishment of salmonella infection, generating localized inflammatory responses and facilitating the systemic spread of the pathogen within the host.

Another Gram-negative bacterium is *Helicobacter pylori*. This microorganism is one of the world's most common infective bacteria, infecting around 50% of the world's population, with its highest prevalence in developing countries. Although most infected individuals have no symptoms, *H. pylori* is a major cause of peptic ulcers and chronic gastritis in adults and children. Moreover, long-term infection may produce gastric atrophy and intestinal metaplasia of the stomach in some individuals and lead to the development of adenocarcinoma and gastric lymphoma in adulthood. *H. pylori* infection is still poorly managed and the eradication failure rate remains as high as 5—20%. Resistance to antibiotics is increasingly recognized as a major contributory factor in eradication failure [18].

Pseudomonas aeruginosa is a noncapsule and nonsporing Gram-negative bacillus that most commonly affects the lower respiratory system, and is a main cause of high morbidity and mortality among immunocompromised individuals. MDR strains have caused an increase rate of nosocomial infections in recent years, which is evolving into a major clinical problem and raising expectations for the development of novel therapeutic drug.

ESSENTIAL OILS FROM THE ASTERACEAE FAMILY

In recent years, there has been considerable interest in assaying the composition and antibacterial activity of EOs from Asteraceae species from around the world [19]. EOs generally have a broad spectrum of bioactivity, owing the presence of several active ingredients or secondary metabolites, which work through various modes of action. Differences that have been observed in the antibacterial activities of EOs between Gram-negative and Gram-positive bacteria may be attributed to differences in the composition and structure of the bacterial outer membrane and cell wall, which are among the primary sites of drug activity in these organisms. The main plant genera investigated are described below and are selected on the basis of their common use as medicinal plants. Reported constituents of Asteraceae species tested for their antibacterial activity are outlined in Table 14-1.

Artemisia Species

The genus *Artemisia* L. is one of the largest genera in the Asteraceae family, consisting of more than 500 species distributed mainly in the northern temperate regions of the world. Many species have been used in various treatments since ancient times as folk remedies.

Artemisia annua L. is a fragrant annual herb widely distributed in Asia, Europe, and North America. The use of this plant (commonly named *sweet wormwood, annual wormwood,* or *qinghaosu*) in Chinese traditional medicine was

TABLE 14-1   Essential Oils from Asteraceae Species Tested for Antibacterial Activity Against Standard Organisms

Compound	Asteraceae Species	Organisms	References
Artemisia ketone	*Artemisia annua*	*Staphylococcus aureus, Streptococcus pneumoniae*	[20]
Camphor	*Artemisia annua*	*S. aureus, S. pneumoniae*	[20]
	Artemisia echegarayi	*Bacillus cereus, Listeria monocytogenes*	[21]
	Tanacetum parthenium	*S. aureus*	[22]
Caryophylladienol	*Achillea cretica*	*B. cereus, S. aureus*	[23]
Chamazulene	*Matricaria recutita*	*Helicobacter pylori, Propionibacterium acnes*	[24,25]
1,8-Cineole	Turkish *Artemisia* spp.	*Escherichia coli, Pseudomonas aeruginosa, S. aureus, Streptococcus pyogenes*	[26]
	Artemisia herba-alba	*E. coli, Salmonella* ser. Typhimurium	[27]
	Helichrysum cymosum subsp. *cymosum*	*B. cereus, E. coli, P. aeruginosa, S. aureus*	[28]
β-Eudesmol	Turkish *Centaurea* spp.	*S. aureus*	[29]
Fraganol	*Achillea ligustica*	*S. aureus*	[30]
Geraniol	*Helichrysum italicum*	*E. coli, P. aeruginosa*	[31]
Limonene	*Tanacetum trifurcatum*	*Staphylococcus epidermidis*	[32]
	Baccharis trinervis	*S. aureus*	[33]
Methyl eugenol	*Centaurea chamaerhaponticum*	*E. coli, Salmonella* ser. Typhimurium, *S. aureus, S. epidermidis*	[34]
Nuciferyl esters	*Artemisia absinthium*	*P. aeruginosa, S. aureus*	[35]
α-Pinene	*Tanacetum parthenium*	*E. coli, Salmonella* ser. Typhimurium	[36]
	Tanacetum argenteum subsp. *argenteum*	*B. cereus*	[37]
	Tanacetum argenteum flabellifolium	*P. aeruginosa*	[38]
Santolina alcohol	*Achillea ligustica*	*E. coli, P. aeruginosa, S. aureus*	[39]
Thujone	*Artemisia herba-alba*	*B. cereus, S. aureus*	[27]
	Artemisia arborescens	*L. monocytogenes*	[40]
	Artemisia echegarayi	*B. cereus, L. monocytogenes*	[21]
	Tanacetum balsamita subsp. *balsamita*	*E. coli, P. aeruginosa, S. aureus, S. epidermidis*	[41]
Vulgarone B	*Artemisia iwayomogi*	*S. aureus*	[42]

recorded before 168 BC. In 1971, artemisinin, a sesquiterpene lactone with antimalarial properties, was isolated from this plant. The search for other active compounds has led to the discovery and isolation of many phytochemicals and EOs with interesting antibacterial activity. Hydrodistilled volatile oil obtained from the aerial parts of *A. annua* cultivated near Sarajevo,

Bosnia, was analyzed by gas chromatography-mass spectroscopy (GC-MS) [20]. More than 100 compounds were identified, representing 95.5% of the total oil. The major constituents of the EO were oxygenated monoterpenes, artemisia ketone (30.7%; Fig. 14-1), and camphor (15.8%). The high percentages of these compounds proved that this EO clearly belongs to the monoterpene chemotype. In contrast, the content of sesquiterpene compounds was relatively low (18.2%). This EO (10 mg/mL) exhibited antimicrobial activity against strains of *S. aureus*, a common species of staphylococci that causes infections, and *S. pneumoniae*. *Artemisia camphorata* L. showed activity against enterotoxigenic, enteropathogenic, and enteroinvasive *E. coli*, with minimum inhibitory concentrations (MICs) of 500–800 μg/mL [43].

Kordali et al. [26] investigated the chemical composition and antibacterial activities of four Turkish *Artemisia* ssp., *Artemisia absinthium* L. (known locally as *pelin otu*), *Artemisia dracunculus* L. (tarragon), *Artemisia santonicum* L. (*deniz tavşani*), and *Artemisia spicigera* K. Koch. (*yavsan*). The inhibitory effects of these EOs at 600, 900, and 1200 μg/disk concentrations on the growth of 64 bacterial strains, including those of plant, food, and clinical origin, were tested. EOs of *A. santonicum* and *A. spicigera* showed the highest antibacterial activities with the broadest spectrums (against clinical isolates of *E. coli*, *P. aeruginosa*, *S. aureus*, and *S. pyogenes*, among others). Both EOs have a similar chemical composition, the major constituents being camphor, 1,8-cineole (Fig. 14-2), borneol, terpinen-4-ol, bornyl acetate, and spathulenol. Thus, their similar antibacterial activities can

FIGURE 14-2 Structure of 1,8-cineole.

be attributed to their similar chemical compositions. Another *Artemisia* EO with antibacterial activity is that of *Artemisia incana* (L.) Druce, which exhibited considerable inhibitory effects against food-borne bacteria [44].

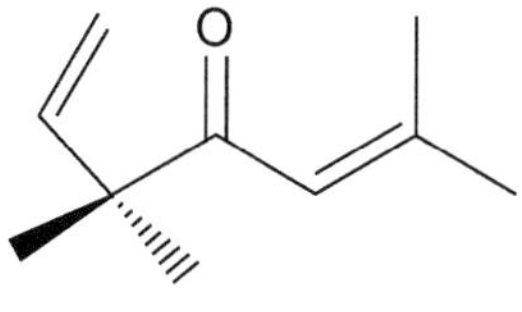

The EO of one of these species growing wild in southeast Serbia, *A. absinthium*, was also active against *S. aureus* and *P. aeruginosa* [35]. GC-MS analyses of the oil revealed that the main constituents were β-thujone and *cis*-β-epoxyocimene. The oil also contained nuciferyl esters (Fig. 14-3). These rather uncommon EO constituents, however, seems to be quite characteristic for species belonging to the *Artemisia* genus. Lopes-Lutz et al. [45] investigated the chemical composition and antibacterial activity of *A. absinthium* EO, together with those of other *Artemisia* species growing wild in the prairies of

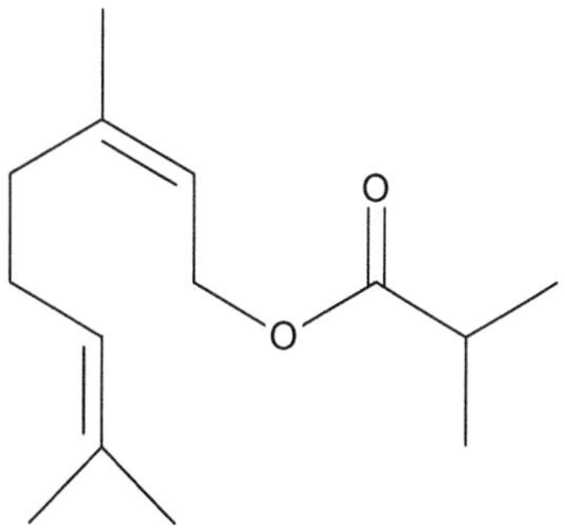

FIGURE 14-1 Structure of artemisia ketone.

FIGURE 14-3 Structure of nuciferyl esters.

central Alberta in western Canada, including *Artemisia biennis* Willd., *Artemisia cana* Pursh, *A. dracunculus*, *Artemisia frigida* Willd., *Artemisia longifolia* Nutt, and *Artemisia ludoviciana* Nutt. All oils had inhibitory effects against *E. coli*, MRSA, and *S. epidermidis*, with the most active being the oil of *A. absinthium*. This oil was characterized by high amounts of myrcene, *trans*-thujone, and *trans*-sabinyl acetate.

Artemisia herba-alba Asso is a greenish-silver perennial dwarf shrub that grows in arid and semiarid climates. Mighri et al. [27] investigated the antibacterial activities of four EO types extracted by hydrodistillation from the aerial parts of *A. herba-alba* cultivated in southern Tunisia. Oil type III, originating from Tunisia and Morocco (with balanced amount of α- and β-thujone), was the most active against *S. aureus* and *B. cereus*, while oil type IV was the most active against *E. coli* and *Salmonella* ser. Typhimurium NRRL 4420 (Northern Regional Research Laboratory, USA). This latter oil is not described in the literature, as this is the first time that such codominance of the four main components [1,8-cineole (Fig. 14-2), β-thujone, camphor and α-thujone) has been reported. Another *Artemisia* EO with antibacterial activity is that of African *Artemisia afra* Jacq. [46].

Artemisia arborescens (Vaill.) L. is a perennial evergreen woody shrub that is widespread in the Mediterranean area (Italy and North Africa) and also in the Pacific areas of North America. Scientific interest in this plant goes back to the early 2000s. The EO of *A. arborescens* growing wild in Sicily and extracted by steam distillation was examined by GC-MS [40]. A total of 43 compounds were identified, accounting for 93.7% of the total oil. Oxygenated monoterpenes constituted the main fraction, with β-thujone being the main compound, followed by the sesquiterpene hydrocarbon chamazulene (Fig. 14-4). The undiluted EO showed a wide spectrum of inhibition against strains of *L. monocytogenes*, while it was ineffective against enterobacteria and salmonella. The highest level of

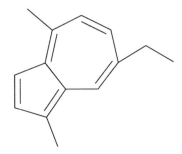

FIGURE 14-4 **Structure of chamazulene.**

inhibition was obtained against *L. monocytogenes* strains DHPS 186 and 7B0 (Department of Health Promotion Sciences, Italy), both of food origin, which survive in food-processing technologies that rely on acidic or salty conditions.

Another *Artemisia* EO with antibacterial activity against *L. monocytogenes* CIP 74904 and 74903 (Collection of Institute Pasteur, France) is *Artemisia echegarayi* Hieron., a plant commonly known in Argentina as *ajenjo* [21]. This oil also inhibited the growth of clinical isolates of *B. cereus*. Two terpenes, thujone and camphor, were identified as the principal constituents of this EO responsible for antibacterial activity. They cause leakage of cell contents due to altering membrane permeability.

Chung et al. [42] investigated the antibacterial activity of vulgarone B, a component of *Artemisia iwayomogi* Kitam. EO, against some antibiotic-susceptible and antibiotic-resistant human pathogens, including *S. aureus*. In checkerboard microtiter tests, vulgarone B and *A. iwayomogi* EO, combined with oxacillin, resulted in synergism or additive effects. Moreover, the safety of *Artemisia* oil and vulgarone B has been confirmed *in vivo*. The antibiotic mechanism may be related to DNA cleavage. Thus, vulgarone B and the EO from *A. iwayomogi* are promising candidates for safe, effective, natural agents against antibiotic-resistant *S. aureus*, especially when combined with oxacillin.

Another example of an antibacterial EO from the *Artemisia* genus is the EO of *Artemisia feddei*

H.Lév. & Vaniot., a perennial plant widely distributed across the mountains and fields of Korea, which had a considerable inhibitory effect on *E. coli*, *S. aureus*, *S. epidermidis*, and *S. pyogenes*, with MIC and minimum bactericidal concentrations (MBCs) of 0.2–3.2 mg/mL and 0.4–3.2 mg/mL, respectively [47]. However, the major compounds in the oil, borneol, α-terpineol, camphor, 1,8-cineole (Fig. 14-2), terpinen-4-ol, and β-caryophyllene (Fig. 14-5), demonstrated different degrees of growth inhibition. Other antibacterial EOs from the *Artemisia* genus include those from species from Iran, such as *Artemisia kulbadica* Boiss. & Buhse [48] and *Artemisia fragrans* Willd. [49].

Centaurea Species

The large genus *Centaurea* L. comprises approximately 500–600 species of annual, biannual, and perennial grassy plants that are distributed all around the world, especially in America, the Mediterranean area, North Africa, and western Asia. Many species of this genus have been used in folk medicine for the treatment of various ailments. Turkey is an important center for endemisms of the *Centaurea* genus. Formisano et al. [50] investigated the antibacterial activity and volatile constituents of aerial parts of three endemic *Centaurea* species from Turkey: *Centaurea amanicola*

Hub.-Mor., *Centaurea consanguinea* DC., and *Centaurea ptosimopappa* Hayek. *S. aureus* was most affected by the oil of *C. amanicola*, with an MIC of 25 µg/mL. The same MIC value was found for *B. cereus*, which was the species most affected by the oil of *C. consanguinea*. Sesquiterpenoids, fatty acids, and carbonylic compounds were the most abundant components of the oils. Other Turkish *Centaurea* species, including *Centaurea sessilis* Willd., *Centaurea armena* Boiss., and *Centaurea aladagensis* Wagenitz, also yielded EOs with antibacterial activity against *S. aureus* [29,51]. The main component of the EOs of these taxa is β-eudesmol, which is present at 12.4% and 19.3% in *C. sessilis* and *C. armena*, respectively.

Centaurea chamaerhaponticum Ball is one of the most conspicuous early spring flowering aromatic plants of the *Centaurea* genus. It is a North African endemic species distributed throughout the north and central area of Tunisia. The chemical composition of the volatile fractions obtained by steam distillation from the capitula and aerial parts of this plant was analyzed by GC-MS [34]. Of the 57 constituents identified, representing 95.5% and 96.3% of the two oils, respectively, the main components were methyl eugenol (Fig. 14-6), epi-13-manool, β-ionone, β-bisabolol, 1-octadecanol, phytol, and farnesyl acetate. Oils from both parts of *C. chamaerhaponticum*, and especially that of capitula, were found to exhibit interesting antibacterial activities against *S. aureus*

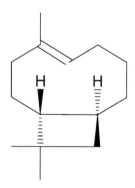

FIGURE 14-5 Structure of β-caryophyllene.

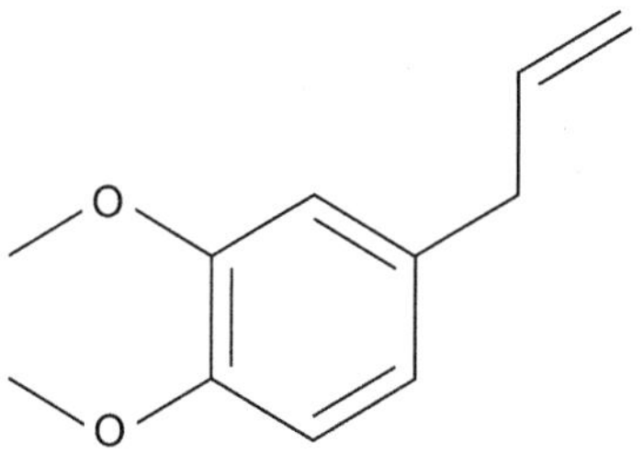

FIGURE 14-6 Structure of methyl eugenol.

(MIC and MBC of 500–800 µg/mL), *S. epidermidis* NCIMB 8853 (National Collection of Industrial Marine & Food Bacteria, Tunisia; MIC and MBC of 800 µg/mL), clinical isolates of *Salmonella* ser. Typhimurium (MIC of 500 µg/mL), and *E. coli* (MIC of 800 µg/mL). The antibacterial properties of the volatile fractions of this plant are thought to be associated with the high content of oxygenated sesquiterpenes. The bacteriostatic activity of the EO is thought to relate to the chemical configuration of its components, the proportions in which they are present, and the interactions between them. The EO from the root of another *Centaurea* species, *Centaurea carthamoides* L., also exhibited antimicrobial activity against *S. aureus* (MIC of 32 µg/mL), *L. monocytogenes* (MIC of 128 µg/mL), and *S. pyogenes* (MIC of 256 µg/mL) [52]. Aplotaxene, cyperene, and the norsesquiterpene, 13-norcypera-1(5),11,(12)-diene, were isolated and identified. Other EOs of the *Centaurea* genus with antimicrobial activity include the oils extracted from *Centaurea pulcherrima* Willd. var. *pulcherrima* by hydrodistillation and microwave distillation [53].

Tanacetum Species

The genus *Tanacetum* L. is one of the largest genera in the Asteraceae family, and is reputed to have excellent medicinal value. This genus comprises about 150 species, compiled mainly from the floras of Europe, Turkey, and Russia. Water-distilled EOs from the herbal parts of *Tanacetum parthenium* (L.) Sch.Bip from two different localities in Turkey were analyzed by GC-MS [22]. The EO of *T. parthenium* collected from Davutpasa, Istanbul was characterized as containing camphor and *trans*-chrysanthenyl acetate. The second plant sample was collected from the remote eastern part of the country, in Şavşat Ardahan, and characterized as containing camphor. The antibacterial activity of these oils was evaluated for five Gram-positive and five Gram-negative bacteria, including *S. aureus*.

The oil from the Istanbul sample showed the highest activity against MRSA (125 µg/mL), representing a twofold increase in concentration compared to the positive control, chloramphenicol (62.5 µg/mL). However, the EO from the same species from Iran showed inhibitory effects on *E. coli* and *Salmonella* ser. Typhimurium, but were not active against *S. aureus* [36]. This oil was characterized by high amounts of camphor (56.4%), α-pinene (50%), *trans*-β-ocimene (23.6%), and germacrene D (15%). Other studies have demonstrated the occurrence of a camphor/bornyl acetate chemotype of *T. parthenium* in the western region of Iran [54].

Water-distilled EO from the aerial parts of *Tanacetum argenteum* (Lam.) Willd. subsp. *argenteum* and *Tanacetum densum* (Lab.) Sch.Bip subsp. *amani* Heywood from Turkey were analyzed by GC-MS [37]. The EO of *T. argenteum* subsp. *argenteum* was characterized by α-pinene, β-pinene, and 1,8-cineole (Fig. 14-2). *T. densum* subsp. *amani* was characterized by β-pinene and 1,8-cineole (Fig. 14-2). The antibacterial activity of the oils was evaluated for five Gram-positive and five Gram-negative bacteria, using a broth microdilution assay. The highest inhibitory activity was observed for *T. argenteum* subsp. *argenteum* oil against *B. cereus* NRRL B-3711 (125 µg/mL), while the oil of *T. densum* subsp. *amani* did not show significant activity against the microorganisms tested. Another *Tanacetum* EO with antibacterial activity against *Bacillus cereus* NRRL B-3711 was the Turkish *Tanacetum argyrophyllum* (K.Koch) Tzvelev var. *argyrophyllum* [55], while the Turkish *Tanacetum cadmeum* (Boiss.) Heywood subsp. *orientale* Grierson also exhibited activity against *Salmonella* ser. Typhimurium [56].

Tabanca et al. [38] reported the chemical composition and antimicrobial activity of the EO of another *T. argenteum* subspecies, *T. argenteum* (Lam.) Willd. subsp. *flabellifolium* (Boiss. & Heldr.) Grierson. This EO showed a good growth inhibitory effect against *P. aeruginosa* (MIC of 125 µg/mL), appearing to be active as the chloramphenicol standard. The oil also demonstrated good to moderate growth inhibition against the pathogenic bacteria *E. coli* (MIC of 250 μg/mL), *S. aureus* (MIC of 125 μg/mL), and *Salmonella* ser. Typhimurium NRRL 4420 (MIC of 250 μg/mL). Its main constituents were found to be α-pinene, (*E*)-sesquilavandulol (Fig. 14-7), and camphor. The antimicrobial activity of *T. argenteum* subsp. *flabellifolium* EO may be directly associated with its major constituents or a consequence of synergy between major and minor components of the oil.

Examples of other *Tanacetum* species with antibacterial activity also include the EO of the Iranian species *Tanacetum balsamita* L. subsp. *balsamita* [41]. This EO was active against *E. coli*, and *P. aeruginosa*, *S. aureus*, and *S. epidermidis*. Carvone and β-thujone were found to be the main components.

Tanacetum chiliophyllum (Fisch. & E.Mey.) Sch.Bip. varieties are native plants from southwest Asia, including Armenia, Azerbaijan, northwest Iran, and Turkey. This species is represented by four varieties in Turkey. One of them, *Tanacetum chiliophyllum* var. *chiliophyllum* is known by the local names *ormadere* and *yavsan* in Eastern Anatolia. A decoction of the flower heads of this plant is used against pulmonary disorders, kidney stones, and as an antipyretic in traditional medicine. The flower and stem oils of *T. chiliophyllum* var. *chiliophyllum* collected from Van-Muradiye in Istanbul were characterized by camphor, 1,8-cineole (Fig. 14-2), and

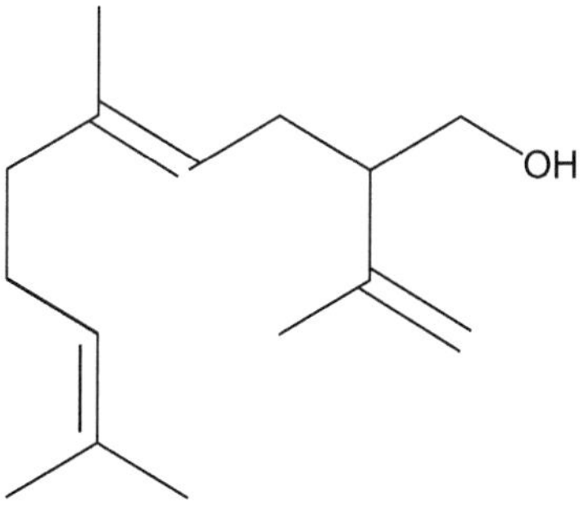

chamazulene (Fig. 14-4) for the first plant sample and 1,8-cineole (Fig. 14-2), terpinene-4-ol, (*E*)-sesquilavandulol (Fig. 14-7), and hexadecanoic acid for the second sample [57]. The antibacterial activities of these oils against five Gram-positive and five Gram-negative bacteria were evaluated using a broth microdilution assay. The highest activity was observed for the stem oil against *E. coli* NRRL B-3008, which gave the same MIC as the positive control, chloramphenicol (62.5 μg/mL).

Ben-Sassi et al. [32] investigated the chemical composition of the EO obtained by hydrodistillation of another *Tanacetum* species growing wild in Tunisia, *Tanacetum trifurcatum* (Desf.) Batt. and Trab. var. *macrocephalum* (viv.) Beg. A total of 56 compounds were identified, representing 97.5% of the oil, of which the most abundant were limonene and γ-terpinene. This oil showed a large antibacterial effect against *S. epidermidis* CIP 53124 (66% inhibition), with an inhibitory concentration 50 (IC50) of 62.5 μg/mL.

Tanacetum indicum (L.) des monl., which is widespread in China, is a well-known herb with small yellow flowers. Traditional oriental medicine has used the aerial parts (stems, leaves, and flowers) of this plant to treat vertigo, hypertensive symptoms, and several infectious diseases. Shunying et al. [58] investigated the chemical composition and antibacterial activity of EOs made from three samples of *T. indicum*: fresh, air-dried, and processed flowers. The major constituents of all three oils were 1,8-cineole (Fig. 14-2), camphor, borneol, and bornyl acetate, although the percentages of these compounds varied greatly according to the type of processing used. Of the three oils, the air-dried flower oil had the most effective bactericidal activity against clinical isolates of *E. coli* (MIC and MBC of 0.39 mg/mL).

Achillea Species

Other aromatic plants from the Asteraceae family belong to the *Achillea* L. genus. This

FIGURE 14-7 Structure of (*E*)-sesquilavandulol.

genus comprises numerous species of wild plants, 50 of which are European species, mainly typical plants of the Mediterranean area. From a wider perspective, *Achillea* species comprise an important biologic resource in Turkish folk medicine. Hydrodistilled volatile oil from the aerial parts of *Achillea cretica* L. was analyzed by GC-MS [23]. Seventy-six components were identified, constituting 86.4% of the oil. The main constituents were caryophylladienol and β-maaliene. Its antimicrobial activity against nine microbial strains was evaluated by the broth dilution method; it proved to be fairly effective against the Gram-positive bacteria *S. aureus* NRRL 1020/06008 (MIC of 125 μg/mL) and *B. cereus* NRRL 869 (MIC of 62.5 μg/mL). The antibacterial properties of *A. cretica* justify its use in traditional medicine for the treatment of bacterially infected wounds.

Achillea ligustica All. (Ligurian yarrow) is a perennial pubescent herbaceous plant. It grows wild throughout the western Mediterranean region, and in Italy is found especially in the Tyrrhenian area from Liguria to Sicily. Tuberoso et al. [39] investigated the chemical composition of the EO made from the flowering tops of this plant, obtained by hydrodistillation using either the Clevenger-type apparatus or simultaneous distillation-extraction. The EO was analyzed by GC-MS and a total of 96 components were detected. The major compounds were santolina alcohol (Fig. 14-8; this was the first time this compound was detected in *A. ligustica*), borneol, sabinol, *trans*-sabinyl acetate, α-thujone, and, among

sesquiterpenes, viridiflorol. No significant differences were detected between EOs extracted by hydrodistillation and simultaneous distillation-extraction. Both oils showed antibacterial activity against *E. coli*, *P. aeruginosa*, and *S. aureus*. Other antibacterial EOs of the *Achillea* genus are those of *Achillea distans* Waldst. & Kit ex Willd. [59] and *Achillea pannonica* Scheele [60].

Achillea umbellata Sibth. & Sm., an endemic Greek plant with white ligules that naturally occurs on limestone cliffs and rocky outcrops in southern and central Greece, is frequently cultivated for its daisy-type flowers and silvery foliage. *A. umbellata* EO exhibits antimicrobial activity in the range of 0.39−6.25 mg/mL, with the highest inhibitory effect against *S. aureus* isolates (MIC of 0.39 mg/mL) [30]. A GC-MS analysis of the oil revealed the presence of a series of fraganyl esters (six new natural products), with the main constituents being the rare monoterpene alcohol, fraganol (Fig. 14-9), fraganyl acetate, and another ester (benzoate). Among the *A. umbellata* EO constituents, fraganol showed the highest antibacterial potential against *S. aureus* isolates (MIC of 0.39 mg/mL).

Matricaria Species

The genus *Matricaria* L. (syn. *Anthemis* L.) comprises approximately 130 species, primarily Mediterranean flowers, although some species are found in southwest Asia and South Africa. *Matricaria recutita* L. (chamomile) is a herbaceous

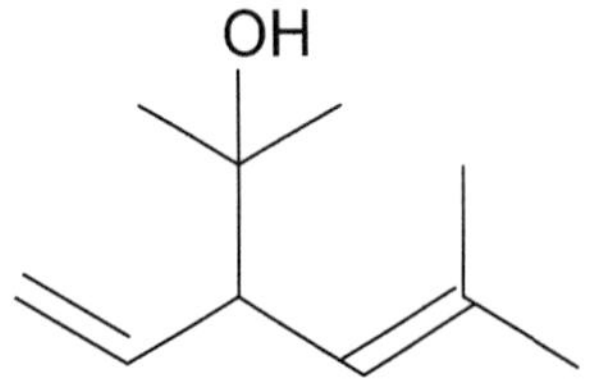

FIGURE 14-8 Structure of santolina alcohol.

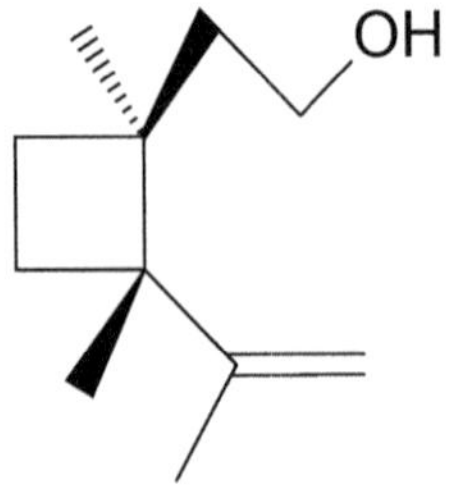

FIGURE 14-9 Structure of fraganol.

plant indigenous to Europe and western Asia. Nowadays, it is mainly cultivated in Europe, South America, and, to a lesser extent, in Africa. This plant is one of the most popular single ingredient herbal teas or tisanes. Chamomile tea, brewed from dried flower heads, has been traditionally used for medicinal purposes. Several chamomile products are commercially available, including beverages, detergents, fragrances, hair products, infusions, lotions, and soaps. There is evidence that the EO of this plant contains antibacterial activity: it was found to be effective against 25 different Gram-positive and Gram-negative bacteria, including *E. coli*, *L. monocytogenes*, *P. aeruginosa*, *Salmonella* ser. Typhimurium, *S. aureus*, and *S. epidermidis* [61,62]. It is also active against *Propionibacterium acnes*, the main causative microorganism of acne vulgaris [24], and *H. pylori* [25]. *H. pylori* can withstand the low pH in the stomach by producing urease, which hydrolyzes urea and converts it into ammonia and carbon dioxide. Thus, an increase in pH disturbs the mechanisms controlling pH homeostasis and leads to the survival of the bacteria within the stomach. Urease production by *H. pylori* is inhibited by EO from *M. recutita*, with an IC50 value of 62.5 mg/mL. The application of chamomile oil extract influences the morphologic and fermentative properties of *H. pylori* and also inhibits colony formation by this microorganism.

The principal components of the EO extracted from chamomile flowers are the terpenoid α-bisabolol and its oxides and azulenes, including chamazulene (Fig. 14-4). Several studies have reported the use of individual compounds detected in chamomile EO. For example, investigations carried out with α-bisabolol showed that this compound has antimicrobial activity against *B. cereus*, *E. coli*, *P. aeruginosa*, and *S. aureus* [63]. It promotes disruption of the bacterial cell membrane, allowing permeation into the cell of exogenous solutes such as ethidium bromide (a membrane-impermeant model drug) and antibiotics.

The EO of the Turkish *Anthemis aciphylla* Boiss. var. *discoidea* Boiss. was obtained by hydrodistillation and analyzed by GC and GC-MS [64]. An unknown component isolated from the EO was characterized as an isofaurinone. The EO showed moderate antibacterial activity against *S. epidermidis*.

Baccharis Species

Baccharis L. is the largest genus of the Asteraceae family, with over 500 species distributed throughout the North and South American continents. The species of this genus are mainly found in the warm temperate and tropical regions of Argentina, Brazil, Chile, Colombia, and Mexico. The many general traditional medicinal uses of *Baccharis* spp. include the treatment of wounds and ulcers, fevers, gastrointestinal illnesses, and bacterial/fungal infections. Rojas et al. [33] investigated the chemical composition and antibacterial activity of the EO of *Baccharis trinervis* (Lam.) Pers. collected in Venezuela, where it is known as *chilca*. The major components were identified as germacrene D and limonene. The EO showed antibacterial activity against the important human pathogenic Gram-positive bacterium, *S. aureus*, with an MIC of 80 μg/mL. Similar activity was also observed with the EOs of *Baccharis latifolia* Pers. and *Baccharis prunifolia* H. B. & K. [31]. The main components in *B. latifolia* were found to be oxygenated sesquiterpenes (69.8%) and sesquiterpenes (20.4%), and in *B. prunifolia* were sesquiterpenes (45.9%) and monoterpenes (43.9%). Another example of an EO from the *Baccharis* genus with antibacterial activity is that of *Baccharis flabellata* Hook. & Arn. from Argentina [66]: *B. cereus* and *S. aureus* were shown to be the most sensitive microorganisms to this EO.

Four *Baccharis* species growing wild in Bolivia, *Baccharis papilosa* Rusby, *Baccharis polycephala* Wedd., *Baccharis buxifolia* (Lam.) Pers., and *Baccharis conwayi* L., also yielded volatile oils with antibacterial activity against the Gram-positive bacterium *S. aureus* [67], while the oil of *Baccharis trimera* DC. from Brazil was active against enterotoxigenic *E. coli* [43].

Helichrysum Species

Members of the genus *Helichrysum* Mill. are usually aromatic perennial shrubs, with dense leaves and hardy flower heads. South Africa harbors approximately 245 species of *Helichrysum*. Extracts of various species have been used to treat tropical infections, respiratory ailments, and as dressings in circumcision rites. Its administration by inhalation suggests that the volatile aromatic compounds may play a role in anti-infective therapies, and several studies have indicated significant antimicrobial properties for *Helichrysum* oils. Van Vuuren et al. [28] investigated the antibacterial activity of the EO of *Helichrysum cymosum* (L.) D.Don subsp. *cymosum* against 10 pathogens. This EO, whose main component is 1,8-cineole (Fig. 14-2), was active against *B. cereus* (MIC of 8 mg/mL), *E. coli* (MIC of 8 mg/mL), *P. aeruginosa* (MIC of 4 mg/mL), and *S. aureus* (MIC of 8 mg/mL). The chemical composition of EOs obtained from both the leaves and flowers of another *Helichrysum* species, *Helichrysum pallasii* (Spreng.) Ledeb., growing wild in the Lebanon were also analyzed [68]. The oils were both

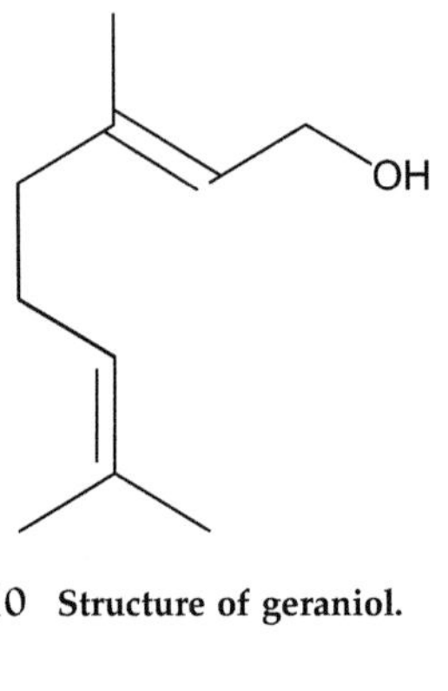

FIGURE 14-10 **Structure of geraniol.**

characterized by the presence of sesquiterpenes, fatty acids, and esters. The *in vitro* activity of these EOs against a number of microorganisms was determined by the broth dilution method. The oils exhibited growth inhibitory activity toward *S. epidermidis* (MIC of 100 μg/mL).

Helichrysum italicum (Roth.) G.Don is widespread throughout the Mediterranean area, where it grows as a small perennial shrub. The ornamental value conferred by its distinctive yellow, scented flowers and the properties of its EO contribute to its popularity. Lorenzi et al. [31] demonstrated that the EO from this plant contains compounds that modulate drug resistance in several Gram-negative bacterial species, including clinical isolates of *E. coli* and *P. aeruginosa*, by targeting their efflux mechanisms. Geraniol (Fig. 14-10), a component of this EO, was shown to significantly increase the efficacy of chloramphenicol, β-lactams, and quinolones. These findings may provide a new source of drugs to augment current therapies; moreover, research into the mode of action of geraniol may improve our understanding of MDR in Gram-negative bacteria.

CONCLUSIONS AND FUTURE PROSPECTS

If modern medicine is to continue in its present form, novel families of antibiotics must enter the marketplace at regular intervals. Although the development of analogs that kill resistant bacteria can prolong the life of each drug family for a number of decades, eventually this well of resources will dry out. Within the next 10 years, our efforts in screening whole bacteria against novel natural and chemical compound libraries may produce new antibiotics. This review summarizes and characterizes the importance of EOs derived from a wide range of Asteraceae species. A number of compounds present in these oils (and the oils themselves) have medicinal and (ethno) pharmacologic properties. The potential for the development of leads from these EOs is continuing to grow, particularly in the area of infectious diseases. The information summarized here is intended to serve as a reference tool to researchers in all fields of ethnopharmacology and natural product chemistry. The chapter also draws attention to some active ingredients that may lead to the development of new antibacterial drugs.

Acknowledgments

The technical assistance of Ms. Brooke-Turner is gratefully acknowledged.

References

[1] Panackai AA. Global climate change and infectious diseases: invasive mycoses. J Earth Sci Climat Change 2011;2:1–5.

[2] Gibbons S. Phytochemicals for bacterial resistance: strengths, weaknesses and opportunities. Planta Med 2008;74:594–602.

[3] Shahid M, Shazad A, Sobia F, Sahai A, Tripathi T, Singh A, Khan HM, Umesh T. Plant natural products as a potential source for antibacterial agents: recent trends. Anti-infect Agents 2009;8:211–25.

[4] Saleem M, Nazir M, Ali MS, Hussain H, Lee YS, Riaz N, Jabbar A. Antimicrobial natural products: an update on future antibiotic drug candidates. Nat Prod Rep 2010;27:238–54.

[5] Ngwoke KG, Odimegwn DC, Esimone CO. Antimicrobial natural products. In: Méndez-Vilas A, editor. Science Against Microbial Pathogens: Communicating Current Research and Technological Advances, vol. 2; 2011. p. 1011–26. Formatex, Badajoz.

[6] Edris AE. Pharmaceutical and therapeutic potentials of essential oils and their individual volatile constituents: a review. Phytother Res 2007;21:308–23.

[7] Williams LAD, Porter RB, Junor GO. Biological activities of selected essential oils. Nat Prod Commun 2007;2:1295–6.

[8] Faleiro ML. The mode of antibacterial action of essential oils. In: Méndez-Vilas A, editor. Science Against Microbial Pathogens: Communicating Current Research and Technological Advances, vol. 2; 2011. p. 1143–56. Formatex, Badajoz.

[9] Zuzarte M, Goncalves MJ, Canhoto J, Salgueiro L. Antidermatophytic activity of essential oils. In: Méndez-Vilas A, editor. Science Against Microbial

Pathogens: Communicating Current Research and Technological Advances, vol. 2; 2011. p. 1167–78. Formatex, Badajoz.

[10] Teixeira da Silva JA. Mining the essential oils of the Anthemideae. Afr J Biotechnol 2004;3:706–20.

[11] Teixeira da Silva JA, Yonekura L, Kaganda J, Mookdasanit J, Arhut DT, Afach G. Important secondary metabolites and essential oils of species within the Anthemideae. Journal of Herbs, Spices & Medicinal Plants 2004;11:1–46.

[12] Murray PR, Baron EJ, Pfaller MA, Tenover FC, Yolken RH. Manual of Clinical Microbiology. American Soc Microbiol 2003. Washington D.C.

[13] Ribeiro de Souza ML, Ustulin DR. Antimicrobial resistance in Staphylococcus spp. In: Méndez-Vilas A, editor. Science Against Microbial Pathogens: Communicating Current Research and Technological Advances, vol. 2; 2011. p. 714–21. Formatex, Badajoz.

[14] Tothpal A, Laub K, Kardos S, Nagy K, Dobay O. Changes in the serotypes of Hungarian pneumococci isolated mainly from invasive infections: a review of all available data between 1988 and 2011. Acta Microbiol Immunol Hung 2012;59:423–33.

[15] Swaminathan B, Gerner-Smidt P. The epidemiology of human listeriasis. Microb Infect 2011;9:1236–43.

[16] Carlin F. Origin of bacterial spores contaminating foods. Food Microbiol 2011;28:177–82.

[17] Karch H, Denamur E, Dobrindt U, Finlay BB, Hengge R, Johannes L, Ron EZ, Tonjun T, Sansonetti PJ, Vicente M. The enemy within us: lessons from the 2011 European Escherichia coli 0104:H4 outbreak. EMBO Mol Med 2012;4:841–8.

[18] Vale FF, Roxo M, Oleastro M. Helicobacter pylori resistance to antibiotics. In: Méndez-Vilas A, editor. Science Against Microbial Pathogens: Communicating Current Research and Technological Advances, vol. 2; 2011. p. 745–56. Formatex, Badajoz.

[19] Baser KHC, Demirci F. Essential Oils, Kirk-Othmer Encyclopedia of Chemical Technology. 4th ed. Sussex: Wiley; 2011.

[20] Cavar S, Maksimovic M, Vidic D, Paric A. Chemical composition and antioxidant and antimicrobial activity of essential oil of Artemisia annua L. from Bosnia. Ind Crops Prod 2012;37:479–85.

[21] Laciar A, Ruiz ML, Flores RC, Saad JR. Antibacterial and antioxidant activities of the essential oil of Artemisia echegaray Hieron (Asteraceae). Rev Argent Microbiol 2009;41:226–31.

[22] Polatoglu K, Demirci F, Demirci B, Gören N, Baser KH. Antibacterial activity and the variation of Tanacetum parthenium (L.) Schultz-Bip. essential oil from Turkey. J Oleo Sci 2010;59:177–84.

[23] Zebra-Kucukbay F, Kuyumcu E, Bilenler T, Yildiz B. Chemical composition and antimicrobial activity of

the essential oil of Achillea cretica L. (Asteraceae) from Turkey. Nat. Prod. Res. 2011;25:1–6.

[24] Zu Y, Yu H, Liang L, Fu Y. Activities of ten essential oils towards Propionibacterium acnes and PC-3, A-549 and MCF-7 cancer cells. Molecules 2010;15:3200–10.

[25] Shikov AN, Pozharitskaya ON, Makarov VG, Kvetnaya A. Antibacterial activity of Chamomilla recutita oil extract against Helicobacter pylori. Phytother Res 2008;22:252–3.

[26] Kordali S, Kotan R, Mavi A, Cakir A, Ala A, Yildirim A. Determination of the chemical composition and antioxidant activity of Artemisia dracunculus, and of the antifungal and antibacterial activities of Turkish Artemisia absinthium, A. dracunculus, Artemisia santonicum and Artemisia spicigera essential oils. J. Agric. Food Chem 2005;53:9452–8.

[27] Mighri H, Hajlaoui H, Akrout A, Najjaa H, Neffati M. Antimicrobial and antioxidant activities of Artemisia herba-alba essential oil cultivated in Tunisian arid zone. C. R. Chimie 2010;13:380–6.

[28] Van Vuuren SF, Viljoen AM, Van Zyl RL, Van Heerden FR, Baser C. The antimicrobial, antimalarial and toxicity profiles of helihumulone, leaf essential oil and extracts of Helichrysum cymosum (L.) D. Don subsp cymosum S Afr J Bot 2006;72:287–90.

[29] Yavli N, Yasar A, Gulec C, Usta A, Kolavli S, Coskuncelebi K, Karaoglu S. Composition and antimicrobial activity of essential oils from Centaurea sessilis and Centaurea armena. Phytochemistry 2005;66:1741–5.

[30] Radulovic NS, Dekic MS, Randelovic PJ, Stojanovic NM, Zarubica AR, Stojanovic-Radic ZZ. Toxic essential oils: anxiolytic, antinociceptive and antimicrobial properties of the yarrow Achillea umbellata Sibth. et Sm. (Asteraceae) volatiles. Food Chem Toxicol 2012;50:2016–26.

[31] Lorenzi V, Muselli A, Bernardini AF, Berti L, Pages JM, Amaral L, Bolla JM. Geraniol restores antibiotic activities against multidrug-resistant isolates from Gram-negative species. Antimicrob Agents Chemother 2009;53:2209–11.

[32] Ben-Sassi A, Harzallah-Skhiri F, Chraief I, Bourgougnon N, Hammami M, Aouni M. Composition and antimicrobial activities of the essential oil of (Tunisian) Chrysanthemum trifurcatum (Desf.) Batt. and Trab. flowerheads. C. R. Chimie 2008;11:324–30.

[33] Rojas J, Velasco J, Morales A, Rojas L, Diaz T, Rondon M, Carmona J. Chemical composition and antibacterial activity of the essential oil of Baccharis trinervis (Lam.) Pers. (Asteraceae) collected in Venezuela. Nat Prod Commun 2008;3:369–72.

[34] Bousaada O, Ammar S, Saidana D, Chriaa J, Chraif I, Daami M, Helal AN, Mighri Z. Chemical composition

and antimicrobial activity of volatile compounds from capitula and aerial parts of Rhaponticum acaule DC. growing wild in Tunisia. Microbiol Res 2008;163: 87–95.

[35] Blagojevic P, Radulovic N, Palic R, Stojanovic G. Chemical composition of the essential oils of Serbian wild-growing Artemisia absinthium and Artemisia vulgaris. J Agric Food Chem 2006;54:4780–9.

[36] Shafagat A, Sadeghi H, Oji K. Composition and antibacterial activity of essential oils from leaf, stem and root of Chrysanthemum parthenium (L.) Bernh. from Iran. Nat Prod Commun 2009;4:859–60.

[37] Polatoglu K, Demirci F, Demirci B, Gören N, Baser KH. Essential oil composition and antibacterial activity of Tanacetum argenteum (Lam.) Willd. ssp. argenteum and T. densum (Lab.) Schultz-Bip. ssp. amani Heywood from Turkey. J Oleo Sci 2010;59: 361–7.

[38] Tabanca N, Demirci F, Demirci B, Wedge DE, Baser KHC. Composition, enantiomeric distribution and antimicrobial activity of Tanacetum argenteum subsp. flabellifolium essential oil. J Pharm Biomed Anal 2007;45:714–9.

[39] Tuberoso CI, Kowalczyk A, Coroneo V, Russo MT, Dessi S, Cabras P. Chemical composition and antioxidant, antimicrobial and antifungal activities of the essential oil of Achillea ligustica All. J Agric Food Chem 2005;53:10148–53.

[40] Militello M, Settanni L, Aleo A, Mammina C, Moschetti G, Giammanco GM, Blazquez MA, Carrubba A. Chemical composition and antibacterial potential of Artemisia arborescens L. essential oil. Curr Microbiol 2011;62:1274–81.

[41] Yousefzadi M, Ebrahimi SN, Sonboli A, Miraghasi F, Ghiasi S, Arman M, Mosaffa N. Cytotoxicity, antimicrobial activity and composition of essential oil from Tanacetum balsamita L. subsp. balsamita. Nat Prod Commun 2009;4:119–22.

[42] Chung EY, Byun YH, Shin EJ, Chung HS, Lee YH, Shin S. Antibacterial effects of vulgarone B from Artemisia iwayomogi alone and in combination with oxacillin. Arch Pharm Res 2009;32:1711–9.

[43] Teixeira MC, Leme EE, Delarmelina C, Almeida A, Figueira GM, Sartoratto A. Activity of essential oils from Brazilian medicinal plants on Escherichia coli. J Ethnopharmacol 2007;111:197–201.

[44] Cetin B, Ozer H, Cakir A, Mete E, Oztürk E, Polat T, Kandemir A. Chemical composition of hydrodistillated essential oil of Artemisia incana (L.) Druce and antimicrobial activity against foodborne microorganisms. Chem Biodivers 2009;6:2302–10.

[45] Lopes-Lutz D, Alviano DS, Alviano CS, Kolodziejczyk PP. Screening of chemical composition,

antimicrobial and antioxidant activities of Artemisia essential oils. Phytochemistry 2008;69:1732−8.

[46] Viljoen AM, van Vuuren SF, Gwebu T, Demirci B, Baser KHC. The geographical variation and antimicrobial activity of African wormwood (Artemisia afra Jacq.) essential oil. J Essent Oil Res 2006;18:19−25.

[47] Cha JD, Jung EK, Kil BS, Lee KY. Chemical composition and antibacterial activity of essential oil from Artemisia feddei. J Microbiol Biotechnol 2007;17:2061−5.

[48] Aghajani Z, Kazemio M, Dakhili M, Rustaiyan A. Composition and antimicrobial activity of the essential oil of Artemisia kulbadica from Iran. Nat Prod Commun 2009;4:1261−6.

[49] Shafaghat A, Noormohammadi Y, Zaifizadeh M. Composition and antibacterial activity of essential oils of Artemisia fragans Willd. leaves and roots from Iran. Nat Prod Commun 2009;4:279−82.

[50] Formisano C, Rigano D, Senatore F, Celik S, Bruno M, Roselli S. Volatiles constituents of aerial parts of three endemic Centaurea species from Turkey: Centaurea amanicola Hub.-Mor., Centaurea consanguinea DC. and Centaurea ptosimopappa Hayek. and their antibacterial activities. Nat Prod Res 2008;22:833−9.

[51] Kose YB, Iscan G, Altintas A, Celik S, Demirci B, Baser KHC. Composition and antimicrobial activity of the essential oil of Centaurea aladagensis Wagenitz. Fitoterapia 2007;78:253−4.

[52] Havlik J, Budesinsky M, Kloucek P, Kokoska L, Valterova I, Vasickova S, Zeleny V. Norsesquiterpene hydrocarbons, chemical composition and antimicrobial activity of Rhaponticum carthamoides root essential oil. Phytochemistry 2009;70:414−8.

[53] Kahriman N, Tosun G, Yilmaz-Iskender N, Alpay-Karaoglu S, Yayli N. Antimicrobial activity and a comparative essential oil analysis of Centaurea pulcherrina var. pulcherrina extracted by hydrodistillation and microwave distillation. Nat Prod Res 2012;26:1−10.

[54] Mohsenzadeh F, Chehregani A, Amiri H. Chemical composition, antibacterial activity and cytotoxicity of essential oils of Tanacetum parthenium in different development stages. Pharm Biol 2011;49:920−6.

[55] Polatoglu K, Demirci F, Demirci B, Gören N, Baser KHC. Antimicrobial activity and essential oil composition of a new T. argyrophyllum (C. Koch) Tvzel var. argyrophyllum chemotype. J Oleo Sci 2010;59:307−13.

[56] Ozek G, Ozek T, Iscan G, Baser KHC, Hamzaoglu E, Duran A. Composition and antimicrobial activity of the essential oil of Tanacetum cadmeum (Boiss.) Heywood subsp. orientale Grierson. J Essent Oil Res 2007;19:392−5.

[57] Polatoglu K, Demirci B, Demirci F, Gören N, Baser KHC. Biological activity and essential oil composition of two new Tanacetum chiliophyllum (Fisch. & Mey.) Schultz-Bip. var. chiliophyllum chemotypes from Turkey. Ind Crops Prod 2012;39:97−105.

[58] Shunying Z, Yang Y, Huaidong Y, Yue Y, Guolin Z. Chemical composition and antimicrobial activity of the essential oils of Chrysanthemum indicum. J Ethnopharmacol 2005;96:151−8.

[59] Konakchiev A, Todorova M, Mikhova B, Vitkova A, Naidenski H. Composition and antimicrobial activity of Achillea distans essential oil. Nat Prod Commun 2011;6:905−6.

[60] Bozin B, Mimica-Dukic N, Bogavac M, Suvajdzic L, Simin N, Samojilik I, Couladis M. Chemical composition, antioxidant and antibacterial properties of Achillea collina Becker ex Heimerl S. I. and A. pannonica Scheele essential oils. Molecules 2008;13:2058−68.

[61] McKay DL, Blumberg JB. A review of the bioactivity and potential health benefits of chamomile tea (Matricaria recutita L.). Phytother Res 2006;20:519−30.

[62] Sokovic M, Glamoclija J, Marin PD, Brkic D, Griensven Van. Antibacterial effects of the essential oils of commonly consumed medicinal herbs using an in vitro model. Molecules 2010;15:7532−46.

[63] Kamatou GPP, Viljoen AM. A review of the application and pharmacological properties of α-bisabolol and α-bisabolol-rich oils. J Am Oil Chem Soc 2010;87:1−7.

[64] Baser KHC, Demirci B, Iscan G, Asakawa Y, Hashimoto T, Noma Y. The essential oil constituents and antimicrobial activity of Anthemis aciphylla Boiss. var discoidea Boiss Chem. Pharm Bull 2006;54:222−5.

[65] Rojas J, Velasco J, Rojas LB, Diaz T, Carmona J, Morales A. Chemical composition and antibacterial activity of the essential oil of Baccharis latifolia Pers. and B. prunifolia H. B. & K. (Asteraceae). Nat Prod Commun 2007;2:1245−8.

[66] Derno M, Oliva MM, Lopez ML, Zunino MP, Zygadlo J. Antimicrobial activity of essential oils obtained from aromatic plants of Argentine. Pharm Biol 2005;43:129−34.

[67] Davila M, Loayza I, Lorenzo D, Dellacasa E. Searching for natural bioactive compounds in four Baccharis species from Bolivia. Nat Prod Commun 2008;3:551−6.

[68] Formisano C, Mignola E, Rigano D, Senatore F, Arnold DA, Bruno M, Rosselli S. Constituents of leaves and flowers essential oils of Helichrysum pallaswii (Spreng.) Ledeb. growing wild in Lebanon. J Med Food 2009;12:203−7.

Harnessing Traditional Knowledge to Treat Existing and Emerging Infectious Diseases in Africa

M. Fawzi Mahomoodally[1], *Ameenah Gurib-Fakim*[2]

[1]Department of Health Sciences, Faculty of Science, University of Mauritius, Réduit, Mauritius and
[2]Center for Phytotherapy Research, 7th Floor, Cybertower 2, Ebene, Mauritius

INTRODUCTION

Worldwide, infectious diseases, such as diarrheal diseases, human immunodeficiency virus (HIV), malaria, measles, pneumonia, and tuberculosis remain some of the major causes of human mortality and morbidity, even after the arrival of modern antimicrobial chemotherapy. Chemotherapeutic agents developed since World War II include drugs effective against bacteria, fungi, parasites, and viruses. Perhaps the most important antibacterial agents in clinical use remain antibiotics, many of which have been derived from natural sources. Natural sources have also yielded numerous substances with insecticidal, antimicrobial, and antiprotozoal potential. Many such compounds, including essential oils, are active against all these classes of organisms. They can be used internally (e.g., for protozoal infections, and those with antiviral polyphenolics for influenza

and colds), as well as externally for skin infections and infestations. In addition, intestinal worms have been treated with herbal materials such as wormwood.

A comprehensive study on natural products carried out between 1981 and 2002 has shown that of the existing 877 small molecules, 67% of new chemical entities are synthetic but the origin of over 16% could be traced to a pharmacophore derived from a natural source. Interestingly, 12% of these molecules have been modeled on a natural product inhibitor of a molecular target of interest or were designed to mimic an endogenous substrate or active site such as ATP. Therefore, from this logic, only 39% of the 877 molecules could be classified as being truly of synthetic origin. As for the anti-infectives (antimicrobial, antifungal, antiparasitic, and antiviral), almost 70% of the active molecules were derived from natural sources. Of those molecules used to treat cancer,

Fighting Multidrug Resistance with Herbal Extracts,
Essential Oils and Their Components
http://dx.doi.org/10.1016/B978-0-12-398539-2.00015-X

67% are in this category [1,2]. With increasing resistance being observed in various pathogens, there has been a renewed interest in "relooking at natural product as a source of leads" [3]. Hence, Nature has and will continue to play a lead role in the discovery of active natural products that have bearing on and shape drug discovery in the medium to long term [4,5].

Resistance toward existing antibiotics is developing [6] and increases in the death toll related to Methicillin-Resistant *Staphylococcus aureus* (MRSA) or antibiotic-resistant *Escherichia coli* have been reported recently [7,8]. Besides MRSA, other pathogens, such as *Candida albicans* and *Pseudomonas aeruginosa*, are posing an impending threat to human health [9]. MRSA infections affect approximately 94,360 individuals in the USA and are linked to around 18,650 deaths annually [10], even in well-regulated health systems like those prevailing in Europe.

Developing countries from Africa are not spared and will bear the brunt of this pandemic if proper and timely measures are not taken. It is also worth pointing out that the death toll from these infectious diseases is more than 11 million worldwide each year, with the majority of deaths occurring in many parts of Africa. Sub-Saharan countries are the worst hit and are finding it difficult to cope, with their limited infrastructure and resources [11,12].

THE BURDEN OF MULTIDRUG RESISTANCE AND EMERGING INFECTIOUS DISEASES

While existing infectious diseases are proving to be a challenge, newly emerging infections are also adding to the burden. These are attributed to mutations in the microorganisms that infect humans, and reemerging infections are also now known to be spreading at a high rate [13]. Examples of emerging infectious diseases (EIDs) in Africa include avian influenza, Ebola, monkeypox, Marburg, and,

more recently, chikungunya. Over and above the human tragedy, EIDs can have devastating economic effects on livestock and the populations dependent on them.

Moreover, multidrug resistance of existing infectious pathogens is currently hampering efforts to advance eradication of diseases. For instance, multidrug-resistant tuberculosis (MDR-TB) is becoming a life-threatening form of tuberculosis, affecting more than half a million people every year, that causes much higher mortality rates than drug-susceptible tuberculosis. MDR-TB is on the rise in some countries, yet only 3% of cases are being treated according to standards set by the WHO. If MDR-TB is not vigorously addressed, it stands to replace the mainly drug-susceptible strains that cause 95% of the world's tuberculosis today. Using surveillance data from the WHO and its partners generated since 1994, it is estimated that about 510,000 cases of MDR-TB occur every year, of which tens of thousands are classified as extensively drug-resistant tuberculosis (XDR-TB). In some countries, MDR-TB rates are rising, while in others they are falling. Among the world's 12 million cases of tuberculosis in 2010, the WHO estimates that 650,000 involved MDR-TB strains and it is projected that the treatment of MDR-TB is "extremely complicated, typically requiring 2 years of medication with toxic and expensive medicines, some of which are in constant short supply. Even with the best of care, only slightly more than 50 percent of these patients will be cured." [14]. For most countries, the data are not yet good enough to predict trends and, according to Dr. Margaret Chan, director general of the WHO, antibiotic resistance could bring about "the end of modern medicine as we know it" [15,16].

EIDs also present a real challenge to research scientists, who are actively looking for substitute drugs to cope with the growing resistance to antibiotics. Halting the trend of increased emerging and resistant infectious diseases will require a multipronged approach that includes the development of new drugs. In this context, traditional herbal remedies from the tropics, in particular those from the African continent, present an untapped potential. Indeed, using plants as the inspiration for new drugs provides an infusion of novel compounds or substances to combat infectious diseases. To this effect, bio-prospection from tropical flora (both medicinal and nonmedicinal) presents a useful route toward the search for new molecules and remedies.

TRADITIONAL MEDICINES AS AN ALTERNATIVE SOURCE OF NOVEL PHARMACOPHORES

The WHO reported that 80% of the emerging world's population relies on traditional medicine for therapy. Since 2000, the developed world has also been witnessing an ascending trend in the utilization of complementary and alternative medicines (CAMs). While 90% of the population in Ethiopia use herbal remedies for their primary health care, surveys carried out in developed countries like Canada and Germany showed that at least 70% of their population have tried CAMs at least once. It is likely that the profound knowledge of herbal remedies in traditional cultures, developed through trial and error over many centuries, along with the most important cures have been passed on orally from one generation to another. Modern allopathic medicine is firmly anchored in this ancient medicine and it is more than likely that important remedies will be found in traditional remedies and will be commercialized in the future. These successes will rest on leads discovered from traditional knowledge and related expertise.

The composition of medicinal plants is known to be very diverse and to consist of different chemical substances that can act individually, additively or synergistically to improve health conditions. By way of example, one plant can contain anti-inflammatory compounds that bring down swelling or reduce pain, as well as a bitter substance that stimulates digestion. Phenolic compounds are known to act as antioxidants and venotonics, while tannins classically act as antimicrobial agents (or natural antibiotics). In addition, compounds that induce diuresis would promote the elimination of waste products and other toxins. Further, alkaloids are known to be mood enhancers and can promote a sense of wellbeing.

TRADITIONAL AFRICAN MEDICINE

Globally, African herbal medicine is perhaps the oldest and the most diverse form of all medicine systems. The African continent is rich in biologic and cultural diversity and is known as the cradle of mankind. Its cultural diversity is marked in geographic terms, and regional differences affect healing practices. The African system of medicine has been transmitted orally and its documentation remains a challenge, especially in the light of rapid biodiversity loss coupled with the loss of habitats through anthropogenic activities. The African continent has one of the highest rates of deforestation in the world. At the same time, the paradox is that it is also a continent with a high rate of endemism. The island of Madagascar, for example, tops the list, with 82% of its flora being endemic. African traditional medicine is also very varied and holistic, involving both the body and the mind. The traditional healer normally diagnoses and treats the underlying psychologic basis of an illness before prescribing medicines to treat the symptoms.

The recent publication of the *African Herbal Pharmacopoeia* has shed light on the potential of the African flora on various diseases [17]. This document brings together important medicinal plants from all parts of the continent: *Acacia senegal* (gum arabic) and *Aloe vera* from

North Africa; *Aloe ferox* (Cape aloe), *Agathosma betulina* (buchu), *Aspalathus linearis* (rooibos tea), *Harpagophytum procumbens* (Devil's claw), and *Hypoxis hemerocallidea* (African potato) from Southern Africa; *Boswellia sacra* (Frankincense), *Catha edulis* (khat), and *Commiphora myrrha* (myrrh) from Eastern Africa; and *Artemisia afra* (African wormwood), *Hibiscus sabdariffa* (Hibiscus, Roselle), and *Prunus africana* (African cherry) from Central and West Africa. The island of Madagascar has contributed *Catharanthus roseus* (rosy periwinkle). This country has the potential to contribute even more by virtue of its unique biodiversity.

SUB-SAHARAN AFRICAN BIODIVERSITY IN THE FIGHT AGAINST INFECTIOUS DISEASES

Malaria

If there was a disease that illustrated the troubled medical history of humans, it would no doubt be malaria. It kills millions of people annually throughout the world, and the majority of victims are children [18]. More than 10% of the US overseas troops in 1943 acquired malaria, and it has been reported that Alexander the Great died of it in June 2323 BC. Untreated, malaria may kill about 1% of those infected, and the survivors are prone to relapse. It is generally accepted nowadays as the most deadly parasitic disease in the world. It was in the eighteenth century that Dr. Francisco Torti coined the name *malaria* by combining the Italian words for *bad* (*mala*) and *air* (*aria*). At the time, it was generally believed that this disease was caused by the unhealthy air found around marshy areas. It was only later, toward the middle eighteenth century, that the connection was made between the transmission of this disease and the mosquito vector. This resulted in a need for mosquito control, leading to the eventual draining of marshes throughout

parts of the world where this disease was prevalent. Dichlorodiphenyltrichloroethane (or DDT), in spite of its bad reputation, has been perhaps the only insecticide to have saved millions of lives.

It was reported that as far back as the fifteenth century, doctors were dreaming of a plant-based medicine against this scourge. They first identified one such plant in Lima, Peru, the capital of New Spain. It was reported that the recovery was very high for a place where malaria is reported to be endemic. Jesuit priests observed that the local Indian population was not affected by this disease. It was later discovered that the secret of their health resided in the bark of a tree, which upon mixing with water, cured the associated fevers. The locals called the tree the *quina* or *fever bark tree*. By the end of the seventeenth century, quinine powder was the standard treatment for malaria. Over that particular period in history, Spain controlled much of the trade, as it had exclusive mandates in Bolivia and Peru. The demand for the quina was so great that soon there were not enough trees to assure supplies, and collectors of this precious remedy had to go further into the forest to find more trees. Many of them never returned, as they got lost and died either from dysentery or from the darts of Jivaro Indians. By the middle of the eighteenth century, French botanists had confirmed that there were in fact four species to this tree genus. Linnaeus confirmed this information and gave the name *Cinchona* to this tree in honor of the Viceroy of Peru, who lived between 1628 and 1639 [18,19].

The recent WHO Malaria Report (2011) [20] estimates that 3.3 billion people were at risk of malaria in 2010, although, of all geographical regions, those populations living in sub-Saharan Africa have the highest risk of acquiring malaria: among 216 million episodes of malaria in 2010, approximately 81% (or 174 million cases) occurred in the African region; and of an estimated 655,000 malaria deaths in 2010,

91% were from Africa. Resurgent vector-borne diseases result in a high burden of disease, estimated as approximately 56 million disability-adjusted life years [21]. Today, malaria has become a critical and widespread disease; one of the main reasons for this is that the efficacy of antimalarial drugs, including chloroquine, has been reduced by the spread of drug-resistant strains. This loss in efficacy is a major barrier to the effective treatment of malaria and has posed an urgent challenge for the discovery new antimalarial drugs. Malaria is caused by four species of the genus *Plasmodium*, namely *P. falciparum*, *P. malariae*, *P. ovale*, and *P. vivax*. Almost all fatalities are due to *P. falciparum* infections and, therefore, it is the most important species, but *P. vivax* also causes significant morbidity. This shocking reality is largely due to the emergence of drug-resistant strains of *P. falciparum*. In the early days, quinine was the curative agent for malaria and, subsequently, quinoline antimalarials and related aryl alcohols were developed based on the quinine prototype. This led to the emergence of drugs such as chloroquine and mefloquine.

With the rise of parasite resistance to these antimalarials, it became necessary to search for other synthetic and natural product-based agents. Another plant long used in the treatment of fevers in Chinese traditional medicine was therefore considered. The idea of investigating the antimalarial activity of wormwood came from Chinese herbal medicine, as this herb was first prescribed for fevers by the Chinese physician Li Shi-zen in 1527 [22].

Artemisia *Species* (*Asteraceae*)

Artemisia annua is a medicinal plant whose use has long been reported in China, where it is locally known as qinghao. It is now grown commercially in many African countries. Also known as Sweet wormwood, *A. annua* yields artemisinin and the derivatives of this compound are potent antimalarial drugs. Artemisinin is an endoperoxide sesquiterpene lactone

that is effective against multidrug-resistant malaria and is also known to act on *P. falciparum*, the *Plasmodium* species that causes cerebral malaria. The clinical efficacy of this drug and its derivatives is demonstrated by an immediate and rapid reduction of parasitemia following treatment [23]. Since the WHO recommended the use of artemisinin-based combination therapies for malaria in 2001, a number of other forms of *A. annua* L. have appeared as antimalarial *remedies*, including tea bags made from the plant's leaves.

Artemisinin was first isolated in 1972 and has served as prototype for many semisynthetic versions such as arteether and artemether. These compounds have increased solubility in vaccines and have improved antimalarial activities. However, although these synthetic and semisynthetic molecules are being tested widely, malaria remains a big threat to poorer countries, where these modern antimalarial drugs are not available to the general public. In these poorer countries, randomized trials have been performed to assess the efficacy of a traditional herbal tea made from the leaves of *A. annua*, especially for the treatment of uncomplicated malaria. It was observed that after 7 days of medication, cure rates were high (74%). Unfortunately, trials also confirmed that recrudescence was high and, hence, monotherapy with *A. annua* could not be recommended as a potential alternative treatment for this disease [24,25]. A combination of these treatments, however, was recommended [26].

Although Asian *A. annua* is now being grown on the African continent, *A. afra*, commonly referred to as African wormwood, is more commonly used in traditional medicine against infections and malarial fever. *A. afra* essential oil is exceptionally variable and its composition depends on its geographical origin. For example, Ethiopian oil yields artemisyl acetate and yomogi alcohol as the dominant constituents, while those of South African origin contain 1,8-cineole, α- and β-thujone, as well as camphor and sesquiterpenoids. Recent *in vitro* and *in vivo* studies have confirmed the pharmacologic efficacy of these plant extracts [17]. The next question to address is how quickly malaria will evolve resistance to artemisinin. Recent observations in Southeast Asia and sub-Saharan Africa have been worrying. For instance, it was reported that malarial parasites from sub-Saharan Africa may be acquiring mutations that make them resistant to artemisinin, the backbone of new antimalarial therapy. A team of researchers from Canada and the United Kingdom studied parasites obtained from travelers who returned to Canada with malaria after trips abroad (11 from Africa, including Angola, Cameroon, Congo, Ghana, Kenya, Liberia, Nigeria, and Tanzania) between April 2008 and January 2011. They found that 11 of the 28 parasites grown in the laboratory had a mutation that made them resistant to artemether. It is also reported that although parasites are showing drug resistance in malaria patients in Southeast Asia, the same strains are not being identified as resistant in laboratory studies, suggesting that the relationship between laboratory studies and patient treatment is not direct. It is therefore suggested that the spread of resistance may be exacerbated by the poor quality of antimalarials, which only kill the weaker parasites and allow the fittest to survive [27–30].

Strychnos myrtoides *(Loganiaceae)*

The reemergence of malaria in the central highlands of Madagascar in the 1980s, coupled with the lack of inappropriate drugs, compelled the indigenous people to explore traditional herbal remedies. A group of plants showing promising activity are *Strychnos* spp. *Strychnos* spp. are regularly used in the local Malagasy Pharmacopoeia and also on mainland Africa. Their roots are used to treat constipation, coughs, and toothache, as well as epilepsy. The aerial parts of these plants are used against malarial fever [31]. In Madagascar, there is a reported prevalence of quinine-resistant

P. falciparum and attention is increasingly being focused on alternative medicinal plants that can treat drug-resistant malaria. Investigations on several plants have led to the isolation of crude alkaloids from the leaves of *S. myrtoides*. These alkaloids have been used locally as adjuvant to chloroquine. When combined with chloroquine at doses less than the IC50, these molecules were shown to markedly enhance the effectiveness of synthetic drugs against chloroquine-resistant *P. falciparum in vitro*. They also enhanced chloroquine activity against a resistant strain of *P. yoelii in vivo*. By countercurrent distribution separation of the crude alkaloid extract, two major bioactive constituents, strychnobrasiline and malagashanine, were isolated from this plant, along with four minor alkaloids [32]. Malagashanine was identified as the parent compound of a new subtype of *Strychnos* alkaloids, the C-21, Nb-secocuran indole alkaloids, which had previously been isolated from Malagasy *Strychnos* [33,34]. *In vitro*, both strychnobrasiline and malagashanine are devoid of both intrinsic antimalarial activity and cytotoxic effects, but exhibit significant chloroquine-potentiating activities. Tests performed *in vivo*, on the other hand, showed that these extracts exhibited cytotoxicity and significant chloroquine-potentiating activity, which would justify the empirical use of *S. myrtoides* (10 mg/kg conferred a 5% suppression of the parasitemia) [34].

Until now, an infusion of the stem bark of *S. myrtoides* in association with chloroquine has been successfully evaluated within a clinical setting. The final aim is to develop a purified standardized extract for use in clinical trials, with a view to developing an efficient and inexpensive drug to combat chloroquine-resistant malaria.

Nauclea latifolia *(Rubiaceae)*

Nauclea latifolia is a savanna shrub commonly found in the Burkina Faso, Democratic Republic of the Congo, Gambia, and the Republic of

Benin, among others. Its medicinal uses are as a tonic and fever medicine; a chewing stick for treating toothaches, dental caries, and septic mouth, and for treating diarrhea, dysentery, and malaria. In most parts of West Africa, the bark is used against fever and malaria; hence, it has been described as *African quinine*. It is sometimes used in combination with *Khaya senegalensis*. Its key constituents are glycoalkaloids, indole-quinolizidine alkaloids, and saponins. Several indoloquinolizidine alkaloids were isolated from the root and include, among others, nauclefidine and naucletine. Root and stem aqueous extracts have been found to be active against *P. falciparum* (FcB1 strain) *in vitro*, mainly at the end of the erythrocytic cycle (after 32–48 h). Nonetheless, a comparative randomized clinical trial using standardized extracts of the roots has been tested against symptomatic, but uncomplicated malaria in human volunteers in Abuja, Nigeria. The results showed that the standardized extract was efficacious against uncomplicated malaria: parasite clearance was better than with chloroquine and there were no serious side effects on organs or tissues [17]. Additionally, studies have shown that the root has antibacterial activity against Gram-positive and Gram-negative bacteria, as well as antifungal activity. It is most effective against *Corynebacterium diphtheriae*, *Neisseria* spp., *P. aeruginosa*, *Salmonella* spp., *Streptobacillus* spp., *Streptococcus* spp. [35,36].

Cryptolepis sanguinolenta *(Asclepiadaceae)*

This plant, commonly known as Ghana quinine, is a thin-stemmed twining and scrambling shrub. Its dried roots are commonly used in West and Central Africa to treat hepatitis, while the entire plant is used to treat malaria. The major alkaloid isolated from this plant is cryptolepine, but it has been reported that other alkaloids present in the plant are responsible for its biologic/pharmacologic activity. Measurement of its antiplasmodial activity by ^{3}H-hypoxanthine incorporation into the malaria parasite indicates that the hydrochloride and hydroxy derivatives, as well as neocryptolepine, are more active than quindoline. *In vitro* results have proved encouraging, with IC50 values of 47, 42, and 54 μM, compared to values of 2.3, 72, and 68 μM for chloroquine. Cryptolepine was the most effective, with IC50 values of 27, 33, and 41 μM for D6-chloroquine-sensitive, K-1 chloroquine-resistant, and W-2 chloroquine-resistant strains, respectively. The WHO carried out in vivo studies to demonstrate the clinical efficacy of the product converted into a tea-bag formulation—Phyto-Laria. Over a 7-day period, the mean parasite clearance time was 82.3 h. The overall cure rate was 93.5%, with only two cases of recrudescence on days 21 and 28. On the evidence of fever clearance and disappearance of parasitemia by day 7, according to WHO criteria, this tea-bag formulation was deemed to be effective in the treatment of acute uncomplicated malaria [37].

Quillaja saponaria *(Soap bark tree; Rosaceae)*

Quillaja saponaria is a South American tree reported to contain triterpenoid saponins [38]. These ingredients have been used for an experimental malaria vaccine [39]. Partial purification of the crude extract yielded QuilA, which has since been renamed Stimulon. Stimulon works as an adjuvant, i.e., a pharmacologic additive that improves the effectiveness of a vaccine by stimulating the production of antibodies [39].

Plants and Acquired Immunodeficiency Syndrome

Across the world millions of people have been and continue to be infected by HIV, the pathogen directly responsible for acquired immunodeficiency syndrome (AIDS). AIDS is a complex array of disorders resulting from the breakdown of the immune system. Globally, AIDS-related diseases remain a leading cause of death. A person infected with HIV becomes highly susceptible to rare forms of cancer and to infections, often from opportunistic pathogens. HIV uses cells of the immune system (helper T cells and macrophages) as sites for reproduction. There, multiple copies of the viral genetic material (RNA) are made and packaged into new viral particles, ready for dispersal into new hosts. Thus, more cells of the host's immune system are killed or damaged with subsequent rounds of infection, in which millions of viral particles are produced every day. Despite the production of antibodies and helper T cells that normally fight disease, eventually the virus prevails and signs of infections and cancer associated with AIDS start to appear. To date, there is no known cure or vaccine against HIV and drugs that can slow the progression of viral infection or halt the onset of AIDS are scarce.

As early as 1989, the WHO had already voiced the need to evaluate ethnomedicines and other natural products for the management of HIV/AIDS: "In this context, there is need to evaluate those elements of traditional medicine, particularly medicinal plants and other natural products that might yield effective and affordable therapeutic agents. This will require a systematic approach," stated a memorandum of the WHO [40]. Plants and other natural products comprise a large repertoire from which to isolate novel anti-HIV compounds. Increasingly, new compounds from natural sources are being reported daily. Currently, around 55 plant families, containing 95 plant species, and other natural products have been found to contain anti-HIV active compounds, including diterpenes, triterpenes, biflavonoids, coumarins, caffeic acid tetramers, curcumins, hypericin, gallotannins, galloylquinic acids, limonoids, and michellamines. These active compounds can inhibit various steps in the HIV life cycle [41]. However, many remain unproven and others have so far only shown promise in *in vitro* studies. Secondary metabolites will continue to play a significant role in combating viral infections, including AIDS infections, that result from a compromised immune system. It has been estimated that over 36,000 extracts have been tested by the American National Cancer Institute and 10% have been reported to exhibit anti-HIV properties [22].

Calophyllum *Species (Garcinia family; Clusiaceae/Guttiferae)*

One of the most promising compounds against AIDS is been reported to be produced by a Malaysian tree that is a member of the Garcinia family (Clusiaceae or Guttiferae). This tree is valued both for its wood and resin. A thorough investigation of African species of the Garcinia family is warranted in the quest for novel anti-HIV compounds. Research into the Malaysian species showed that the latex of *Calophyllum lanigerum* and related species, such as *C. teysmannii*, manifests significant anti-HIV activity. The active constituent was found to be (−)-calanolide B, which could be isolated to provide yields of 20–30%. Of the eight compounds been isolated from *C. lanigerum*, calanolide A has shown anti-HIV activity; moreover, *C. teysmannii* has yielded calanolide B, which was found to be slightly less active than (+)-calanolide A, but has the advantage of being readily available from latex, which can be tapped in a sustainable manner by making small slash wounds in the bark of mature trees. Calanolide A is a type of coumarin and is now being tested in clinical trials. These drugs are being developed by Sarawak MediChem Pharmaceuticals, a joint venture company formed between the Sarawak State Government and MediChem Research, Inc.: (+)-Calanolide A (which has been synthesized by MediChem chemists) is currently in Phase II clinical trials, while (−)-calanolide B is in preclinical development. Both these calanolides can also be isolated from another *Calophyllum* species, specifically from the leaves of *C. brasiliensis* [42], and exhibit more or less the same pattern of activity. Eventually, these compounds may form part of the antiviral ingredients included in an AIDS *cocktail* to slow the rate of AIDS progression and extend the lives of HIV-infected patients.

Another potential anti-HIV drug originating in Africa comes from the *Ancistrocladus* spp. of woody vines. Three new atropisomeric naphthylisoquinoline alkaloid dimers, michellamines A, B, and C, have been isolated from a newly described species of tropical liana, *A. korupensis*, found in the rainforests of Cameroon. These compounds are capable of completely inhibiting the cytopathic effects of HIV-1 and HIV-2 on human lymphoblastoid target cells *in vitro* [3]. Crude extracts from this plant have yielded michellamine B, a new alkaloid that has been shown to have activity against HIV in initial trials. Based on the observed activity and the efficient synthesis of the di-acetate salt, the National Cancer Institute (NCI) of the United States committed michellamine B to advanced preclinical development, but continuous infusion studies in dogs indicated that effective anti-HIV concentrations *in vivo* could only be achieved at close to neurotoxic dose levels. Thus, despite showing *in vitro* activity against an impressive range of HIV-1 and HIV-2 strains, the difference between the toxic dose level and the level anticipated to be required for effective antiviral activity was small, and NCI decided to discontinue further studies aimed at clinical development. However, the discovery of novel antimalarial agents, the korupensamines, from the same species [43], holds further promise.

Sutherlandia frutescens (Fabaceae)

Sutherlandia frutescens is also known as cancer bush in South African and the southern African region. It is mainly used locally as a bitter tonic and an adaptogen.

The herb is known to be exceptionally variable and contains a large number of triterpenoid saponins. L-canavanine has been adopted as the marker molecule because it is a potent L-arginine antagonist with documented anticancer and antiviral activities, including activity against the influenza virus and retroviruses. Recent observations have shown that significant clinical benefits can be obtained in the treatment of wasting in cancer and AIDS, which is supported by a US patent. Convergent clinical observations by health professionals and community workers suggest that daily treatment with *Sutherlandia* can improve appetite, facilitate weight gain, and improve CD4 counts in HIV-positive patients. However, these observations need to be verified by a controlled clinical study [22].

Catharanthus roseus (Rosy Periwinkle; Apocynaceae)

Patients suffering from AIDS usually find themselves at risk of a range of diseases, including cancers, that would normally be controlled by the immune system. *Catharanthus roseus* has given medicine two very important anticancer drugs. One of these, a semisynthetic version of the anticancer alkaloid, vinorelbine, is known to disrupt the spindle fibers responsible for separating chromosomes during mitosis. It is effective at lower concentrations and has fewer side effects than alkaloids derived directly from the plant material. This new drug could also be useful in combating Kaposi sarcoma, a rare skin cancer usually associated with AIDS [22].

Chikungunya Virus

Chikungunya virus (CHIKV) is an arbovirus belonging to the family Togaviridae and the genus *Alphavirus*, which can be further classified into encephalitic and arthritic viruses. Of the 29 viruses belonging to the genus *Alphavirus*, six are arthritic viruses: CHIKV, Mayaro virus, o'nyong-nyong virus, Ross River virus, Semliki Forest virus, and Sindbis virus. Examples of encephalitic viruses are the western equine encephalitis and Venezuelan equine encephalitis viruses. A recent outbreak of Chikungunya fever in the islands of the Indian Ocean has drawn attention to CHIKV, which was first identified in the 1950s in Africa. Intriguingly, it was initially classified as a neglected tropical disease, and it was only the sheer magnitude of the 2005–2007 CHIKV outbreaks that brought this virus to the attention of both the scientific community and the general public [42]. CHIKV has since then been associated with the urban *Aedes aegypti* mosquito (possibly supplemented by *Aedes albopictus*) in an epidemiologic cycle resembling that of dengue and characterized by the absence of an animal reservoir, direct human-to-human transmission by urban mosquitoes, and the potential for major epidemics [44,45]. *A. albopictus* is considered to be the vector in Reunion Island and other islands of the Indian Ocean.

To date, neither a vaccine nor a selective antiviral drug is available for the prevention or treatment of this debilitating viral infection, and treatment is mainly supportive. The majority of cases are relatively mild, although more significant sequelae are now known. Thus, an antiviral treatment is most useful for prophylaxis in vulnerable groups, such as the immunocompromised, and for management of severe cases [46,47]. Currently, chloroquine use is not justified as there is no conclusive evidence for its effectiveness. The antiviral effects of chloroquine were first described in 1969. Subsequently, in the early 1980s, it was shown to have an inhibitory effect against replication of the Sindbis and Semliki Forest viruses. Recent *in vitro* experiments using chloroquine have led to a successful reduction in CHIKV growth, and use of chloroquine phosphate solution has been shown to provide relief to patients. Chloroquine is active in cell culture and may alleviate the symptoms of arthritis by acting as an anti-inflammatory agent, although this latter activity is still under investigation. However, in a 2006 double-blind, placebo-controlled trial with 54 participants, no statistical difference in the mean duration of febrile arthralgia between the placebo and chloroquine group was found [46,47].

Currently, there is therefore a need to identify new, potential drugs and many investigators have turned toward indigenous biodiversity for this. Interestingly, several Indian Ocean islands (Madagascar, Mauritius, and Reunion Island) have combined forces under an umbrella project—PHYTOCHICK—to combat this emerging virus threat via selecting natural drug candidates from locally available medicinal plants. So far, a number of promising leads have been discovered, and currently several attempts at bioassay-guided purification/fractionation of pure substances are underway and have yielded promising preliminary results. Concomitantly, enzyme assays are being developed to evaluate and provide a detailed characterization of the selective inhibitory effects of these phytocompounds. Overall, more than 1554 crude and filtered extracts and 22 pure compounds have been evaluated for cytotoxicity and evaluation against CHIKV. A total of 13 and 8 hit extracts were recorded for the Madagascar and La Reunion partners, respectively. Interestingly, 12 extracts have proven to be potent (i.e., providing a superhit against CHIKV) from Mauritius; these belong to the Celastraceae, Ebenaceae, Meliaceae, Rubiaceae, Sapindaceae, Sapotaceae, and Sterculiaceae families. Additionally, five plants from Mauritius were initially selected for further fractionation, phytochemical analysis, and anti-CHIKV evaluation. Promising leads have been found *in vitro* from four of these fractions; they have shown maximum inhibition of 88.8% at 20 µg/mL; 3.9% at 4 µg/mL; 100% at 20 µg/mL, and 95.3% at 20 µg/mL against the CHIKV virus, respectively.

CONCLUSIONS AND FUTURE PERSPECTIVES

Undeniably, drugs resistance has created resurgence and insurgence of a panoply of infectious diseases, mainly CHIKV, HIV, and malaria. The major victims for these killers are developing countries with the poorest resources, such as African and Asian countries. Many investigators now strongly believe that studying traditional medicines may offer new template molecules to combat these diseases. Evaluating plants from the traditional African system of medicine can provide us with clues about how these plants can be used in the treatment and management of diseases. Many of the plants presented in this chapter show very promising activity as antimicrobial agents, thus warranting their further investigation. Nevertheless, the discovery of compounds with antimicrobial activities from traditional medicinal plant remedies remains a challenging task. Indeed, to be successful in such an endeavor, more highly reproducible and robust innovative bioassays are needed as our understanding of the multifactorial pathogenicity of microbial infection evolves. Therefore, it is of the utmost importance that investigators should devise new automated bioassays, with a special emphasis on high through-put procedures, for screening and processing data from a large number of phytochemicals within shorter time periods. Additionally, these procedures should be able to rule out false-positive hits and incorporate dereplication methods to remove duplicate compounds. The ultimate goal will be to establish structure-activity phytochemical libraries to boost new antimicrobial drug discovery.

On the other hand, one of the main constraints to the growth of a modern African phytomedicine industry has been identified as a lack of proper validation of traditional knowledge and of technical specifications and quality control standards. This makes it extremely difficult for buyers, whether national or international, to evaluate the safety and efficacy of plants and extracts, or to compare batches of products from different places or from year to year. This stands in marked contrast to Europe and Asia, where traditional methods and formulations are recorded and evaluated at both the local and national levels. This could explain why the level of trade in Asia and Europe is higher than in Africa. Other issues that need to be addressed are those of Access and Benefit Sharing following the Nagoya Protocol. Local laws need to be TRIPS compliant if trade is to increase and, at the same time, issues of sustainable development need to be addressed. Nonetheless, despite the continuous, comprehensive, and mechanism-orientated evaluation of medicinal plants worldwide, there is still a dearth of literature since 2000 from investigations addressing procedures to be adopted for quality assurance, authentication, and standardization of crude medicinal plant products. Finally, above and beyond simple *in vitro* and *in vivo* assays, randomized, controlled trials must be carried out and reported for each claim and the data amassed should be provided to traditional healers.

References

[1] Newman D, Cragg MG, Snader KM. Natural products as sources of new drugs over the period 1981−2002. J Nat Prod 2002;66:1022−37.

[2] Okem A, Finnie JF, Van Staden J. Pharmacological, genotoxic and phytochemical properties of selected South African medicinal plants used in treating stomach-related ailments. J Ethnopharmacol 2012;139:712−20.

[3] Boyd M, Hallock Y, Cardellina II J, Manfredi K, Blunt J, McMahon J, et al. Anti-HIV michellamines from Ancistrocladus korupensis. Med Chem 1994;37: 1740−5.

[4] Cragg GM, Newman DJ. Plants as source of anti-cancer agents. J Ethnopharmacol 2005;100:72−9.

[5] Chintamunnee V, Mahomoodally MF. Herbal medicine commonly used against infectious diseases in the tropical island of Mauritius. J Herbal Med 2012;. http://dx. doi.org/10.1016/j.hermed.2012.06.001 (in press).

[6] Madoff LC, Kasper DL. Introduction to infectious diseases: Host pathogen interactions. In: Kasper DL, Fauci AS, editors. Harrison's Infectious Diseases. United States: The McGraw Hill Companies; 2010. p. 2−8.

[7] World Health Organization. The World Medicines Situation. Geneva: WHO Press; 2011 (WHO/EMP/ MIE/2011.2.3).

[8] WHO. Containing Antimicrobial Resistance: A Renewed Effort. Bulletin of the World Health Organization 2010;88(12):877−953, http://www.who.int/bulletin/volumes/88/12/10; [accessed 2.01.13].

[9] Obeidat M. Antimicrobial activity of some medicinal plants against multiresistant skin pathogens. J Med Plants Res 2011;5(16):3856−60.

[10] Quave CL, Pieroni A, Bennett BC. Dermatological remedies in the traditional pharmacopoeia of Vulture-Alto Bradano, inland southern Italy. J Ethnobio Ethnomed 2008:1−10.

[11] Hotez PJ, Molyneux DH, Fenwick A, Ottesen E, Ehrlich-Sachs S, Sachs J. Incorporating a rapid-impact package for neglected tropical diseases with programs for HIV/AIDS, tuberculosis, and malaria. PLoS Med 2007;3(5):e277.

[12] Hotez PJ, Molyneux DH, Fenwick A, Kumaresan MB, Ehrlich-Sachs S, Sachs J, et al. Control of Neglected Tropical Diseases. New Eng J Med 2007;357:1018−27.

[13] Morens DM, Folkers GK, Fauci AS. The challenge of emerging and re-emerging infectious diseases. Nature 2004;430:242−9.

[14] Jain A, Dixit P, Prasad R. Pre-XDR & XDR in MDR and Ofloxacin and Kanamycin resistance in non-MDR Mycobacterium tuberculosis isolates. Tuberculosis 2012;92:404−6.

[15] Zager EM, McNerney R. Multidrug-resistant tuberculosis. BMC Infec Dis 2008;8:10.

[16] World Health Organization (2007) XDR-TB: extensively drug-resistant tuberculosis March 2007. http://www.who.int/tb/xdr/news_mar07.pdf. Geneva. [accessed 08.01.13].

[17] Brendler T, Eloff JN, Gurib-Fakim A, Philips LD. "African herbal pharmacopoeia." Graphic press limited. Mauritius 2010.

[18] Meshnick S. Artemisinin: mechanisms of action, resistance and toxicity. Inten J Parasitol. 2002;32:1655−60.

[19] World Health Organization. Assessment of therapeutic efficacy of antimalarial drugs for uncomplicated falciparum malaria in areas with intense transmission. WHO/MAL/96:1077, http://www.who.int/drugresistance/malaria/en/Assessment_malaria_96.pdf 1996; [accessed 8.01.13].

[20] WHO Malaria Report. World Health Organization. Geneva: Switzerland; 2011.

[21] Karunamoorthi K, Tsehaye E. Ethnomedicinal knowledge, belief and self-reported practice of local inhabitants on traditional antimalarial plants and phytotherapy. J Ethnopharmacol 2012;141:143−50.

[22] Gurib-Fakim A. Medicinal plants: Traditions of yesterday and drugs of tomorrow. Mol Asp Med 2006;27:1−93.

[23] Eckstein-Ludwig U, Webb RJ, van Goethem ID, East JM, Lee AG, et al. Artemisinins target the SERCA of Plasmodium falciparum. Nature 2003;424:957−61.

[24] Mueller MS, Runyambo N, Wagner I, Borrmann S, Dietz K, Heidi L. Randomized controlled trial of a traditional preparation of Artemisia annua L. (Annual wormwood) in the treatment of malaria. Trans R Soc Med Hyg 2004;98:318−21.

[25] Willcox M, Rasoanaivo P, Sharma VP, Bodeker G. Comments on: Randomised controlled trial of a traditional preparation of Artemisia annua L. (Annual Wormwood) in the treatment of malaria. Trans R Soc Trop Med Hyg 2004;98:755−6.

[26] Balint GA. Artemisinin and its derivatives: an important new class of antimalarial agents. Pharmacol Ther 2001;90:261−5.

[27] Pillai DR, Lau R, Khairnar K, Lepore R, Via A, Staines HM, et al. Artemether resistance *in vitro* is linked to mutations in PfATP6 that also interact with mutations in PfMDR1 in travellers returning with Plasmodium falciparum infections. Malaria J 2012;11:131.

[28] Basco LK, Le Bras J, Rhoades Z, Wilson CM. Analysis of pfmdr1 and drug susceptibility in fresh isolates of Plasmodium falciparum from sub-Saharan Africa. Mol. Biochem Parasitol 1995;74:157−66.

[29] Wang Z, Parker D, Meng H, Wu L, Li J, Zhao Z, et al. In vitro sensitivity of Plasmodium falciparum from China-Myanmar border area to major ACT drugs and polymorphisms in potential target genes. PLoS ONE 2012;7(5):e30927.

[30] Nayyar G, Breman J, Newton P, Herrington J. Poor-quality antimalarial drugs in southeast Asia and sub-Saharan Africa. Lancet Infect Dis 2012;12(6):488−96.

[31] Neuwinger HD. African Traditional Medicine. Stuttgart, Germany: MedPharm Scientific Publishers; 2000.

[32] Rasoanaivo P, Ratsimamanga-Urverg S, Milijaona R, Rafatro H, Galeffi C, Nicoletti M. *In vitro* and *in vivo* chloroquine-potentiating action of Strychnos myrtoides alkaloids against chloroquine-resistant strains of Plasmodium malaria. Planta Med 1994;60:13−6.

[33] Rafatro H, Verbeeck RK, Gougnard TY, De Jonghe PJ, Rasoanaivo P, et al. Isolation from rat urine and human liver microsomes and identification by electrospray and nanospray tandem mass spectrometry or new malagashanine metabolites. J Mass Spectro 2000;35:1112−20.

[34] Rasoanaivo P, Ratsimamanga-Urverg S, Frappier F. Recent results on the pharmacodynamics of Strychnos malgaches alkaloids. Sante 1996;6:249−53.

[35] Iwu MW, Duncan AR, Okunji CO. New antimicrobials of plant origin. p. 457−462. In: Janick J, editor. Perspectives on new crops and new uses. Alexandria: ASHS Press; 1999.

[36] Deeni Y, Hussain H. Screening for antimicrobial activity and for alkaloids of Nauclea latifolia. J Ethnopharmacol 1991;35:91–6.

[37] Boye GL, Ampofo O. Proceedings of the First International Seminar on Cryptolepine. Kumasi, Ghana: University of Science and Technology. Clinical uses of Cryptolepis sanguinolenta (Asclepidaceae); 1983. p. 37.

[38] Guo S, Kenne L. Structural studies of triterpenoid saponins with new acyl components from Quillaja saponaria Molina. Phytochem 2000;55:419–28.

[39] Kirk DD, Rempel R, Pinkhasov J, Walmsley AM. Application of Quillaja saponaria extracts as oral adjuvants for plant-made vaccines. Expert Opin Biol Ther 2004;4:947–58.

[40] World Health Organisation (WHO). *In vitro* screening of traditional medicines for anti-HIV activity: memorandum from a WHO meeting. Bull World Health Organization 1989;87:613–8.

[41] Chinsembu KC, Hedimbi M. Ethnomedicinal plants and other natural products with anti-HIV active compounds and their putative modes of action. Intern J Biotechn Mol Bio Res 2010;1(6):74–91.

[42] Huerta-Reyes M, Basualdo Mdel C, Abe F, Jimenez-Estrada M, Soler C, Reyes-Chilpa R. HIV-1 inhibitory compounds from Calophyllum brasiliensis leaves. Biol Pharm Bull 2004;27:1471–5.

[43] Hallock YF, Manfredi KP, Blunt JW, Cardellina II JH, Schaffer M, Gulden KP, et al. Korupensamines A-D, novel antimalarial alkaloids from Ancistrocladus korupensis. J Org Chem 1994;59:6349–55.

[44] Powers AM, Logue CH. Changing patterns of Chikungunya virus: re-emergence of a zoonotic arbovirus. J General Virol 2007;88:2363–77.

[45] Leyssen P, Litaudon M, Guillemot J, Rasoanaivo P, Smadja J, Gurib-Fakim A, et al. PHYTOCHIK: Biodiversity as a source of selective inhibitors of CHIKV replication. Antiviral Res 2011;90:A1–A20.

[46] Mohan A. Chikungunya fever: clinical manifestations & management. Indian J Med Res 2006;124: 471–4.

[47] Coombs K, Mann E, Edwards J, Brown DT. Effects of chloroquine and cytochalasin B on the infection of cells by Sindbis virus and vesicular stomatitis virus. J Virol 1981;37:1060–5.

Natural Products with Activity against Multidrug-Resistant Tumor Cells

Camila Bernardes de Andrade Carli, Marcela Bassi Quilles,
Iracilda Zeppone Carlos

São Paulo State University, Faculdade de Ciências Farmacêuticas,
C. Postal 502, 14801−902 Araraquara, São Paula, Brazil

MULTIDRUG RESISTANCE

Multidrug resistance (MDR) is a significant challenge in the treatment of infectious diseases and cancer. The antitumor treatments currently in use often fail at some stage of the sickness. This is because many types of cancers develop resistance to chemotherapeutic drugs, rendering them unresponsive to treatment. For metastatic cancers, approximately 90% of patients are unresponsive to therapy because of MDR [1]. MDR affects several types of blood and solid tumors, including breast, gastrointestinal, lung, and ovarian tumors. Because tumors are composed of heterogeneous cells with structural and biochemical differences, drug-sensitive and drug-resistant cells often coexist in the same tumor microenvironment.

Antitumor therapy can select drug-resistant cells because the drug-sensitive cells die in the early stage of treatment, leaving a large population of drug-resistant cells. Without competition

from the sensitive cells, the drug-resistant cells grow and the tumor becomes resistant to the initial treatment [2].

RESISTANCE MECHANISMS

Several studies have been performed to analyze drug resistance mechanisms. There are three major drug resistance mechanisms in cells. The first involves decreased uptake of water-soluble drugs, including foliate antagonists, nucleoside analogs, and cisplatin, which require transporters to enter cells. This decrease can be attributed to upregulation of drug transporters, such as P-glycoprotein. These transporters can export a range of anticancer drugs from the cell, thus lowering the concentration to below the level required to provide a cytotoxic effect. A second mechanism involves changes to cellular processes that affect the capacity of cytotoxic drugs to kill these cells, including

Fighting Multidrug Resistance with Herbal Extracts,
Essential Oils and Their Components
http://dx.doi.org/10.1016/B978-0-12-398539-2.00016-1

alterations to the cell cycle, increased repair of DNA damage, reduced apoptosis, and altered drug metabolism. One hallmark of human cancers is their ability to evade apoptosis. In principle, cell death signaling can be blocked by an increase in antiapoptotic molecules and/or decreased function of or defects in proapoptotic proteins. Bcl-2 is a mitochondrial surface molecule that inhibits activation of the apoptosis cascade and is overexpressed in many types of cancer, thereby contributing to tumor initiation and progression [3–5]. This molecule inhibits the formation of pores in the mitochondrial membrane and thus inhibits the release of cytoplasmic molecules that signal cell death. Through this mechanism, reduced Bcl-2 expression can enhance the sensitivity of tumors to antineoplastic drugs and consequently reduce their survival. For this reason, Bcl-2 inhibitory substances have been extensively studied [5–7]. Finally, the third mechanism that mediates drug resistance involves an increase in the energy-dependent efflux of hydrophobic drugs that can easily enter cells by diffusion across the plasma membrane [2].

Because tumor angiogenesis is inhibitory to tumor growth, intrinsic resistance to antiangiogenic therapy must be considerable. An evolving hypothesis is that angiogenic tumors can adapt to the presence of angiogenesis inhibitors, thus acquiring means to functionally evade the therapeutic blockade of angiogenesis [8,9]. In contrast to the traditional concept of drug resistance, i.e., that it is acquired by mutational alterations to genes encoding drug targets or by alterations in drug uptake and efflux, evasive resistance is largely indirect: alternative ways of sustaining tumor growth are activated although the specific therapeutic target of the antiangiogenic drug remains inhibited [10–12]. Evidence suggests that at least four distinct adaptive mechanisms can result in evasive resistance to antiangiogenic therapies: (1) activation and/or up regulation of alternative proangiogenic signaling pathways within the tumors

and (2) recruitment of bone marrow-derived proangiogenic cells, both of which rely on vascular endothelial growth factor (VEGF) signaling effect reinitiation and continued tumor angiogenesis; (3) increased pericyte coverage of the tumor vasculature, which supports its integrity and decreases the need for VEGF-mediated survival signaling; and (4) activation and enhancement of invasion and metastasis to provide access to normal tissue vasculature without obligatory neovascularization [13].

Since the development of resistance to multiple chemotherapeutic drugs occurs frequently during the treatment of several types of cancer, drug resistance remains one of the most important issues in the management of cancer patients because most patients die when their disseminated cancer has become resistant to all available drugs.

PLANTS WITH ANTITUMOR ACTIVITY AGAINST MULTIDRUG-RESISTANT CELLS

The use of medicinal plants is one of the oldest human practices and remains widespread around the world. During the late twentieth century, *herbal* remedies regained popularity in the West because of recognition that traditional pharmacopeias are important sources of affordable treatments and the perception that these treatments would be safer and more efficient than those based on allopathic drugs. This recognition is primarily derived from an awareness that many allopathic drugs that are currently produced and consumed on a large scale are based on natural products, and that most are derived from plant species.

The search for antitumor drugs derived from plants began in the early 1950s with the discovery and development of the vinca alkaloids, vincristine and vinblastine. As a result, the National Cancer Institute of the United States initiated an extensive sampling program in the 1960s, focused mainly on temperate regions. This program led to the discovery of several molecules that exhibited a range of cytotoxic activities, including camptothecins and taxanes. Although the program ended in 1982, the development of new technologies to conduct screens of biologic activities led to a renewed interest in collecting plants and other organisms, with a focus on the subtropical and tropical regions of the world. With this in mind, some natural products have been studied to find new drugs against tumors. Unfortunately, most tumors become resistant to several anticancer drugs. These limitations have stimulated efforts to search for new, effective compounds without adverse side effects. In this regard, recent research shows natural products to be potential MDR modulators.

Thus, some natural products have been well studied. For example, species of the Simaroubaceae family exhibit a striking feature; their cortical substance has a very bitter taste called *quassina*. Several quassinoids are known to have antitumor and cytotoxic activities [14,15]. Among the quassinoids that exhibit significant *in vitro* activity against lymphocytic leukemia P388 cells, which are resistant to multiple anticancer drugs, cedronolactone and 14-hydroxychaparrinone are particularly effective [15]. Others quassinoids, like bruceanols D, E, and F, showed potent cytotoxicity *in vitro* against five multidrug-resistant (MDR) tumor cell lines (A549, TE-671, HCT-8, KB, and RPMI-7951) and lymphocytic leukemia (P338) [16]. The compounds euricomalactone, 14,15 b-dihydroxyclaineanone, 6-dihydroxylongilactone, and pasakbumina C also showed significant activity against lung cancer cells (A549), which was lower than the activity of the quassinoids bruceanol D and F against A-549 cells and the MCF-7 multidrug-resistant human breast cancer cell line [14].

In another study, betulinic, oleanolic, and pomolic acids were obtained from *Licania tomentosa* fruits and *Chrysobalanus icaco* leaves. These triterpenoids inhibited growth and induced apoptosis in an erythroleukemic cell line (K562). They also inhibited the proliferation of Lucena 1, a vincristine-resistant derivative of K562 that displays several MDR characteristics [15].

In addition to their cytotoxic effects, natural products may exhibit other activities against MDR cells. Some plants can change the resistance of tumor cells to antitumor drugs [16,17]. Limtrakul et al. [18] reported the reversal of MDR against colchicine, paclitaxel, and vinblastine in KB-V1 cells [MDR human cervical carcinoma cells with high P-glycoprotein (P-gp) expression] by treatment with a root extract of *Stemona curtisii*. In this study, the chemosensitizing property of *S. curtisii* extract was demonstrated using an MTT [3-(4,5-dimethythiazol-2-yl)-2,5-diphenyl tetrazolium bromide] assay. The *S. curtisii* extract was shown to selectively inhibit P-gp: it downregulated anticancer drug efflux *in vitro* by acting as a competitive or noncompetitive blocker and may therefore be effective in the treatment of MDR cancers.

Fractions of *Curcuma wenyujin* (C10 and E10) and *Chrysanthemum indicum* (E10) were tested for their ability to modulate the MDR phenotype and P-gp function in MCF-7/ADR and A549/Taxol cells *in vitro*. The function of P-gp was assessed by measuring the intracellular accumulation and efflux of rhodamine 123 in MCF-7/ADR cells by flow cytometry. Of the three fractions, *C. wenyujin* C10 induced the greatest decrease in P-gp expression in MCF-7/ADR cells at nontoxic concentrations, which might be due to the presence of curcumin in this fraction, as curcumin has been shown to downregulate P-gp in MDR cells. Moreover, all three fractions enhanced the level of apoptosis induced by doxorubicin in MCF-7/ADR cells and restore docetaxel induction of G2/M arrest in A549/Taxol cells [19].

Flavonoids are constituents of fruits, nuts, vegetables, plant-derived beverages, traditional eastern medicines, and herb-containing dietary supplements. They are considered to form a new class of chemosensitizers, which interacts with both the ATP-binding site and the vicinal steroid-interacting hydrophobic sequence of P-gp [20,21]. Chung et al. investigated the effects of various flavonoids, including biochanin A, diadzein, fisetin, morin, naringenin, quercetin, and silymarin, on P-gp function in the human breast cancer cell lines, MCF-7 (sensitive) and MCF-7/ADR (resistant) [22]. Using a sulforhodamine B staining assay, they showed that biochanin A and silymarin significantly decrease the IC50 value of the standard drug, daunomycin, thus potentiating the cytotoxicity of this drug. Biochanin A induced the greatest increase in daunomycin accumulation, while the effect of silymarin on daunomycin accumulation was similar to that of the P-gp inhibitor, verapamil. This study suggests that biochanin A and silymarin are potent and safe P-gp inhibitors that can increase the efficacy of chemotherapeutic agents when administered concomitantly [22].

Resveratrol (a flavonoid-like molecule), which is found in the skins of red grapes, mulberries, and other plants, has been shown to be effective against some tumor cells. The anticancer activities of resveratrol are mediated through modulation of several cell signaling molecules that regulate cell cycle progression, inflammation, proliferation, apoptosis, invasion, metastasis, and angiogenesis in tumor cells. Resveratrol has been shown to sensitize resistant cells to chemotherapeutic agents by decreasing the activity of membrane transporters such as P-gp and MDR-associated protein [23,24]. One study showed that resveratrol induces the accumulation of daunorubicin in MDR human oral epidermoid carcinoma cells (KBv200) in a concentration-dependent manner [23]. The researchers examined the potential of resveratrol to overcome KBv200 cell chemoresistance to doxorubicin, paclitaxel, and vincristine [23]. Resveratrol produced a synergistic effect when combined with these chemotherapeutic agents and reversed the MDR phenotype of KBv200 cells by downregulating P-gp and inducing apoptosis [23,24].

Several diterpenes, diterpenic lactones and other polycyclic diterpenes, steroids, and a triterpene isolated from methanol extracts of *Euphorbia* spp. have shown antineoplastic activity in various human MDR cancer cell lines, such as colon (HT-29), gastric (EPG85-257), and pancreatic (EPP85-181) cells [25]. For assessing the cytotoxicity of the test compounds in each cell line, the IC50 values of all agents were determined by cell proliferation assays. Most of the drug-resistant cancer sublines showed increased sensitivities to the test compounds compared to the parental lines. The most active compounds were the lathyrane diterpenes, latilagascenes C and D, and the diterpenic lactones, 3β-acetoxy-helioscopinolide B and helioscopinolide E, which exhibited high antineoplastic activities against the drug-resistant subline EPG85-257RDB, derived from gastric carcinoma. In addition, the macrocyclic lathyrane diterpene, jolkinol B, was found to be highly effective against the MDR variant HT-29RNOV [25].

QUALEA SPECIES AND THEIR MEDICAL APPLICATIONS

The Vochysiaceae family includes eight genera and approximately 200 tropical species of trees and shrubs [26,27]. Six genera are found in South America and these species are well represented in the Brazilian flora. The dispersal centers are located in the Guyana-Amazonian region and the Central Brazilian Plateau. From these centers, they seem to have spread to other neotropical and paleotropical areas [28–30]. This family is traditionally divided into two tribes: (1) Vochysiaceae, with five genera, *Callisthene*, *Qualea*, *Ruizteranea*, *Salvertia*, and *Vochysia* [31,32]; and (2) Erismeae, with two genera, *Erisma* and *Erismadelphus*, the latter of which is found in tropical Africa. A new monospecific genus, *Korupodendron*, was recently described in western Africa [27]. All of these genera are well defined and are clearly separated by their floral features.

The *Qualea* genus occurs throughout tropical America, from Mexico, through Peru and the Guianas, and south to Brazil, where there is a strong presence in the Cerrado. The species are popularly known as cassava or *quaruba* [33]. Species include *Q. calantha, Q. cordata* Spreng., *Q. dichotoma* (Mart.) Warm., *Q. glauca* Warm., *Q. grandiflora* Mart., *Q. grandiflora* Mart., *Q. impexa, Q. ingens* Warm.(known as *arrayán* in Colombia), *Q. multiflora* Mart., and *Q. parviflora* Mart. Research conducted on plants of this genus has demonstrated their versatility in the treatment of various diseases. Potts and Potts [34] demonstrated the presence of tannin in *Q. parviflora*; its shell and tea powder are used as an external antiseptic, while its leaves are used as a proton pump inhibitor. In addition, the sheets and shells of the inner part of *Q grandiflora* are used as an astringent and antidiarrheal, and used to clean sores and treat external inflammation [34]. Studies have shown that *Q. grandiflora* also has antimicrobial and antiulcer activities [35,36]. Another report indicates that *Q. grandiflora* is active in the central nervous system as an analgesic and anticonvulsant [37]. Grandi et al. [38] reported that a decoction of *Qualea* spp. sheets can be used for treating amebiasis, bloody diarrhea, gastritis, intestinal cramps, and ulcers. Recent studies identified the presence of ellagic acid derivatives in *Qualea* spp. and cytotoxic activity against tumor cells in *Q grandiflora* and *Q. multiflora* extracts [39–41].

ANTITUMOR ACTIVITY OF *QUALEA GRANDIFLORA* AND *QUALEA MULTIFLORA*

Q. multiflora Mart. and *Q. grandiflora* Mart. are found in the Brazilian Cerrado and are traditionally used in folk medicine.

Recent studies in our group by Carli et al. [40] and Quilles et al. [41] evaluated the antitumor activity of these two plant species against the breast cancer cell lines, Ehrlich (MDR) and LM3. Experiments were performed to assess the cytotoxic activity of a hexane extract and a terpenoid fraction, and of β-sitosterol isolated from the bark of *Q. grandiflora,* and of a terpenoid fraction isolated from the chloroform

TABLE 16-1 Viability of Different Cell Types Treated with a Terpenoid Fraction of *Qualea multiflora*

Concentration	Macrophage survival (%)	Lymphocyte survival (%)	LM3 cell survival (%)
2 mg/mL	97.98 ± 7.95	99.32 ± 8.45	0.00 ± 0.00^c
1 mg/mL	98.31 ± 5.24	98.99 ± 3.96	8.22 ± 1.35^c
0.5 mg/mL	85.12 ± 6.47^a	99.31 ± 6.14	34.34 ± 2.67^c
0.25 mg/mL	83.15 ± 5.77^a	99.97 ± 2.48	48.34 ± 2.15^c
0.125 mg/mL	83.15 ± 4.91^a	98.64 ± 5.44	54.51 ± 3.47^c
0.0625 mg/mL	80.47 ± 3.83^b	99.14 ± 7.37	64.37 ± 3.33^c
0 (NC)	100	100	100

[a] *P > 0.05;*
[b] *P < 0.01; and*
[c] *P < 0.001, compared with negative control.*

Negative control (NC) corresponds to 100% viability. Each value represents the mean ± SD for at least four independent experiments carried out in triplicate.

TABLE 16-2 Effects of *Qualea grandiflora* Hexane Extract, Terpenoid Fraction, and B-Sitosterol on Cell Viability

Treatment	Macrophage survival (%)	Ehrlich cell survival (%)
HE, 2 mg/mL	41.46 ± 6.89^c	2.60 ± 1.67^c
HE, 1 mg/mL	93.58 ± 5.98	24.7 ± 4.66^c
HE, 0,5 mg/mL	96.32 ± 8.61	60.40 ± 8.56^a
TF, 2 mg/mL	76.73 ± 2.88^a	60.17 ± 2.56^b
TF, 1 mg/mL	75.27 ± 3.77^a	56.60 ± 4.78^b
TF, 0,5 mg/mL	63.68 ± 5.99^b	51.54 ± 4.65^c
BS, 2 mg/mL	44.81 ± 3.90^c	37.33 ± 3.89^c
BS, 1 mg/mL	49.13 ± 4.44^c	45.53 ± 5.77^c
BS, 0.5 mg/mL	52.67 ± 5.89^c	48.52 ± 3.99^c
NC	100	100

[a] $P > 0.05$;
[b] $P < 0.01$; and
[c] $P < 0.001$, compared with negative control

Negative control (NC) corresponds to 100% viability. Each value represents the mean ± SD for at least four independent experiments carried out in triplicate.

BS, β-sitosterol; HE, hexane extract; TF, terpenoid fraction.

extract of *Q. multiflora* against tumor cells using the MTT assay. In addition, the toxic effect of these natural products on immune cells was assessed, since the immune system is very important for tumor growth inhibition.

In these studies, a terpenoid fraction of a *Q. multiflora* chloroform extract showed cytotoxic activity (Table 16-1) against the LM3 cell line and lower toxicity against immune cells (macrophages and lymphocytes). For *Q. grandiflora*, we observed a cytotoxic effect of the hexane extract and terpenoid fraction, as well as of β-sitosterol (isolated from the bark) against Ehrlich cells and lower toxicity against macrophages (Table 16-2). It should be emphasized that the *Q. multiflora* terpenoid fraction induced a very high selective toxicity against LM3 cells: following treatment with the *Q. multiflora* terpenoid fraction at a concentration of 2 mg/mL, viability of the macrophages and lymphocytes was almost 100%, while LM3 tumor cell viability was 0%.

The standard drug, taxol, was used as a positive control (Table 16-3). Although taxol has been extensively used for cancer treatment, its cytotoxicity was not superior to that of the *Q. multiflora* derivatives against LM3 cells or to

TABLE 16-3 Effect of Taxol on the Viability of Different Cell Types

Taxol concentration	Macrophage survival (%)	Lymphocyte survival (%)	Erlich Cell survival (%)	LM3 cell survival (%)
2 mg/mL	22.87 ± 3.93^a	98.65 ± 6.81	16.02 ± 4.05^a	54.39 ± 3.84^a
1 mg/mL	30.41 ± 2.67^a	98.67 ± 8.66	18.42 ± 3.2^a	54.43 ± 4.27^a
0.5 mg/mL	85.64 ± 8.37	98.99 ± 8.47	16.56 ± 2.22^a	56.13 ± 3.99^a
0.25 mg/mL	98.91 ± 5.70	98.95 ± 8.19	34.36 ± 6.88^a	54.61 ± 3.76^a
0.125 mg/mL	98.76 ± 8.33	99.34 ± 7.58	33.21 ± 3.49^a	56.20 ± 5.80^a
0.0625 mg/mL	98.54 ± 7.91	99.86 ± 5.84	35.42 ± 1.78^a	56.63 ± 2.87^a
0 (NC)	100	100	100	100

[a] $P < 0.001$, compared with negative control.

Negative control (NC) corresponds to 100% of viability. Each value represents the mean ± SD for at least four independent experiments carried out in triplicate.

that of the *Q. grandiflora* hexane extract *in vitro*. Furthermore, taxol caused a considerable decrease in macrophage viability and was more toxic against this immune cell than against LM3 cells at higher concentrations. Thus, comparing the results from *Q. multiflora* and taxol reveals that the *Q. multiflora* terpenoid fraction is more selectively cytotoxic than the standard drug, taxol.

CONCLUSIONS AND FUTURE PERSPECTIVES

Since tumors can become resistant to most chemotherapeutic agents, it is necessary to discover novel treatments for combating MDR cancer cells. Understanding the mechanisms of resistance to chemotherapeutic agents is necessary to develop strategies for more effective treatments.

Based on this knowledge and the results presented in this review, we conclude that the constituents of several plants have potential as antitumor drugs against MDR cells. Research efforts should be focused on *Qualea* spp. that exhibit selective activity against breast tumor cells. These studies show the great potential of plant biodiversity to provide treatments for diseases such as cancer and highlight the importance of conducting in-depth studies on promising species. Based on these promising results, future studies should be conducted to establish alternative forms of cancer treatment.

References

[1] Türk D, Hall MD, Chu BF, Ludwig JA, Fales HM, Gottesman MM, et al. Identification of Compounds Selectively Killing Multidrug-Resistant Cancer Cells. Cancer Resm 2009;69:8293—301.

[2] Szakács G, Paterson JK, Ludwig JA, Genthe CB, Gottesman MM. Targeting multidrug resistance in cancer. Nature 2006;5:219—34.

[3] Danial N, Korsmeyer SJ. Cell death: critical control points. Cell. 2004;116:205—19.

[4] Youle RJ, Strasser A. The BCL-2 protein family: opposing activities that mediate cell death. Nat Rev Mol Cell Biol 2008;9:47—59.

[5] Lai F, Chen J, Farrelly C, Margaret L, Zhang X, Hersey P. Evidence for upregulation of Bim and the splicing factor SRp55 in melanoma cells from patients treated with selective BRAF inhibitors. Melanoma Res 2012;22:244—51.

[6] Letai A. Puma strikes Bax. Journal of Cell Biology 2009;185:189—91.

[7] Becattini B, Kitada S, Leone M, Monosov E, Chandler S, Zhai D, et al. Rational design and real time, in-cell detection of the proapoptotic activity of a novel compound targeting Bcl-X(L). Chem Biol 2004;11:389—95.

[8] Miller KD, Sweeney CJ, Sledge Jr GW. Can tumor angiogenesis be inhibited without resistance? Exp Suppl 2005;2:95—112.

[9] Kerbe 1RS, Yu J, Tran J, Man S, Viloria-Petit A, Klement G, et al. Possible mechanisms of acquired resistance to anti-angiogenic drugs: implications for the use of combination therapy approaches. Cancer Metastasis Rev 2001;20:79—86.

[10] Casanovas O, Hicklin DJ, Bergers G, Hanahan D. Drug resistance by evasion of antiangiogenic targeting of VEGF signaling in late-stage pancreatic islet tumors. Cancer Cell 2005;8:299—309.

[11] Blouw B, Song H, Tihan T, Bosze J, Ferrara N, Gerber H, et al. The hypoxic response of tumors is dependent on their microenvironment. Cancer Cell 2003;4:133—46.

[12] Carmeliet P. Angiogenesis in life, disease and medicine. Nature 2005;438:932—6.

[13] Bergers G, Hanahan D. Modes of resistance to anti-angiogenic therapy. Nat Rev Cancer 2008;8: 592—603.

[14] Imamura K, Okano M, Fukamiya N, Tagahara K, Lee K. Three new cytotoxic quassinoids from Brucea antidysenteric.a J. Nat Prod 1993;56:2091—7.

[15] Fernandes J, Castilho RO, Costa MR, Wagner-Souza K, Kaplan MAC, Gattass CR. Pentacyclic triterpenes from Chrysobalanaceae species: cytotoxicity on multidrug resistant and sensitive leukemia cell lines. Cancer Let 2003;190:165—9.

[16] Murakami C, Fukamya N, Tamura S, Okano M, Bostow KF, Tokuda H, et al. Multidrug-resistant cancer cell susceptibility to cytotoxic quassinoids and cancer chemopreventive effects os quassinoids and canthin alkaloids. Bioorg Med Chem 2004;12:4963—8.

[17] Almeida MMB, Arriaga AMC, Santos AKL, Lemos TLG, Filho RB, Vieira IJC. Occurrence and biological activity of quassinoids in the last decade. Quim Nova 2007;30:935—51.

[18] Limtrakul P, Siwanon S, Yodkeereea S, Duangrat C. Effect of Stemona curtisii root extract on P-glycoprotein and MRP-1 function in multidrug-resistant cancer cells. Phytomed 2007;14:381—9.

[19] Yang L, Wei D-D, Chen Z, Wang J-S, Kong L-Y. Reversal of multidrug resistance in human breast cancer cells by Curcuma wenyujin and Chrysanthemum indicum. Phytomed 2011;18:710—8.

[20] Suttana W, Mankhetkorn S, Poompimon W, Palagani A, Zhokhov S, Gerlo S, et al. Differential chemosensitization of P-glycoprotein overexpressing K562/Adr cells by withaferin A and Siamois polyphenols. Mol Cancer 2010;3:9—99.

[21] Zand H. Chemopreventive and chemosensitization potential of flavonoids in acute lymphoblastic leukemia. J Pediatric Biochem 2012;2:15—21.

[22] Chung SY, Sung MK, Kim NH, Jang JO, Go EJ, Lee HJ. Inhibition of P-glycoprotein by natural products in human breast cancer cells. Arch Pharmacol Res 2005;28:823—8.

[23] Quan F, Pan CE, Zhang SQ, Yan LY, Yu L. Reversal effect of resveratrol on multidrug resistance in KBv200 cell line. Biomed Pharmacother 2008;62:622—9.

[24] Gupta SC, Kannappan R, Reuter S, Kim JH, Aggarwal BB. Chemosensitization of tumors by resveratrol. Ann N Y Acad Sci 2011;12:150—60.

[25] Lage H, Duarte N, Coburger C, Hilgeroth A, Ferreira MJU. Antitumor activity of terpenoids against classical and atypical multidrug resistant cancer cells. Phytomed 2010;17:441—8.

[26] Heywood V. Flowering Plants of the World. London: Croom Helm; 1985.

[27] Litt A, Cheek M. Korupodendron songweanum, a new genus and species of Vochysiaceae from West-Central Africa. Brittonia 2002;54:13—7.

[28] Barroso GM, Peixoto AL, Ichaso CLF, Costa CG, Guimarães EF, Lima HC. Sistemática de Angiospermas do Brasil. Viçosa: Academic press; 1984.

[29] Kawasaki ML. Systematics of Erisma (Vochysiaceae). Memoirs of the New York Botanical Garden 1998;81:1—40.

[30] Litt A. Floral morphology and phylogeny of Vochysiaceae. PhD Thesis. New York, City: University of New York; 1999.

[31] Stafleu FA. A monography of the Vochysiaceae. I. Salvertia and Vochysia. Mededelingen van het Botanich Museum en Herbarium van de Rijksuniversiteit te Utrecht 1948;95:397—540.

[32] Marcano-Berti L. Ruizterania. Pittieria 1969;2:6—27.

[33] Corrêa MP. Dicionário das plantas úteis do Brasil e das exóticas cultivadas. Rio de Janeiro: Ministério da Agricultura; 1974.

[34] Almeida SP, Proença CEB, Sano SM, Ribeiro JF. Cerrado: espécies vegetais úteis. Brasília: EMBRAPA; 1998.

[35] Hiruma-lima CA, Santos LC, Kushima H, Pellizzon CH, Silveira GG, Vasconcelos PC, et al. Qualea grandiflora, a Brazilian "Cerrado" medicinal plant presents an important antiulcer activity. J Ethopharmacol 2006;104:207—14.

[36] Costa ES, Hiruma-Lima CA, Lima EO, Sucupira GC, Bertolin AO, Lolis SF, et al. Antimicrobial activity of some medicinal plants of the Cerrado. Brazil Phytoter Res 2008;22:705—7.

[37] Gaspi FO, Foglio MA, Carvalho JE, Moreno RA. Pharmacological activities investigation of crude extracts and fractions from Qualea grandiflora Mart. J Ethopharmacol 2006;107:19—24.

[38] Grandi TSM, Trindade JA, Pinto MJF, Ferreira LL, Catella AC. Plantas Medicinais de Minas Gerais, Brasil. Acta Bot Bras 1986;3:185—224.

[39] Nasser ALM, Carli CBA, Rodrigues CM, Maia DCG, Carlos IZ, Eberlin MN, et al. Identification of ellagic acid derivatives in methanolic extracts from Qualea Species. Z Naturforsch 2008;63c. 794—80.

[40] Carli CBA, Quilles MB, Vilegas W, Carlos IZ. Efeito dos componentes obtidos de Qualea multiflora sobre modelo tumoral mamário murino e sua influência no sistema imunológico. Araraquara: PhD Thesis, UNESP; 2012.

[41] Quilles MB, Carli CBA, Vilegas W, Carlos IZ. Qualea grandiflora e sua interface com o sistema imunológico em modelo experimental de adenocarcinoma murino. Araraquara.: PhD Thesis, UNESP; 2012.

Development of New Antiherpetic Drugs Based on Plant Compounds

Adil M. Allahverdiyev[1], Melahat Bagirova[1], Serkan Yaman[1], Rabia Cakir Koc[2], Emrah Sefik Abamor[1], Sezen Canim Ates[1], Serap Yesilkir Baydar[1], Serhat Elcicek[3], Olga Nehir Oztel[1]

[1]Department of Bioengineering, Yildiz Technical University, Istanbul, Turkey,
[2]Department of Biomedical Engineering, Yeni Yuzil University, Istanbul, Turkey and
[3]Department of Bioengineering, Firat University, Elazig, Turkey

INTRODUCTION

Herpes simplex viruses (HSVs) are double-stranded, enveloped DNA viruses of the Herpesviridae family. HSVs are characterized by their icosahedral symmetry and possess 162 capsomers. These viruses consist of four basic structures: an envelope, a nucleocapsid, a DNA containing core, and a tegument. The virion contains 11 envelope glycoproteins, which are found as bulges on the outer surface. Four glycoproteins (gB, gD, gH, and gL) play important roles as receptors for determining virus infectivity. Approximately 100 species of herpes virus have been identified as pathogens both for humans and animals, and 25 of these have been shown to cause diseases affecting humans [1]. HSV infections usually give rise to painful lesion in tissues originating from the embryonic ectoderm [2]. Hence, these infections are frequently seen in skin, mouth, oral cavity, genital region, and central nerve system. HSV-1 and HSV-2 have been characterized as the mostly common virus species, responsible for a variety of mild to severe diseases such as herpes corneae, herpes labialis, herpes genitalis, herpetic whitlow, gingivostomatitis, Kaposi varicelliform eruption, and neonatal herpes. Among these diseases, herpes labialis, gingivostomatitis, and herpetic whitlow, which are seen in patients infected by HSV-1, causes painful cold sores and blisters at facial sites such as the mouth, tongue, and skin. On the other hand, HSV-2 infects genital sites of the body and induces ulcers in the genital mucosa.

Primary HSV infections are caused by penetration of the virus into the skin and mucosa of susceptible individuals from scratches and wounds on the skin or mucosal surface. Generally, intact skin or mucosa is resistant to HSV

Fighting Multidrug Resistance with Herbal Extracts,
Essential Oils and Their Components
http://dx.doi.org/10.1016/B978-0-12-398539-2.00017-3

infections. Broadly, HSV-1 infections are transmitted into individuals via aerosols or saliva, while HSV-2 infections are frequently transmitted sexually. HSV diseases occur by one of two mechanisms: a new infection or reactivation of latent viruses. HSV is known to reside in nerve ganglia after initial infection and can survive at these sites as a latent virus for many years. Viral reactivation inside nerves can be induced by fever, ultraviolet rays, stress, fatigue, menstruation, sexual intercourse, or immune suppression; after reactivation, herpetic blisters can be observed at different sites of the human body [3].

According to the World Health Organization, an estimated 536 million people all over the world were infected with HSV-2 in 2003 [4]. In addition, more than 50 million people are currently infected with HSV-2 in the USA and an estimated 0.5 million new cases occur each year [5]. It is known that HSV-1 infections are more commonly seen in humans than are HSV-2 infections. For example, 57% of the population of the United States is reported to be seropositive for HSV-1, while 20% are seropositive for HSV-2 infection [6]. Worse news is that the prevalence of both types of the disease has been gradually increasing and has reached life-threatening levels, especially for immune-compromised patients. On this subject, it was reported that the prevalence of HSV-2 has been increasing at a high rate within HIV-positive patients, indicating that the herpetic infection is a major cause of morbidity, especially in immune-suppressed patients. Acyclovir (Zovirax) and other nucleoside derivatives, including famciclovir, ganciclovir, penciclovir, and valacyclovir, have been approved for the treatment of HSV (both HSV-1 and HSV-2) infections worldwide [7–11]. The basic efficacy mechanisms of these drugs involve inhibition of thymidine kinase and DNA polymerase enzymes specific to HSV. However, it was recently reported that resistance can develop against acyclovir, especially in individuals that are immune compromised, undergoing cancer

chemotherapy, or have had bone-marrow transplants. Treatment failure can also be caused by the recurrence of latent viruses. Considering the constant increase on HSV infections worldwide and recently developed resistance to current therapies, it is clear that there is an urgent need to identify new, more effective antiviral agents to combat HSV infections [12].

Medicinal plants have been used for the treatment of human diseases since ancient times and their use offers an alternative to traditional antibiotics as antimicrobial agents. In contrast to synthetic drugs that may contain several harmful components, medicinal plants have the advantage of being 100% natural and biologically sourced. Recently, the acceptance of traditional medicine as an alternative form of health care and the development of microbial resistance to the current antibiotics have encouraged researchers to investigate the antimicrobial activities of medicinal plants. In several studies, medicinal plant products, such as extracts and essential oils (EOs), have been proven to possess significant antimicrobial properties [13].

Additionally, plant products have been recently marketed as pharmaceuticals in developed and developing countries, since they include biologically active substances and have been indicated to be safe and without side effects. According to the literature, many plant products obtained from traditional medicinal plants also have strong antiviral activity, and some have already been used in the treatment of viral infections in humans and animals. One type of viral infection for which researchers are investigating plant products is herpes simplex infection. There are severally reports of studies on antiherpetic plants. Therefore, we will review the information on various herbal compounds available in the current literature, compare the antiherpetic efficacy of these compounds, and determine the potential of these plants for developing new antiherpetic therapies. In this chapter, the antiherpetic activity of herbal extracts and EOs will be presented. Such plant extracts and EOs could be important for the discovery of new and effective antiviral drugs.

PLANT EXTRACTS WITH ANTIHERPETIC ACTIVITY

Carissa edulis (Forssk.) Vahl (Apocynaceae)

Carissa edulis is a medicinal plant that grows particularly in Kenya. *C. edulis* has been used as a traditional medicine in Kenyan for the treatment of various ailments, without any reported side effects [14]. The roots of this plant have been also used for antiherpetic purposes. Briefly, roots are washed, dried, boiled, and then lyophilized for the preparation of extracts. The antiherpetic effects of *C. edulis* were investigated in Vero E6 cells infected with 100 plaque forming units of wild-type or resistant strains of HSV; under these conditions, 50 g/mL of the plant extract showed antiviral activity. *In vivo* animal tests with Balb/c mice showed that the extract at an oral dose of 250 mg/kg significantly delayed the onset of HSV infections by > 50% and increased the survival time of treated animals. *C. edulis* toxicity was low [50% cytotoxic concentration (CC50) value of 480 g/mL] and it did not cause acute toxicity in BALB/c mice at the oral therapeutic dose of 250 mg/kg [15].

Rhubarb and Sage

The antiviral activity of rhubarb root extract has been observed and confirmed by several *in vitro* studies [16,17]. The antiviral activity of sage leaf extract was also confirmed, although is showed a low efficiency. In a screening study, the antiherpetic effect of rhubarb-sage cream, sage cream, and acyclovir cream were compared in an intention to treat analysis. Aqueous-ethanol extracts of rhubarb (23 mg/g) and dried sage (23 mg/g) were used, and the healing time of treated patients was 6−7 days for sage cream, 6−7 days for rhubarb-sage cream, and 5−6 days for acyclovir cream. The results of this study showed that sage was less effective than rhubarb against HSV, while a combined topical preparation of these plant extracts was as effective as topical acyclovir [18].

Hypericum connatum (St. John's Wort)

The Guttiferae family produces several antiviral compounds, and species of the *Hypericum* genus have long been used in traditional medicine. Extracts and isolated compounds from plants of the *Hypericum* genus are commonly used for the treatment of viral infections. For example, *H. connatum* is utilized for the treatment of mouth lesions in southern Brazil. In a study, ethanol extracts of *H. connatum* compounds were tested for antiherpetic effects *in vitro* using the Vero cell line infected with HSV-1. This study demonstrated different rates of antiviral activity against HSV for the crude methanol extract, fractions, and isolated compounds. Therefore, compounds obtained from *H. connatum* have inhibitory effects against HSV. The mechanisms of the antiherpetic effects of these compounds may involve a direct virucidal effect; interference with virus entry across the cell membrane; or interference with viral genome replication or other early steps in intracellular viral replication [19].

Clinacanthus nutans (Snake Plant)

Clinacanthus nutans extract is traditionally used in Thailand for the topical treatment of HSV and varicella-zoster virus infection [20]. The active ingredient was purified from the heartwood of *Artocarpus lakoocha* Roxb. (Moraceae), is extracted in methanol and then purified by vacuum liquid chromatography. The purified compound was identified as 2,3′, 4,5′-tetrahydroxy-*trans*-stilbene (oxyresveratrol) by mass spectroscopy, and high performance liquid chromatography analysis confirmed it to be > 99% pure. Antiviral activities of oxyresveratrol were demonstrated against HSV-1, HSV-2, measles virus (Tanabe strain), and poliovirus type 1 (Sabin strain). There were no significant differences among the IC50 values ($P > 0.05$) for various strains of HSV-1. Acyclovir-resistant HSV-1 strains were also susceptible to oxyresveratrol, suggesting that the mode of anti-HSV activity of oxyresveratrol differs from that of acyclovir. Concentrations of 25–100 µg/mL had no effect on viral infectivity. However, treatment of mice with three applications a day showed a significant therapeutic efficacy ($P = 0.004$), although two applications a day did not show significant efficacy ($P = 0.10$), compared to the control group. An ointment containing 30% oxyresveratrol showed no significant difference in efficacy ($P = 0.62$) compared to 5% acyclovir cream [21]. Therefore, topical administration of oxyresveratrol is suitable as an anti-HSV therapeutic route for cutaneous HSV infections in mice. Oxyresveratrol also demonstrated efficacy on acyclovir-resistant mutant strains, and may therefore be effective against acyclovir-resistant virus strains. Oxyresveratrol showed synergy with the anti-HSV-1 activity of acyclovir. The IC50 for combined acyclovir and oxyresveratrol was reduced by two- to four-fold, compared to the IC50 values when these drugs were used alone. This provides further supporting evidence that the mode of oxyresveratrol anti-HSV-1 activity differs from that of acyclovir: acyclovir is a nucleoside analog that exhibits antiherpetic activity after its phosphorylation by viral thymidine kinase; in contrast, oxyresveratrol affects viral replication by inhibiting late viral protein synthesis [21].

According to the results of topical treatment experiments, ointment containing 30% oxyresveratrol applied 4–5 times a day is most effective in delaying the skin lesion development and protecting against death. Further, topical treatment of oxyresveratrol is reported to inhibit

HSV in animal experiments. Cutaneous application of oxyresveratrol was found to be more effective than oral administration. Therefore, oxyresveratrol may be an appropriate topical treatment for HSV-1 [21–23].

In conclusion, oxyresveratrol has been shown to be effective against both wild-type and acyclovir-resistant HSV strains. These results provided evidence that topical rather than oral applications of oxyresveratrol provide a greater therapeutic efficacy against cutaneous HSV infection in mice.

Coptidis rhizoma (Coptis Root or Goldenthread)

Three extracts from the *Coptidis rhizoma* plant are berberine, *C. rhizome* extract, and Ching Wei San. Berberine is an antimicrobial herbal extract with activity against a variety of organisms, including human immunodeficiency virus [24], human cytomegalovirus (HCMV) [25], and hepatitis B virus [26]. This compound intercalates DNA, thus inhibiting DNA synthesis and reverse transcription [24,27–29]. Berberine does not affect membrane permeability, but does affect protein biosynthesis [30]. A study compared the antiviral activity of all three *C. rhizome* extracts against HSV and demonstrated that all exhibited antiherpetic effects, with IC50 values of 8.2, 2.4, and 24.3 mg/mL for berberine, *C. rhizome* extract, and Ching Wei San, respectively. Levels of cytotoxicity induced by these extracts were evaluated in Vero cells; the toxicity of berberine was found to be lower than those of the other two extracts. Thus, the selectivity of berberine appears to be 1.2–1.5-fold higher than that of the *C. rhizome* extract and Ching Wei San.

Ilex paraguariensis (Yerba Mate)

Ilex paraguariensis (yerba mate) is an evergreen tree from the holly family that grows in the subtropical forests of Paraguay, Uruguay, and the Parana state of Brazil. In a study, an aqueous extract of *I. paraguariensis* leaves showed antiviral activity against HSV-1 *in vitro* [31]. Caffeoylquinic acid derivatives and triterpenoid saponins are widely distributed among plants, including yerba mate, which shows an antiviral effect that may be linked to the presence of these compounds. The 50% effective concentrations (EC50) of caffeoylquinic acid derivatives and triterpenoid saponins derived from *I. paraguariensis* were 80 µg/mL and the CC50 was 1260 µg/mL [selectivity index (SI = CC50/EC50) value of 15.8] [31]. Therefore, these compounds may play important roles in the drug discovery process in the future.

Lafoensia pacari

Lafoensia pacari is a plant of the Lythraceae family that grows in Brazil and Paraguay. In a study, a methanol extract of *L. pacari* leaves showed antiviral activity against HSV-1 *in vitro*. A high SI value (19.0) was obtained for tannins isolated from *L. pacari*, indicating that these compounds are very effective against the HSV-1 virus and that extracts containing tannins may play an important role against HSV [31].

Passiflora edulis (Passion Flower)

Passiflora edulis is a vine species of passion flower that grows in Brazil, Northern Argentina, and Paraguay. In a study, an aqueous extract of *P. edulis* roots showed antiviral activity against HSV-1 *in vitro* [31]. The EC50 of *P. edulis* extracts (containing saponins and flavonoids) was 89.9 µg/mL and the CC50 was 1600 µg/mL. Therefore, saponins and flavonoids may exhibit antiherpetic activity, consistent with previous reports [32].

Rubus imperialis

Rubus imperialis grows abundantly in southern Brazil. In a study, a methanol extract of *R. imperialis* leaves showed antiviral activity against HSV-1 *in vitro* [31]. The EC50 of *R. imperialis* extracts was 70 µg/mL, while the CC50 was 1390 µg/mL (SI value of 19.8). As a result, the presence of tannins and flavonoids may be associated with the high antiviral activity of this extract against HSV-1 [31].

Slonea guianensis

In a study, the methanol extract of *Slonea guianensis* leaves showed antiviral activity against HSV-1 *in vitro*. The EC50 of the *S. guianensis* methanol extract was 140 µg/mL and the CC50 was 1400 µg/mL [31].

Pongamia pinnata (Indian Beech)

Pongamia pinnata belongs to the Papilionaceae family. Seeds were collected from dried fruits and, in an *in vitro* study, a seed extract completely inhibited HSV-1 and HSV-1 multiplication at concentrations of 1 mg/mL (w/v) and 20 mg/mL (w/v), respectively [33].

Other Important Plant Extracts

In contrast to the other studies, a study performed in Thailand identified specific activity against HSV *in vitro* in extracts of *Aglaia edulis*, *Centella asiatica*, *Glyptopetalum sclerocarpum*, *Maclura cochinchinensis*, and *Mangifera indica* [34]. Additionally, extracts of *Azadirachta indica*, *Rinorea anguifera*, and *Protium serratum* were also demonstrated to be effective against HSV-1 and poliovirus *in vitro* [34]. Moreover, extracts of *Hura crepitans*, *Moringa oleifera*, *Schima wallichii*, and *Willughbeia edulis* specifically inhibited plaque formation by HSV-1; in addition, none of these plant extracts were cytotoxic [34]. The *Moringa oleifera* extract was more effective against a phosphono acetate-resistant (APr) strain and *Aglaia odorata* was less effective against a thymidine kinase-deficient strain compared to their activities against the wild-type strain (7401H HSV-1) [34]. *In vivo* studies revealed that *A. odorata, Cerbera odollam, M. oleifera, Ventilago denticulata*, and *W. edulis* either inhibit plaque formation or exhibit therapeutic levels of antiviral efficacy against HSV-1 following oral application [34]. The anti-HSV activity of *M. oleifera* in mice does not involve a direct antiviral effect, because this extract can also be used as an immune modulator in mice. The three plant extracts containing active compounds (*A. odorata, M. oleifera*, and *V. denticulata*) were also nontoxic. Therefore, all may be used as prophylactic or therapeutic anti-HSV medicines for the treatment of HSV-1 infections. The *M. oleifera* extract is particularly promising because of its immunomodulatory and anti-HSV features.

Combinations of acyclovir and herbal extracts have also been used to improve the treatment of HSV-1 infection in humans. Some studies have shown that *Geum japonicum, Rhus javanica, Syzygium aromaticum*, and *Terminalia chebula* can augment the anti-HSV-1 activity of acyclovir in the brain, rather than the skin [35]. Controversially, the anti-HSV-1 activity of acyclovir appeared to be weaker in the brain than in the skin. The researchers also suggested that the combined use of acyclovir with herbal extracts at various concentrations showed strong synergistic anti-HSV-1 activity, without a toxic effect. Additionally, combined treatment with acyclovir and *T. chebula* was more efficient in the brain than the use of *T. chebula* alone [35]. Thus, these extracts may be beneficial for prophylactic use against central nervous system complications.

ANTIHERPETIC ACTIVITY OF PLANT ESSENTIAL OILS

Melissa officinalis (Lemon Balm)

Melissa officinalis L. contains the flavonoids, quercitrin and rhamnocitrin; the 7-glucosides, apigenin, kaempferol, quercetin, and luteolin; phenolic acids and tannins; rosmarinic acid and glycosidically bound caffeic and chlorogenic acids; and the triterpenes, ursolic and oleanolic acids [36—38]. This plant has been used in a variety of practical applications in medical science. *M. officinalis* extract has been reported to inhibit protein synthesis and to exhibit antiviral activity against HSV-1 [39—41].

Allahverdiyev et al. showed the antiviral effect of volatile oils of *M. officinalis* against HSV-1 and HSV-2 viruses. They showed that concentrations up to 100 mg/mL are not toxic to cells, although concentrations > 100 mg/mL showed toxic effects on HEp-2 cells. Volatile oil concentrations up to 100 mg/mL exhibited antiviral activity against HSV-2. Since the maximal nontoxic concentration was determined to be 100 mg/mL, they used this concentration in all experiments to test the effects of volatile oils on virus replication. Their *in vivo* findings indicate that *M. officinalis* L. extract inhibit the HSV-2 replication at nontoxic doses [42].

This research indicates that *M. officinalis* L. volatile oils combined with synthetic materials may serve as an effective alternative to the currently available antiviral drugs by producing novel, more effective drugs or using the volatile oils to increase the efficacy of available drugs.

Illicium verum (Star Anise)

Illicium verum is also known as star anise, star aniseed, or Chinese star anise, and is a member of Schisandraceae family. *I. verum* is a small native evergreen tree of northeast Vietnam and southwest China. In a study, EOs obtained from *I. verum* were investigated for their antiherpetic effect on HSV-2 [43]. The pharmaceutical quality and identity of the EOs and their chemical composition was confirmed by quantitative and qualitative analysis using gas chromatography (GC) and gas chromatography-mass spectrometry (GC-MS) [43]. In this study, EO derived from star anise showed antiviral activity prior to viral adsorption. The antiviral activity of star anise EO against HSV-2 is shown in Table 17-1 [43]. EOs derived from *I. verum* seed showed a much higher SI value than those of other extracts and EOs examined. The 50% EC50 was 1 μg/mL and the CC50 was 160 μg/mL (SI value of 160). They concluded that the star anise EO interferes with the virion envelope structure or may mask viral compounds that are necessary for viral adsorption and thus inhibits the virulence of the virus. In another study, the inhibitory effect of star anise oil against HSV-1 infection was compared with the antiviral potential of selected phenylpropanoids and sesquiterpenes, important constituents of EOs, and acyclovir. Star anise EO and most isolated compounds exhibited high levels of antiviral activity against HSV-1 in viral suspension tests [44].

Hyssopus officinalis (Hyssop)

Hyssopus officinalis is a species native to southern Europe, the Middle East, and the region surrounding the Caspian Sea. An EO extract of *H. officinalis* showed antiviral activity against HSV-2 *in vitro* [43]. In this study, the EC50 was 6 ng/mL, the CC50 was 7.5 ng/mL, and the SI value of this oil was calculated as 13.

Thymus vulgaris (Thyme)

Thymus vulgaris is a gramineous plant that grows widely in southern Europe. This plant is generally known as thyme. The characterization of thyme oil by GC and GC-MS methods indicated that thymol (40.5%), *p*-cymene (23.6%), carvacrol (3.2%), linalool (5.4%), β-caryphyllene (2.6%), and terpinen-4-ol (0.7%) are present in thyme EO. Investigations into the effects of thyme EO on HSV-2 indicated that pretreatment of cells with the EO has a stronger effect on HSV virulence than does EO pretreatment of the virus. In this study, the EC50 of *T. vulgaris* was

0.7 ng/mL, and the CC50 was 7 ng/mL, making the SI of thyme oil 10 [43].

Zingiber officinale (Ginger)

Culinary ginger is the root of the *Zingiber officinale* plant. Ginger is a member of Zingiberaceae family and is generally cultivated in southern Asia and East Africa. The EO of ginger is very effective against HSV-2, but is relatively cytotoxic (CC50 = 0.004%), compared to the other oils tested [43]. However, the IC50 of ginger EO (containing zingiberene, limonene/cineol, β-sesquiphellandrene, camphene, and pinocamphene) was found to be 0.0001% and the SI value was 40 [43].

Matricaria chamomilla (German Chamomile)

Matricaria chamomilla, also known as German chamomile, is a member of a very wide range of different species belonging to Asteraceae family. Chamomile originated from Europe and western Asia, and is also found in Australia and North America. The toxic concentration of chamomile in RC-37 cells was found to be very low compared to those of other oils. In addition, the SI of this plant is 20, which indicates an average antiviral effect compared to controls [43]. Pretreatment of the virus with this oil showed a relatively strong effective compared to acyclovir. The IC50 was 0.15 ng/mL and the oils exhibited low toxicity (EC50 = 0.003 μg/mL) [43].

Geum japonicum

Geum japonicum is a yellow flowered perennial native to North America and eastern Asia. *G. japonicum* has been investigated for its antiherpetic effects and was reported as being potentially galenic *in vivo*. Statistically significant symptom alleviation by *G. japonicum* has been reported in HSV-infected

TABLE 17-1 Antiviral Activities of Plant Essential Oils and Purified Compounds Against Herpes Simplex Virus

Plant	Plant part used	Active compound	In vivo/ In vitro	Extract or purified compounds	Type of herpes	CC50 (μg/mL)	EC50 (μg/ mL)	SI	Reference(s)
Melissa officinalis	Leaves	Volatile oils	*In vitro*	Extracted by vapor distillation method	HSV-2	[a]	[a]	[a]	[42]
Illicium verum	Seed	Sesquiterpene	*In vitro*	Ethanol	HSV-2	160 ± 30.71	± 0.1	160	[44]
Hyssopus officinalis	Flowering tops and leaves	Iso-pinocamphene, pinocamphene, spathulenol, linalool, sandalwood oil, α-santalol, *trans-α*-santalol, cis-lanceol, α-santalene	*In vitro*	Ethanol	HSV-2	0.0075	0.0006	13	[43]
Thymus vulgaris	Flowering tops and leaves	Thymol, *p*-cymene, carvacrol, linalool, β-caryphyllene, terpinen-4-ol	*In vitro*	Ethanol	HSV-2	0.007	0.0007	10	
Zingiber officinale	Root	Zingiberene, limonene/cineol, β-sesquiphellandrene, camphene, pinocamphene	*In vitro*	Ethanol	HSV-2	0.004	0.0001	40	
Matricaria chamomilla	Flowers	bisabolol, β-farnesene, bisabolol oxide A, bisabolol oxide B	*In vitro*	Ethanol	HSV-2	0.003	0.00015	20	
Geum japonicum	Whole plant	—	*In vivo*	Dry-boil filtered aqueous extract	HSV-1	[a]	[a]	[a]	[45]
Rhus javanica	Gall	—	*In vivo*	Dry-boil filtered aqueous extract	HSV-1	[a]	[a]	[a]	
Syzygium aromaticum	Flower buds	—	*In vivo*	Dry-boil filtered aqueous extract	HSV-1	[a]	[a]	[a]	
Terminalia chebula	Fruit	—	*In vivo*	Dry-boil filtered aqueous extract	HSV-1	[a]	[a]	[a]	

Plant	Source	Active compounds	Assay	Extract	Virus	CC_{50}	EC_{50}	SI	Ref.
Santalum album	Commercially obtained	—	—	—	HSV-1, HSV-2	c	25	c	[46]
Minthostachys verticillata	Leaves and Stems	—	In vitro	—	HSV-1	613	20.25	30.3	[47]
Aloysia gratissima	Fruits and leaves	—	In vitro	—	HSV-1	150	65	2.31	[48]
Artemisia douglasiana	Leaves	—	In vitro	—	HSV-1	313	83	3,77	
Eupatorium patens	Flowers, fruits, and leaves	—	In vitro	—	HSV-1	294	125	2,35	
Tessaria absinthioides	Leaves	—	In vitro	—	HSV-1	263	105	2,50	
Melaleuca alternifolia	Commercially obtained	—			b	b	b	b	[49]
Menthae piperitae aetheroleum	Commercially obtained	Menthol, menthone, isomenthone, menthyl acetate, cineole	In vitro	Ethanol	HSV-1, HSV-2	~0.003, ~0.0009	~0.014	~0.215, ~0,06	[50]
Rosmarinus officinalis L.	Leaves (in spice form)	—	In vitro	Aqueous	HSV-1	1800	—	—	[51]
Artemisia arborescens	Whole plant	—	In vitro	Ethanol, aqueous	HSV-1, HSV-2	2.4, 4.1	—	—	[52]
Santolina insularis	Full blossom	—	In vitro	—	HSV-1	112	—	127	[53]
Santolina insularis	Full blossom	—	In vitro	—	HSV-2	112	—	160	

[a] Evaluated tissue culture infective dose decreasing method at three different concentrations.
[b] Statistically significant difference compared to placebo.
[d] Statistically evaluated (analysis of variance, Fisher's exact test, or Chi-square test and Student's t-test).
[e] Inhibition of virus replication was measured by the plaque reduction assay.
CC_{50}, 50% cytotoxic concentration; EC_{50}, 50% effective concentration; HSV, herpes simplex virus; SI, selectivity index.
$SI = CC_{50}/EC_{50}$.
[c] Determined as the viral titer reduction factor (log10), compared with untreated controls.

mice [45]. However, other parameters were not investigated.

Rhus javanica (Sumac)

Rhus javanica is a member of the Anacardiaceae family, which contains a large number of species in the *Rhus* genus. *Rhus* spp. are commonly known as sumac and grow in temperate and subtropical regions throughout the world. *R. javanica* is native to Asia. For investigating the potential antiherpetic activity of *R. javanica*, researchers have studied compounds present in the whole or separate parts of the plant. In an *in vivo* animal study, *R. javanica* promoted significant healing when applied in combination with acyclovir onto the pinna and ganglia of HSV-1-infected mice [45].

Syzygium aromaticum (Cloves)

S. aromaticum is commonly known as cloves and has a wide range of uses in both medicine and cooking. Cloves are the flower buds of the *S. aromaticum* tree, a member of the Myrtaceae family. In general, cloves are harvested in India, Indonesia, Madagascar, Pakistan, and Sri Lanka and are used as a spice all over the world. The antiherpetic activity of different EOs of clove was investigated and satisfactory results were obtained. The recurrence of HSV infection was investigated in mice infected with HSV-1. Ultraviolet irradiation of skin was also applied during the prophylactic treatment with clove EO. Of the seven mice investigated, no HSV DNA was detected in samples taken from the pinna of one HSV-infected and clove oil-treated mouse, indicating that *S. aromaticum* may be capable of preventing recurrence of the disease [45].

Terminalia chebula (Black Myrobalan)

The common name for *Terminalia chebula* is black myrobalan and it is native to southern Asia, from India and Nepal to Southwestern China. In an antiherpetic treatment study, *T. chebula* was examined with acyclovir; in two of the six mice treated, disease recurrence was observed. However, *T. chebula* was able to induce a statistically significant improvement in acyclovir efficiency *in vivo* [45].

Santalum album L. (Sandalwood)

Santalum album L. is a small tropical tree known as sandalwood, which grows abundantly in China, India, Indonesia, Malaysia, northwestern Australia, the Philippines, and Sri Lanka,. In a study, sandalwood EO showed antiviral activity against HSV-1 and HSV-2 *in vitro*. However, this effect was not due to a virucidal activity of the EO, but rather to inhibition of herpes virus replication [46].

Minthostachys verticillata (Peperina)

Minthostachys verticillata, known as peperina, is the only species of the genus *Minthostachys* that grows in Argentina. Its EO has pharmacologic importance. In a study, pulegone, a component of this EO, showed antiviral activity against HSV-1 *in vitro*. In this study, the EC50 of *M. verticillata* was 20.25 µg/mL and the CC50 was 613 µg/mL (Table 17-2). The SI value of *M. verticillata* EO was calculated as 30.3, while the SI values of its components were: pulegone, 17.6; menthone and limonene were both < 1 [47]. Thus, the antiviral activity of pulegone and the EO may play a future role in the drug discovery process.

Melaleuca alternifolia (Tea Tree)

In a clinical trial, *Melaleuca alternifolia* oil gel (6%) was used for the treatment of recurrent herpes labialis. The oil gel was applied to patients five times a day, and assessed for antiherpetic (evaluated by HSV DNA analysis)

TABLE 17-2 Antiviral Activity of Plant Crude Extracts Against Herpes Simplex Virus

Plant	Plant part used	Active compound	In vivo/ In vitro	Extract/purified compounds	Herpes subtype	CC50 (µg/mL)	EC50 (µg/mL)	SI	Reference(s)
Ilex paraguariensis	Leaves	Caffeoylquinic acid, triterpenoid	*In vitro*	Aqueous	HSV-1	1260	80.0	15.8	[31]
Lafoensia pacari	Leaves	Tannin	*In vitro*	Methanol	HSV-1	1140	60.0	19.0	
Passiflora edulis	Roots	Saponin, flavonoid	*In vitro*	Aqueous	HSV-1	1600	89.9	17.8	
Rubus imperialis	Leaves	Tannin, flavonoid	*In vitro*	Methanol	HSV-1	1390	70.0	19.8	
Sloanea guianensis	Leaves	–	*In vitro*	Methanol	HSV-1	1400	140.0	10.0	
Clinacanthus nutans	Heartwood	Oxyresveratrol	*In vitro*	Commercially produced	HSV-1	237,6	24.0	9.9	[21]
Artocarpus lakoocha	Heartwood	Oxyresveratrol	*In vitro*	Methanol	HSV-1, HSV-2	237.5	18,70	12.7	
Coptidis rhizoma	Whole plant	Berberine, *Coptidis rhizoma* extract Ching Wei San extract	*In vitro*	Dried plant	HSV-1, HSV-2	13.2, 321, 2500	8.2, 2.4, 24.3	147, 111, 99	[54]
Pongamia pinnata	Seed	–	*In vitro*	Ethanol-aqueous	HSV-1, HSV-2	1000 20000	–	–	[33]
Combination of *Salvia officinalis* and *Rheum palmatum*	Leaf, Root	Tannin, anthraquinones	*In vivo*	Aqueous extract, aqueous-Ethanol extract	HSV-1	–	–	–	[55]
Carissa edulis	Root	Unknown	*In vivo* *In vitro*	Boiled-lyophilized aqueous	HSV-1, HSV-2	480	15.1 ± 0.57, 6.9 ± 1.27	31.79, 69.57	[15]
Hypericum connatum	Aerial parts	Hexane extract, flavonoid	*In vitro*	Ethanol	HSV-1	–	–	–	[19]
Aglaia odorata	Leaf	–	*In vivo* and *In vitro*	Ethanol	HSV-1	312	9.5	32.9	[34]
Moringa oleifera	Leaf	–	*In vivo* and *In vitro*	Ethanol	HSV-1	875	100.0	8.8	[34]
Ventilago denticulata	Leaf	–	*In vitro*	Ethanol	HSV-1	838	46.3	18.1	

CC50, 50% cytotoxic concentration; EC50, 50% effective concentration; HSV, herpes simplex virus; SI, selectivity index. SI = CC50/EC50.

and adverse effects. Their findings indicated some benefit from tea tree oil treatment; however, none of the differences between groups reached statistical significance. The SI of this compound was determined to be 0.215 against HSV-1 and approximately 0.06 against HSV-2 (Table 17-1) [49].

Menthae piperitae aetheroleum (Peppermint Oil)

Menthae piperitae aetheroleum (peppermint oil) is an EO derived from leaves of *M. piperitae aetheroleum* L. Peppermint oil is used as a traditional treatment for different human diseases and various pain conditions, such as headaches [54], postherpetic neuralgia [55], or mild bacterial or fungal infections of the skin. Recently, the antiviral properties of peppermint oil against HSV-1 and HSV-2 were reported. Viral inhibitory activity was tested *in vitro* on RC-37 cells using a plaque reduction assay, and the findings showed that peppermint oil exhibits high levels of virucidal activity against both HSV-1 and HSV-2 in viral suspension tests at noncytotoxic concentrations of the oil. The lipophilic nature of this oil provides penetration into the skin and it may therefore have topical therapeutic use as a virucidal agent against recurrent herpes infections [50].

Other Important Plant Essential Oils

Other plant extracts, chemical compounds, and products have been studied for activity against HSV-2 both *in vitro* and *in vivo*. The extracts contained a very wide range of chemical compounds (including borneol, caffeic acid, carrageenan polysaccharide, cineole, cinnamon oil, citral, curcumin) and the effects of these against HSV-2 were investigated. Following *in vitro* analysis, the compounds with the highest antiviral activities were tested in mice. Four compounds, carrageenan lambda type IV, cineole, curcumin, and eugenol, provided significant protection in a mouse model of intravaginal HSV-2 [56].

HIGHLY EFFECTIVE PLANT EXTRACTS AND ESSENTIAL OILS

The most promising herbs, such as *A. odorata* (SI value of 32.9) [34], *C. edulis* (SI value of 69.57) [15], *I. verum* (SI value of 160), *M. verticillata* (SI value of 30.3) [47], *Santolina insularis* (SI value of 127−160), and *Z. officinale* (SI value of 40) [53] are shown as very active in Tables 17-1 and 17-2. In addition, most of the other promising herbals showed SI values of > 2. In conclusion, natural treatments comprising the different herbal compounds described in this chapter can be effective, safe, and without unpleasant side effects. However, information on the antiviral activities and identities of the active substances present in these extracts and EOs is not complete. The information provided in this chapter should enable further studies should be conducted to enable the development of effective antiherpetic drugs based on herbal compounds. In addition, the literature review indicates that herbal extracts are more effective when combined with acyclovir. Hence, it is possible that through its combination with herbs, the increasing resistance to acyclovir may be prevented and therapeutic dose of acyclovir may be reduced.

REVIEW CRITERIA

The Science Citation Index of the Web of Science and the PubMed database were searched using various combinations of the terms *Herpes simplex*, *HSV-1*, *HSV-2*, *plant compounds* (and derivative words), *natural products*, *antiherpetic drugs*, *clinical trial*, *in vivo*, *mouse*, *in vitro*, and *therapeutic index* to identify relevant articles published between 1954 and 2010. Full-length articles published in English were selected. For *in vitro* and *in vivo* studies, the reference lists of recent research papers and reviews were used to identify further articles.

CONCLUSIONS AND FUTURE PROSPECTS

Several drugs such as acyclovir are available for the treatment of HSV infections. Acyclovir is a nucleoside analog and selective antiherpetic agent. It inhibits viral DNA replication and DNA synthesis by targeting viral thymidine kinase. However, acyclovir resistance is being increasingly described, caused by mutations in either the thymidine kinase or the DNA polymerase genes [59]. Acyclovir-resistant herpes viruses have been isolated from the patients with AIDS or cancer, as well as recipients of bone marrow or organ transplants [60,61].

Therefore new anti-HSV drugs are urgently needed, and many studies have focused on herbals. Medicinal and aromatic plants are widely used today in modern phytotherapy [62]. Recently, anti-HSV activity has been reported in some plant extracts and EOs. Several herbal components with activity against HSV-1 and HSV-2 have been proposed as promising alternative therapeutic tools [44]. These herbals are often also effective against acyclovir-resistant strains. Many herbal extracts and oils have been investigated against HSV; however, there was no comprehensive review of the knowledge gleaned from these studies. Therefore, for the first time, we have tried to evaluate and review the anti-HSV effects of different herbal extracts and oils. Our aim was to bring together the information about various herbal compounds available in the literature, investigate the antiherpetic effectiveness of these compounds, and identify the potential of these plants for developing novel herpetic treatments.

In this chapter, plants investigated for their antiherpetic effectiveness are presented as herbal extracts and EOs. We have described in detail the data obtained from several antiherpetic studies in the tables. Herbal extracts and EOs were evaluated for their antiherpetic effects according to different parameters. Therefore, we reviewed the data according to their SI values. Generally, for most of these *in vitro* studies, the SI value was calculated to determine whether the extracts contained sufficient antiviral activity. The SI measures the antiviral efficiency and should exceed the level of toxicity. Herbal components with an SI of ≥ 2 are considered active and those with an SI of ≥ 10 are considered very active.

Consequently, this chapter reveals that plant extracts and EOs show great potential for use in the treatment of HSV infections. However, we found that many studies are restricted to *in vitro* investigations and that there are insufficient numbers of *in vivo* studies and clinical trials. However, considering recent developments in this field, we believe that the applications of plant compounds for treating HSV infections in humans will begin shortly and that the promising results described in this chapter will guide these applications.

References

[1] Snoeck R. Antiviral therapy of herpes simplex. Int J Antimicrob Agents 2000;16:157–9.

[2] Ablin J, Symon Z, Mevorach D. Sacral herpes-zoster infection presenting as sciatic pain. J Clin Rheumatol 1996;2:167–9.

[3] Aurelian L, Specter S, Hodinka R, Young S, Wiedbrauk D. Herpes simplex viruses. Clin Virol Manual 2009:424–53.

[4] Looker KJ, Garnett GP, Schmid GP. An estimate of the global prevalence and incidence of herpes simplex virus type 2 infection. Bull WHO 2008;86:805–12.

[5] Fatahzadeh M, Schwartz RA. Human herpes simplex virus infections: epidemiology, pathogenesis, symptomatology, diagnosis, and management. J Am Acad Dermatol 2007;57:737–63. quiz 764–36.

[6] Whitley RJ, Kimberlin DW, Roizman B. Herpes simplex viruses. Clin Infectious Diseases 1998;26:541–53.

[7] Brady RC, Bernstein DI. Treatment of herpes simplex virus infections. Antivir Res. 2004;61:73–81.

[8] De Clercq E. Discovery and development of BVDU (brivudin) as a therapeutic for the treatment of herpes zoster. Biochem Pharmacol 2004;68:2301−15.

[9] De Clercq E. Antiviral drugs in current clinical use. J Clin Virol 2004;30:115−33.

[10] De Clercq E, Andrei G, Snoeck R, De Bolle L, Naesens L, Degreve B, et al. Acyclic/carbocyclic guanosine analogues as anti-herpesvirus agents. Nucleosides Nucleotides Nucleic Acids 2001;20:271−85.

[11] Leung DT, Sacks SL. Current recommendations for the treatment of genital herpes. Drugs 2000;60:1329−52.

[12] Collins P, Darby G. Laboratory studies of herpes simplex virus strains resistant to acyclovir. Rev Med Virol 1991;1:19−28.

[13] Khan MTH, Ather A, Thompson KD, Gambari R. Extracts and molecules from medicinal plants against herpes simplex viruses. Antivir Res 2005;67:107−19.

[14] Gachathi FN. Kikuyu botanical dictionary of plant names and uses. Nairobi, Kenya: AMREF; 1989.

[15] Tolo FM, Rukunga GM, Muli FW, Njagi ENM, Njue W, Kumon K, et al. Anti-viral activity of the extracts of a Kenyan medicinal plant Carissa edulis against herpes simplex virus. J Ethnopharmacol 2006;104:92−9.

[16] Taylor A, Mc KG, Burlage HM, Stokes DM. Plant extracts tested against egg cultivated viruses. Tex Rep Biol Med 1954;12:551−7.

[17] Sydiskis R, Owen D, Lohr J, Rosler K, Blomster R. Inactivation of enveloped viruses by anthraquinones extracted from plants. Antimicrob Agents Chemother 1991;35:2463.

[18] Saller R, Buechi S, Meyrat R, Schmidhauser C. Combined herbal preparation for topical treatment of Herpes labialis. Forsch Komplementarmed Klass Naturheilkd. 2001;8:373−82.

[19] Fritz D, Venturi CR, Cargnin S, Schripsema J, Roehe PM, Montanha JA, et al. Herpes virus inhibitory substances from Hypericum connatum Lam., a plant used in southern Brazil to treat oral lesions. J Ethnopharmacol 2007;113:517−20.

[20] Sangkitporn S, Chaiwat S, Balachandra K, Na-Ayudhaya TD, Bunjob M, Jayavasu C. Treatment of herpes zoster with Clinacanthus nutans (bi phaya yaw) extract. J Med Assoc Thai 1995;78:624−7.

[21] Chuanasa T, Phromjai J, Lipipun V, Likhitwitayawuid K, Suzuki M, Pramyothin P, et al. Anti-herpes simplex virus (HSV-1) activity of oxy-resveratrol derived from Thai medicinal plant: mechanism of action and therapeutic efficacy on cutaneous HSV-1 infection in mice. Antivira Res 2008;80:62−70.

[22] Docherty JJ, Fu MM, Stiffler BS, Limperos RJ, Pokabla CM, DeLucia AL. Resveratrol inhibition of

herpes simplex virus replication. Antivir Res 1999;43:145−55.

[23] Docherty JJ, Smith JS, Fu MM, Stoner T, Booth T. Effect of topically applied resveratrol on cutaneous herpes simplex virus infections in hairless mice. Antivir Res 2004;61:19−26.

[24] Gudima S, Memelova L, Borodulin V, Pokholok D, Mednikov B, Tolkachev O, et al. [Kinetic analysis of interaction of human immunodeficiency virus reverse transcriptase with alkaloids]. Molekuliarnaia Biologiia 1994;28:1308.

[25] Hayashi K, Minoda K, Nagaoka Y, Hayashi T, Uesato S. Antiviral activity of berberine and related compounds against human cytomegalovirus. Bioorganic & medicinal chem let 2007;17:1562−4.

[26] Li HL, Han T, Liu RH, Zhang C, Chen HS, Zhang WD. Alkaloids from Corydalis saxicola and their anti-hepatitis B virus activity. Chem & Biodiversity 2008;5:777−83.

[27] Schmeller T, Latz-Brüning B, Wink M. Biochemical activities of berberine, palmatine and sanguinarine mediating chemical defence against microorganisms and herbivores. Phytochemistry 1997;44:257−66.

[28] Sethi ML. Enzyme inhibition VI: inhibition of reverse transcriptase activity by protoberberine alkaloids and structure−activity relationships. J Pharm Sci 1983;72:538−41.

[29] Sethi MI. Comparison of inhibition of reverse transcriptase and antileukemic activities exhibited by protoberberine and benzophenanthridine alkaloids and structure-activity relationships. Phytochemistry 1985;24:447−54.

[30] Roizman B, Sears A. Herpes simplex viruses and their replication. In: The human herpesviruses. New York, NY: Raven Press; 1993. p. 11−68.

[31] Müller V, Chávez JH, Reginatto FH, Zucolotto SM, Niero R, Navarro D, et al. Evaluation of antiviral activity of South American plant extracts against herpes simplex virus type 1 and rabies virus. Phytother Res 2007;21:970−4.

[32] Goncalves JL, Leitao SG, Monache FD, Miranda MM, Santos MG, Romanos MT, et al. In vitro antiviral effect of flavonoid-rich extracts of Vitex polygama (Verbenaceae) against acyclovir-resistant herpes simplex virus type 1. Phytomedicine 2001;8:477−80.

[33] Elanchezhiyan M, Rajarajan S, Rajendran P, Subramanian S, Thyagarajan S. Antiviral properties of the seed extract of an Indian medicinal plant, Pongamia pinnata, Linn., against herpes simplex viruses: in vitro studies on Vero cells. J Med Microbiol 1993;38:262−4.

[34] Lipipun V, Kurokawa M, Suttisri R, Taweechotipatr P, Pramyothin P, Hattori M, et al. Efficacy of Thai medicinal

plant extracts against herpes simplex virus type 1 infection in vitro and in vivo. Antivir Res 2003;60:175–80.

[35] Kurokawa M, Nagasaka K, Hirabayashi T, Uyama S, Sato H, Kageyama T, et al. Efficacy of traditional herbal medicines in combination with acyclovir against herpes simplex virus type 1 infection in vitro and in vivo. Antivir Res 1995;27:19–37.

[36] Bruneton J, Pharmacognosy P. Medicinal Plants. Intercept, Hampshire 1995.

[37] Khan IA, Abourashed EA. Leung's Encyclopedia of Common Natural Ingredients: Used in Food, Drugs and Cosmetics. NY, USA: John Wiley & Sons; 2010.

[38] Bisset NG, Czygan FC. Herbal drugs and phytopharmaceuticals: a handbook for practice on a scientific basis; [with reference to German Commission E monographs]. Germany: Medpharm GmbH; 2001.

[39] Cohen RA, Kucera LS, Herrmann Jr EC. Antiviral activity of Melissa officinalis (lemon balm) extract. Exp Biol Med 1964;117:431.

[40] Kucera LS, Herrmann Jr EC. Antiviral substances in plants of the mint family (Labiatae). I. Tannin of Melissa officinalis. Exp Biol Med 1967;124:865.

[41] May G, Willuhn G. Antiviral effect of aqueous plant extracts in tissue culture]. Arzneimittel-Forschung 1978;28:1.

[42] Allahverdiyev A, Duran N, Ozguven M, Koltas S. Antiviral activity of the volatile oils of Melissa officinalis L. against Herpes simplex virus type-2. Phytomedicine 2004;11:657–61.

[43] Koch C, Reichling J, Schneele J, Schnitzler P. Inhibitory effect of essential oils against herpes simplex virus type 2. Phytomedicine 2008;15:71–8.

[44] Astani A, Reichling J, Schnitzler P. Screening for antiviral activities of isolated compounds from essential oils. Evidence-based Complement. Alternat Med 2011;2011:253643.

[45] Kurokawa M, Nakano M, Ohyama H, Hozumi T, Kageyama S, Namba T, et al. Prophylactic efficacy of traditional herbal medicines against recurrent herpes simplex virus type 1 infection from latently infected ganglia in mice. J Dermatol Sci 1997;14:76–84.

[46] Benencia F, Courreges MC. Antiviral activity of sandalwood oil against herpes simplex viruses-1 and -2. Phytomedicine 1999;6:119–23.

[47] María V, Vogt SBSFME, Sabini María C, Cariddi Laura N, Torres Cristina V, Zanon Silvia M, et al. Minthostachys verticillata essentials oil and its major components: antiherpetic selective action in HEp-2 cells. Mol Med Chem 2010;21:117–20.

[48] Garcia C, Talarico L, Almeida N, Colombres S, Duschatzky C, Damonte E. Virucidal activity of essential oils from aromatic plants of San Luis, 17. Argentina: Phytother. Res 2003; 1073–1075.

[49] Carson C, Ashton L, Dry L, Smith D, Riley T. Melaleuca alternifolia (tea tree) oil gel (6%) for the treatment of recurrent herpes labialis. J Antimicrob Chemoth 2001;48:450.

[50] Schuhmacher A, Reichling J, Schnitzler P. Virucidal effect of peppermint oil on the enveloped viruses herpes simplex virus type 1 and type 2 in vitro. Phytomedicine 2003;10:504–10.

[51] Mancini DAP, Torres RP, Pinto JR, Mancini-Filho J. Inhibition of DNA virus: Herpes-1 (HSV-1) in cellular culture replication, through an antioxidant treatment extracted from rosemary spice. Braz J Pharm Sci 2009;45:127–33.

[52] Saddi M, Sanna A, Cottiglia F, Chisu L, Casu L, Bonsignore L, et al. Antiherpevirus activity of Artemisia arborescens essential oil and inhibition of lateral diffusion in Vero cells. Ann Clin Microbiol Antimicrob 2007;6:10.

[53] De Logu A, Loy G, Pellerano ML, Bonsignore L, Schivo ML. Inactivation of HSV-1 and HSV-2 and prevention of cell-to-cell virus spread by Santolina insularis essential oil. Antivir Res 2000;48:177–85.

[54] Göbel H, Schmidt G, Dworschak M, Stolze H, Heuss D. Essential plant oils and headache mechanisms. Phytomedicine 1995;2:93–102.

[55] Davies SJ, Harding LM, Baranowski AP. A novel treatment of postherpetic neuralgia using peppermint oil. Clin. J Pain 2002;18:200.

[56] Bourne KZ, Bourne N, Reising SF, Stanberry LR. Plant products as topical microbicide candidates: assessment of in vitro and in vivo activity against herpes simplex virus type 2. Antivir Res 1999;42:219–26.

[57] Chin LW, Cheng YW, Lin SS, Lai YY, Lin LY, Chou MY, et al. Anti-herpes simplex virus effects of berberine from Coptidis rhizoma, a major component of a Chinese herbal medicine, Ching-Wei-San. Arch Virol 2010;155:1933–41.

[58] Reuter J, Wolfle U, Weckesser S, Schempp C. Which plant for which skin disease? Part 1: Atopic dermatitis, psoriasis, acne, condyloma and herpes simplex. J Dtsch Dermatol Ges 2010;8:788–96.

[59] Pottage Jr J, Kessler H. Herpes simplex virus resistance to acyclovir: clinical relevance. Infect Agent Dis 1995;4:115.

[60] Bacon TH, Levin MJ, Leary JJ, Sarisky RT, Sutton D. Herpes simplex virus resistance to acyclovir and penciclovir after two decades of antiviral therapy. Clin Microbiol Rev 2003;16:114.

[61] Morfin F, Thouvenot D. Herpes simplex virus resistance to antiviral drugs. J Clin Virol 2003;26:29–37.

[62] Demirci B, Demirci F. Essential oil of Betula pendula Roth. Buds. Evidence-based complement. Alternat Med 2004;1:301–3.

Index

Note: Page numbers followed by "f" and "t" indicate figures and tables, respectively.

Printed and bound by CPI Group (UK) Ltd, Croydon, CR0 4YY

08/05/2025

01864981-0001